Lecture Notes in Computer Science 14039

Founding Editors

Gerhard Goos
Juris Hartmanis

Editorial Board Members

The series Lecture Notes in Computer Science (LNCS), including its subseries Lecture Notes in Artificial Intelligence (LNAI) and Lecture Notes in Bioinformatics (LNBI), has established itself as a medium for the publication of new developments in computer science and information technology research, teaching, and education.

LNCS enjoys close cooperation with the computer science R & D community, the series counts many renowned academics among its volume editors and paper authors, and collaborates with prestigious societies. Its mission is to serve this international community by providing an invaluable service, mainly focused on the publication of conference and workshop proceedings and postproceedings. LNCS commenced publication in 1973.

Fiona Nah · Keng Siau
Editors

HCI in Business, Government and Organizations

10th International Conference, HCIBGO 2023
Held as Part of the 25th HCI International Conference, HCII 2023
Copenhagen, Denmark, July 23–28, 2023
Proceedings, Part II

 Springer

Editors
Fiona Nah
City University of Hong Kong
Kowloon, Hong Kong

Keng Siau
City University of Hong Kong
Kowloon, Hong Kong

ISSN 0302-9743   ISSN 1611-3349 (electronic)
Lecture Notes in Computer Science
ISBN 978-3-031-36048-0   ISBN 978-3-031-36049-7 (eBook)
https://doi.org/10.1007/978-3-031-36049-7

This Springer imprint is published by the registered company Springer Nature Switzerland AG
The registered company address is: Gewerbestrasse 11, 6330 Cham, Switzerland

Foreword

Human-computer interaction (HCI) is acquiring an ever-increasing scientific and industrial importance, as well as having more impact on people's everyday lives, as an ever-growing number of human activities are progressively moving from the physical to the digital world. This process, which has been ongoing for some time now, was further accelerated during the acute period of the COVID-19 pandemic. The HCI International (HCII) conference series, held annually, aims to respond to the compelling need to advance the exchange of knowledge and research and development efforts on the human aspects of design and use of computing systems.

The 25th International Conference on Human-Computer Interaction, HCI International 2023 (HCII 2023), was held in the emerging post-pandemic era as a 'hybrid' event at the AC Bella Sky Hotel and Bella Center, Copenhagen, Denmark, during July 23–28, 2023. It incorporated the 21 thematic areas and affiliated conferences listed below.

A total of 7472 individuals from academia, research institutes, industry, and government agencies from 85 countries submitted contributions, and 1578 papers and 396 posters were included in the volumes of the proceedings that were published just before the start of the conference, these are listed below. The contributions thoroughly cover the entire field of human-computer interaction, addressing major advances in knowledge and effective use of computers in a variety of application areas. These papers provide academics, researchers, engineers, scientists, practitioners and students with state-of-the-art information on the most recent advances in HCI.

The HCI International (HCII) conference also offers the option of presenting 'Late Breaking Work', and this applies both for papers and posters, with corresponding volumes of proceedings that will be published after the conference. Full papers will be included in the 'HCII 2023 - Late Breaking Work - Papers' volumes of the proceedings to be published in the Springer LNCS series, while 'Poster Extended Abstracts' will be included as short research papers in the 'HCII 2023 - Late Breaking Work - Posters' volumes to be published in the Springer CCIS series.

I would like to thank the Program Board Chairs and the members of the Program Boards of all thematic areas and affiliated conferences for their contribution towards the high scientific quality and overall success of the HCI International 2023 conference. Their manifold support in terms of paper reviewing (single-blind review process, with a minimum of two reviews per submission), session organization and their willingness to act as goodwill ambassadors for the conference is most highly appreciated.

This conference would not have been possible without the continuous and unwavering support and advice of Gavriel Salvendy, founder, General Chair Emeritus, and Scientific Advisor. For his outstanding efforts, I would like to express my sincere appreciation to Abbas Moallem, Communications Chair and Editor of HCI International News.

July 2023   Constantine Stephanidis

HCI International 2023 Thematic Areas and Affiliated Conferences

Thematic Areas

- HCI: Human-Computer Interaction
- HIMI: Human Interface and the Management of Information

Affiliated Conferences

- EPCE: 20th International Conference on Engineering Psychology and Cognitive Ergonomics
- AC: 17th International Conference on Augmented Cognition
- UAHCI: 17th International Conference on Universal Access in Human-Computer Interaction
- CCD: 15th International Conference on Cross-Cultural Design
- SCSM: 15th International Conference on Social Computing and Social Media
- VAMR: 15th International Conference on Virtual, Augmented and Mixed Reality
- DHM: 14th International Conference on Digital Human Modeling and Applications in Health, Safety, Ergonomics and Risk Management
- DUXU: 12th International Conference on Design, User Experience and Usability
- C&C: 11th International Conference on Culture and Computing
- DAPI: 11th International Conference on Distributed, Ambient and Pervasive Interactions
- HCIBGO: 10th International Conference on HCI in Business, Government and Organizations
- LCT: 10th International Conference on Learning and Collaboration Technologies
- ITAP: 9th International Conference on Human Aspects of IT for the Aged Population
- AIS: 5th International Conference on Adaptive Instructional Systems
- HCI-CPT: 5th International Conference on HCI for Cybersecurity, Privacy and Trust
- HCI-Games: 5th International Conference on HCI in Games
- MobiTAS: 5th International Conference on HCI in Mobility, Transport and Automotive Systems
- AI-HCI: 4th International Conference on Artificial Intelligence in HCI
- MOBILE: 4th International Conference on Design, Operation and Evaluation of Mobile Communications

List of Conference Proceedings Volumes Appearing Before the Conference

1. LNCS 14011, Human-Computer Interaction: Part I, edited by Masaaki Kurosu and Ayako Hashizume
2. LNCS 14012, Human-Computer Interaction: Part II, edited by Masaaki Kurosu and Ayako Hashizume
3. LNCS 14013, Human-Computer Interaction: Part III, edited by Masaaki Kurosu and Ayako Hashizume
4. LNCS 14014, Human-Computer Interaction: Part IV, edited by Masaaki Kurosu and Ayako Hashizume
5. LNCS 14015, Human Interface and the Management of Information: Part I, edited by Hirohiko Mori and Yumi Asahi
6. LNCS 14016, Human Interface and the Management of Information: Part II, edited by Hirohiko Mori and Yumi Asahi
7. LNAI 14017, Engineering Psychology and Cognitive Ergonomics: Part I, edited by Don Harris and Wen-Chin Li
8. LNAI 14018, Engineering Psychology and Cognitive Ergonomics: Part II, edited by Don Harris and Wen-Chin Li
9. LNAI 14019, Augmented Cognition, edited by Dylan D. Schmorrow and Cali M. Fidopiastis
10. LNCS 14020, Universal Access in Human-Computer Interaction: Part I, edited by Margherita Antona and Constantine Stephanidis
11. LNCS 14021, Universal Access in Human-Computer Interaction: Part II, edited by Margherita Antona and Constantine Stephanidis
12. LNCS 14022, Cross-Cultural Design: Part I, edited by Pei-Luen Patrick Rau
13. LNCS 14023, Cross-Cultural Design: Part II, edited by Pei-Luen Patrick Rau
14. LNCS 14024, Cross-Cultural Design: Part III, edited by Pei-Luen Patrick Rau
15. LNCS 14025, Social Computing and Social Media: Part I, edited by Adela Coman and Simona Vasilache
16. LNCS 14026, Social Computing and Social Media: Part II, edited by Adela Coman and Simona Vasilache
17. LNCS 14027, Virtual, Augmented and Mixed Reality, edited by Jessie Y. C. Chen and Gino Fragomeni
18. LNCS 14028, Digital Human Modeling and Applications in Health, Safety, Ergonomics and Risk Management: Part I, edited by Vincent G. Duffy
19. LNCS 14029, Digital Human Modeling and Applications in Health, Safety, Ergonomics and Risk Management: Part II, edited by Vincent G. Duffy
20. LNCS 14030, Design, User Experience, and Usability: Part I, edited by Aaron Marcus, Elizabeth Rosenzweig and Marcelo Soares
21. LNCS 14031, Design, User Experience, and Usability: Part II, edited by Aaron Marcus, Elizabeth Rosenzweig and Marcelo Soares

22. LNCS 14032, Design, User Experience, and Usability: Part III, edited by Aaron Marcus, Elizabeth Rosenzweig and Marcelo Soares
23. LNCS 14033, Design, User Experience, and Usability: Part IV, edited by Aaron Marcus, Elizabeth Rosenzweig and Marcelo Soares
24. LNCS 14034, Design, User Experience, and Usability: Part V, edited by Aaron Marcus, Elizabeth Rosenzweig and Marcelo Soares
25. LNCS 14035, Culture and Computing, edited by Matthias Rauterberg
26. LNCS 14036, Distributed, Ambient and Pervasive Interactions: Part I, edited by Norbert Streitz and Shin'ichi Konomi
27. LNCS 14037, Distributed, Ambient and Pervasive Interactions: Part II, edited by Norbert Streitz and Shin'ichi Konomi
28. LNCS 14038, HCI in Business, Government and Organizations: Part I, edited by Fiona Nah and Keng Siau
29. LNCS 14039, HCI in Business, Government and Organizations: Part II, edited by Fiona Nah and Keng Siau
30. LNCS 14040, Learning and Collaboration Technologies: Part I, edited by Panayiotis Zaphiris and Andri Ioannou
31. LNCS 14041, Learning and Collaboration Technologies: Part II, edited by Panayiotis Zaphiris and Andri Ioannou
32. LNCS 14042, Human Aspects of IT for the Aged Population: Part I, edited by Qin Gao and Jia Zhou
33. LNCS 14043, Human Aspects of IT for the Aged Population: Part II, edited by Qin Gao and Jia Zhou
34. LNCS 14044, Adaptive Instructional Systems, edited by Robert A. Sottilare and Jessica Schwarz
35. LNCS 14045, HCI for Cybersecurity, Privacy and Trust, edited by Abbas Moallem
36. LNCS 14046, HCI in Games: Part I, edited by Xiaowen Fang
37. LNCS 14047, HCI in Games: Part II, edited by Xiaowen Fang
38. LNCS 14048, HCI in Mobility, Transport and Automotive Systems: Part I, edited by Heidi Krömker
39. LNCS 14049, HCI in Mobility, Transport and Automotive Systems: Part II, edited by Heidi Krömker
40. LNAI 14050, Artificial Intelligence in HCI: Part I, edited by Helmut Degen and Stavroula Ntoa
41. LNAI 14051, Artificial Intelligence in HCI: Part II, edited by Helmut Degen and Stavroula Ntoa
42. LNCS 14052, Design, Operation and Evaluation of Mobile Communications, edited by Gavriel Salvendy and June Wei
43. CCIS 1832, HCI International 2023 Posters - Part I, edited by Constantine Stephanidis, Margherita Antona, Stavroula Ntoa and Gavriel Salvendy
44. CCIS 1833, HCI International 2023 Posters - Part II, edited by Constantine Stephanidis, Margherita Antona, Stavroula Ntoa and Gavriel Salvendy
45. CCIS 1834, HCI International 2023 Posters - Part III, edited by Constantine Stephanidis, Margherita Antona, Stavroula Ntoa and Gavriel Salvendy
46. CCIS 1835, HCI International 2023 Posters - Part IV, edited by Constantine Stephanidis, Margherita Antona, Stavroula Ntoa and Gavriel Salvendy

47. CCIS 1836, HCI International 2023 Posters - Part V, edited by Constantine Stephanidis, Margherita Antona, Stavroula Ntoa and Gavriel Salvendy

https://2023.hci.international/proceedings

Preface

The use and role of technology in the business and organizational context have always been at the heart of human-computer interaction (HCI) since the start of management information systems. In general, HCI research in such a context is concerned with the ways humans interact with information, technologies, and tasks in the business, managerial, and organizational contexts. Hence, the focus lies in understanding the relationships and interactions between people (e.g., management, users, implementers, designers, developers, senior executives, and vendors), tasks, contexts, information, and technology. Today, with the explosion of the metaverse, social media, big data, and the Internet of Things, new pathways are opening in this direction, which need to be investigated and exploited.

The 10th International Conference on HCI in Business, Government and Organizations (HCIBGO 2023), an affiliated conference of the HCI International (HCII) conference, promoted and supported multidisciplinary dialogue, cross-fertilization of ideas, and greater synergies between research, academia, and stakeholders in the business, managerial, and organizational domain.

HCI in business, government, and organizations ranges across a broad spectrum of topics from digital transformation to customer engagement. The HCIBGO conference facilitates the advancement of HCI research and practice for individuals, groups, enterprises, and society at large. The topics covered include emerging areas such as artificial intelligence and machine learning, blockchain, service design, live streaming in electronic commerce, visualization, and workplace design.

Two volumes of the HCII 2023 proceedings are dedicated to this year's edition of the HCIBGO conference. The first volume covers topics related to advancing technology and management in public sector organizations and governance, user experience and business perspectives in the context of mobile commerce and e-commerce, as well as the use of disruptive technologies to enhance customer experience. The second volume focuses on topics related to exploring the intersection of robotics and autonomous agents in business and industry, applications of AI in business and society, as well as case studies and empirical research in the domain of exploring human behavior and communication in business.

Papers of this volume are included for publication after a minimum of two single-blind reviews from the members of the HCIBGO Program Board or, in some cases, from members of the Program Boards of other affiliated conferences. We would like to thank all of them for their invaluable contribution, support, and efforts.

July 2023

Fiona Nah
Keng Siau

10th International Conference on HCI in Business, Government and Organizations (HCIBGO)

Program Board Chairs: **Fiona Nah,** *City University of Hong Kong, China* and **Keng Siau**, *City University of Hong Kong, China*

Program Board:

- Kaveh Abhari, *San Diego State University, USA*
- Andreas Auinger, *University of Applied Sciences Upper Austria, Austria*
- Denise Baker, *Missouri University of Science and Technology, USA*
- Gaurav Bansal, *University of Wisconsin - Green Bay, USA*
- Valerie Bartelt, *University of Denver, USA*
- Kaveh Bazargan, *Allameh Tabataba'i University, Iran*
- Langtao Chen, *Missouri University of Science and Technology, USA*
- Constantinos K. Coursaris, *HEC Montréal, Canada*
- Brenda Eschenbrenner, *University of Nebraska at Kearney, USA*
- J. M. Goh, *Simon Fraser University, Canada*
- Netta Iivari, *University of Oulu, Finland*
- Qiqi Jiang, *Copenhagen Business School, Denmark*
- Yi-Cheng Ku, *Fu Jen Catholic University, Taiwan*
- Murad Moqbel, *University of Texas Rio Grande Valley, USA*
- Norman Shaw, *Toronto Metropolitan University, Canada*
- Martin Stabauer, *Johannes Kepler University Linz, Austria*
- Werner Wetzlinger, *University of Applied Sciences Upper Austria, Austria*
- I-Chin Wu, *National Taiwan Normal University, Taiwan*
- Dezhi Wu, *University of South Carolina, USA*
- Jie Yu, *University of Nottingham Ningbo China, P.R. China*

The full list with the Program Board Chairs and the members of the Program Boards of all thematic areas and affiliated conferences of HCII2023 is available online at:

http://www.hci.international/board-members-2023.php

HCI International 2024 Conference

The 26th International Conference on Human-Computer Interaction, HCI International 2024, will be held jointly with the affiliated conferences at the Washington Hilton Hotel, Washington, DC, USA, June 29 – July 4, 2024. It will cover a broad spectrum of themes related to Human-Computer Interaction, including theoretical issues, methods, tools, processes, and case studies in HCI design, as well as novel interaction techniques, interfaces, and applications. The proceedings will be published by Springer. More information will be made available on the conference website: http://2024.hci.international/.

General Chair
Prof. Constantine Stephanidis
University of Crete and ICS-FORTH
Heraklion, Crete, Greece
Email: general_chair@hcii2024.org

https://2024.hci.international/

Contents – Part II

Exploring the Intersection of Robotics and Autonomous Agents in Business and Industry

Evaluation of Risk Factors in the Armed Forces Physical Fitness Test 3
    Yu-Jing Chiu, Chien-Cheng Li, So-Ra Song, and Yi-Chung Hu

Chatbots in Academic Advising: Evaluating the Acceptance and Effects
of Chatbots in German Student-University Communication 18
    Annebeth Demaeght, Natalie Walz, and Andrea Müller

Study on the Impact of Service Robot Autonomy on Customer Satisfaction 30
    Keli Li and Guoxin Li

Interactive Robot-Aided Diagnosis System for Children with Autism
Spectrum Disorder 41
    *Szu-Yin Lin, Yi-Pei Lai, Hao-Chun Chiang, Yawei Cheng,
    and Shih-Yi Chien*

The Role of Technology-Ethical Leadership Interaction in Minimising
Unethical Acts: Implications for Research and Practice 53
    Majd Megheirkouni and David Weir

Systematic Literature Review on the User Evaluation of Teleoperation
Interfaces for Professional Service Robots 66
    *Gaayathri Sankar, Soussan Djamasbi, Zhi Li, Jing Xiao,
    and Norbou Buchler*

Safety of Human-Robot Collaboration within the Internet of Production 86
    *Minh Trinh, Hannah Dammers, Mohamed Behery, Ralph Baier,
    Thomas Henn, Daniel Gossen, Burkhard Corves, Stefan Kowalewski,
    Verena Nitsch, Gerhard Lakemeyer, Thomas Gries, and Christian Brecher*

Fuzzy-Set Qualitative Comparative Analysis (fsQCA) in Xiamen's
High-Quality Industrial Development 104
    Yin Zheng, Hang Jiang, and Yuyang Lin

Applications of AI in Business and Society

Effect of AI Generated Content Advertising on Consumer Engagement 121
  Duo Du, Yanling Zhang, and Jiao Ge

AI-Enabled Smart Healthcare Ecosystem Model and Its Empirical Research ... 130
  Qianrui Du, Changlin Cao, Qichen Liao, and Qiongwei Ye

Using Co-word Network Community Detection and LDA Topic Modeling
to Extract Topics in TED Talks 140
  Li-Ting Hung, Muh-Chyun Tang, and Sung-Chien Lin

Emotional Analysis through EEG on In-Store Journey: Pilot Methodology
for Evocation of Emotions Through Video Stimuli to Measure Performance
Metrics Using EEG Emotiv EPOC+ on In-Store Experiences 155
  Fernando U. Osornio García, Gilberto A. Fragoso González,
  Mayté V. Martínez Pérez, Fernando Báez Martínez,
  Mario H. Salas Barraza, and Víctor M. González

A Machine Learning Model for Predicting a Movie Sequel's Revenue
Based on the Sentiment Analysis of Consumers' Reviews 170
  Suyanee Polsri, Ya-Wen Chang Chien, and Li-Chen Cheng

Participative Process Model for the Introduction of AI-Based Knowledge
Management in Production 181
  Tobias Rusch, Nicole Ottersböck, and Sascha Stowasser

Automating Data Personas for Designing Health Interventions 192
  Gaayathri Sankar, Soussan Djamasbi, Daniel J. Amante,
  Adarsha S. Bajracharya, Qiming Shi, Yunus Dogan Telliel,
  and Torumoy Ghoshal

Change Management Process and People's Involvement when Introducing
AI Systems in Companies 202
  Sascha Stowasser, Oliver Suchy, Sebastian Terstegen,
  and Alexander Mihatsch

Option Pricing Using Machine Learning with Intraday Data of TAIEX
Option 214
  Chou-Wen Wang, Chin-Wen Wu, and Po-Lin Chen

Portfolio Performance Evaluation with Leptokurtic Asset Returns 225
  Chin-Wen Wu, Chou-Wen Wang, and Yang-Cheng Chen

Distinguishing Good from Bad:
Distributed-Collaborative-Representation-Based Data Fraud
Detection in Federated Learning 242
  Zongxiang Zhang, Chenghong Zhang, Gang Chen, Shuaiyong Xiao,
  and Lihua Huang

**Exploring Human Behavior and Communication in Business: Case
Studies and Empirical Research**

Why Do People Donate for Live Streaming? Examining the S-O-R
Process of Sponsors in the Live Streaming Context 259
  Tsai-Hsin Chu, Wei-Hsin Chu, and Yen-Hsien Lee

The Socioemotional Wealth Effect on Intra-family Succession in Family
Businesses 274
  Pi Hui Chung

An Experimental Study of the Relationship Between Static Advertisements
Effectiveness and Personality Traits: Using Big-Five, Eye-Tracking,
and Interviews 285
  Semira Maria Evangelou and Michalis Xenos

User Study on a Multi-view Environment to Identify Differences Between
Biological Taxonomies 300
  Manuel Figueroa-Montero and Lilliana Sancho-Chavarría

Analysis and Application of a Batch Arrival Queueing Model
with the Second Optional Service and Randomized Vacation Policy 320
  Kai-Bin Huang

Communicating Sustainability Online: A Soft Systems Methodology
and Corpus Linguistics Approach in the Example of Norwegian Seafood
Companies 334
  Nataliya Berbyuk Lindström and Cheryl Marie Cordeiro

Comparative Evaluation of User Interfaces for Preventing Wasteful
Spending in Cashless Payment 352
  Daisuke Mashiro, Yoko Nishihara, and Junjie Shan

Emotional Communication and Interaction with Target Groups
Exemplified by Public Educational Institutions 364
  Christina Miclau, Josef Nerb, and Bernhard Denne

A New Perspective on the Prediction of the Innovation Performance:
A Data-Driven Methodology to Identify Innovation Indicators Through
a Comparative Study of Boston's Neighborhoods 377
  Eleni Oikonomaki and Dimitris Belivanis

Interruptions in the Workplace: An Exploratory Study Among Digital
Business Professionals 400
  Fabian J. Stangl and René Riedl

Author Index 423

Contents – Part I

Advancing Technology and Management in Public Sector Organizations and Governance

Virtual Reality for Smart Government – Requirements, Opportunities,
and Challenges 3
   Matthias Baldauf, Hans-Dieter Zimmermann, Pascale Baer-Baldauf,
   and Valmir Bekiri

Measuring the Effectiveness of U.S. Government Security Awareness
Programs: A Mixed-Methods Study 14
   Jody L. Jacobs, Julie M. Haney, and Susanne M. Furman

Stakeholder-in-the-Loop Fair Decisions: A Framework to Design Decision
Support Systems in Public and Private Organizations 34
   Yuri Nakao and Takuya Yokota

Theoretical Model of Electronic Management for the Development
of Human Potential in a Local Government. Peru Case 47
   Moisés David Reyes-Perez, Jhoselit Lisset Facho-Cornejo,
   Carmen Graciela Arbulú-Pérez Vargas,
   Danicsa Karina Carrasco-Espino, and Luis Eden Rojas-Palacios

Enhancing Transparency for Benefit Payments in the Digital Age:
Perspectives from Government Officials and Citizens in Thailand 57
   Saiphit Satjawisate and Mark Perry

An Assessment of the Green Innovation, Environmental Regulation,
Energy Consumption, and CO_2 Emissions Dynamic Nexus in China 74
   Taipeng Sun, Hang Jiang, and Xijie Zhang

Introduction to Ontologies for Defense Business Analytics 87
   Bethany Taylor, Christianne Izumigawa, and Jonathan Sato

Learning by Reasoning: An Explainable Hierarchical Association
Regularized Deep Learning Method for Disease Prediction 102
   Shuaiyong Xiao, Gang Chen, Zongxiang Zhang, Chenghong Zhang,
   and Jie Lin

Government Initiative to Reduce the Failed or Unsuccessful Delivery
Orders Attempts in the Last Mile Logistics Operation 114
   Muhammad Younus, Achmad Nurmandi, Misran, and Abdul Rehman

**Mobile Commerce and e-Commerce: User Experience and Business
Perspectives**

The Impact of Country-of-Origin Images on Online Customer Reviews:
A Case Study of a Cross-Border E-Commerce Platform 141
   Wen-Hsin Chen and Yi-Cheng Ku

Key Successful Factors of E-commerce Platform Operations 150
   Yu-Jing Chiu, Hsi-His Chen, and Chin-Yi Chen

On the Role of User Interface Elements in the Hotel Booking Intention:
Analyzing a Gap in State-of-The-Art Research 170
   Stefan Eibl and Andreas Auinger

The Mediation Role of Compatible Advantage in Mobile Wallet Usage 190
   Brenda Eschenbrenner and Norman Shaw

Research on the Optimization Design of Mobile Vending Cart Service
Process 202
   Cui Hangrui

E-Commerce and Covid-19: An Analysis of Payment Transactions
and Consumer Preferences 219
   Sarah Krennhuber and Martin Stabauer

Consumers' Intentions to Use Mobile Food Applications 230
   Ralston Kwan and Norman Shaw

Influence of Artificial Intelligence Recommendation on Consumers'
Purchase Intention Under the Information Cocoon Effect 249
   Siyi Liang, Nurzat Alimu, Hanchi Si, Hong Li, and Chuanmin Mi

The Dynamic Update of Mobile Apps: A Research Design with HMM
Method 260
   Xinhui Liu, Kaiwen Bao, Lele Kang, Jianjun Sun, and Yanqing Shi

An Analysis of Survey Results on the User Interface Experiences
of E-wallet Services 271
   Kwan Panyawanich, Martin Maguire, and Patrick Pradel

Acceptance of Mobile Payment: A Cross-Cultural Examination Between
Mainland China, Taiwan, and Germany 293
    Vipin Saini, Julian Reckter, Yu-Chen Yang, and Yong Jin

Types of Mobile Retail Consumers' Shopping Behaviors
from the Perspective of Time 302
    I-Chin Wu, Hsin-Kai Yu, and Shao-I. Lien

The Study of Different Types of Menus Layout Design on the E-Commerce
Platform via Eye-Tracking 314
    Ya-Chun Yang and Tseng-Ping Chiu

Use of Disruptive Technologies to Enhance Customer Experience

Gamification in Organizational Contexts: A Systematic Literature Review 331
    Luciana S. Assis and Sergio A. A. Freitas

The Impact of Gender and Visual Presentation of Advertising on User
Experience in Mobile Shopping Apps 353
    Yan Cao, Weimin Zhai, and Weiren Zhai

What Do User Experience Professionals Discuss Online? Topic Modeling
of a User Experience Q&A Community 365
    Langtao Chen

The Study of User Experience Within Advertising in Virtual Reality 381
    Sara Dieter, Ben Mark, Matt Childress, Anthony Anderson,
    Andrea Mower, and Max Harberg

Increasing Customer Interaction of an Online Magazine for Beauty
and Fashion Articles Within a Media and Tech Company 401
    Christina Miclau, Veronika Peuker, Carolin Gailer, Adrian Panitz,
    and Andrea Müller

Achieve Your Goal Without Dying in the Attempt: Developing
an Area-Based Support for Nomadic Work 421
    Guillermo Monroy-Rodríguez, Sonia Mendoza,
    Luis Martín Sánchez-Adame, Ivan Giovanni Valdespin-Garcia,
    and Dominique Decouchant

Digital Showroom in 3DWeb, the Scene Effect on Object Placement 439
    Giorgio Olivas Martinez, Valeria Orso, and Luciano Gamberini

Exploiting 3D Web to Enhance Online Shopping: Toward an Update
of Usability Heuristics 450
Valeria Orso, Maria Luisa Campanini, Leonardo Pierobon,
Giovanni Portello, Merylin Monaro, Alice Bettelli,
and Luciano Gamberini

Booking Shore Excursions for Cruises. The Role of Virtual 360-Degree
Presentations 461
Jenny Wagner, Christopher Zerres, and Kai Israel

Analyzing Customer Experience and Willingness to Use Towards Virtual
Human Products: Real Person Generated vs. Computer Program Generated 476
Mingling Wu, Michael Xu, and Jiao Ge

When Virtual Influencers are Used as Endorsers: Will Match-Up
and Attractiveness Affect Consumer Purchase Intention? 487
Yanling Zhang, Duo Du, and Jiao Ge

Author Index 497

Exploring the Intersection of Robotics and Autonomous Agents in Business and Industry

Evaluation of Risk Factors in the Armed Forces Physical Fitness Test

Yu-Jing Chiu[✉], Chien-Cheng Li, So-Ra Song, and Yi-Chung Hu

Department of Business Administration, Chung Yuan Christian University, Chung Li District, Taoyuan, Taiwan

{yujing,ychu}@cycu.edu.tw, ossa082@naver.com

Abstract. For the past few years, some officers and soldiers have remained with the Armed Forces for training purposes and for identifying injuries and deaths caused by accidents. This study constructed a research framework through literature review and expert interviews, and then used the decision laboratory method as the basis for network analysis procedures to determine the key factors for assessing risk factors in the national military fitness assessment. Results showed that there were five main influencing criteria, which included 'potential risk factors of physical and mental conditions of officers and enlisted personnel', 'body monitoring equipment', 'medical preparation and knowledge', 'first aid equipment' and 'hazard prevention equipment'. Among them, "Identifying potential risk factors related to the physical and psychological state of officers and soldiers" can be used to improve other key factors. The research results and analyses will be useful for the establishment of a risk assessment mechanism for the effective prevention of accident risks through the national military fitness assessment in combination with new technological devices and systems.

Keywords: Physical fitness · Body Mass Index · Delphi method · DEMATEL-based Analytic Network Process(DANP) · Importance-Performance Analysis

1 Introduction

With regard to physical fitness training and the detection of accidental injuries and fatalities, this study collects statistics from national news reports from the years 2013 to 2016, as shown in Table 1.

According to US military reports, two to three people have died each year since 2010 during physical training (Caitlin 2020). This study uses news media broadcasts to collect statistics from 2017 to 2020, as shown in Table 2.

Especially recently, events such as marathon and triathlon have gradually become national sports, and runners and ordinary people can monitor and adjust their running schedule and training strategy anytime and anywhere (Xu and Zhuang 2018).The data is then transmitted to cloud computing via wired or wireless communication, and the cloud feeds the results back to personal mobile devices, allowing users to obtain necessary information (Chen 2015). To improve the overall effectiveness of the unit's physical

© Springer Nature Switzerland AG 2023

F. Fui-Hoon Nah and K. Siau (Eds.): HCII 2023, LNCS 14039, pp. 3–17, 2023.

https://doi.org/10.1007/978-3-031-36049-7_1

Table 1. Accidents caused by physical fitness assessment and training in the National Army.

Item	Reporting time	Incident	Result
1	August 7, 2013(Jane Zhengfeng, Li Guanzhi)	A sergeant surnamed Lu from the Kinmen area went to Taoyuan Army Academy for training at the end of July. After classes ended at 5 o'clock in the evening on 7 August, he went jogging with the chief sergeant. After 40 min, he was found lying on the ground, unable to get up, and was rushed to hospital, but he didn't recover	training casualties
2	October 27, 2015 (Wen Yude)	A male major in the Air Force took part in the 3,000 m barehanded running test, but fainted at the end and his heart stopped beating for a while. Fortunately, he was rescued and survived	Detect damage
3	January 7, 2016 (Cai Qinghua)	Colonel Gu, the captain of a certain unit of the Marine Corps, did a 3,000 m unarmed running test. He was sent to hospital after the test when he suddenly fainted during a relaxation exercise	Detect casualties
4	October 11, 2016 (Yang Jiyu)	An NCO surnamed He, who served in a certain naval command, was doing an unarmed 3,000-m run in the camp in the afternoon. When he had finished running and was returning to his dormitory, he suddenly fainted on the stairs. He was pronounced dead	Detect casualties

training and prevent training accidents, the US Marines have developed a physical training APP (2019) to improve the physical fitness level of members and units, as well as to keep track of the injury rate of the entire Marines and which training is more likely to lead to injury (Jiang 2020).

The aim of this study is to examine the existing process of the National Army's physical fitness inspection at the station, reduce the labour cost and inspection time consumption, and establish and improve the risk factor assessment method to effectively monitor the physiological monitoring data generated by officers and soldiers during the inspection. In addition, it effectively analyses and evaluates the accident risk during the

Table 2. US military physical fitness assessment and training accidents.

Item	Reporting time	Incident	Result
1	June 2017(Andy)	The Air Force Times reported that the director of the Air Force Service Center in Denver suddenly collapsed due to physical discomfort during a physical fitness test at Buckley Air Force Base in Colorado, was taken to hospital and died the next day	Detect casualties
2	May 13, 2019 (Meghann)	Likat Ahan, a 51-year-old officer at the U.S. Army War College in New York, reportedly collapsed while running on the campus and was taken to Keller Army Community Hospital for treatment, where he was pronounced dead	Detect casualties
3	October 17, 2019 (Stephen)	Three pilots reportedly died following fitness tests in 2019. Senior pilot Amalia Joseph died on 26 May after suffering a medical emergency while she was running on a track at an Air Force base in South Carolina. Senior pilot Aaron Hall died on 1 June in a similar emergency on another track at the air base. Captain Tranai Rashawn Tanner died on 17 August after being taken to hospital with a medical condition following a physical fitness test at Eglin Air Force Base in Florida on 16 August	Detect casualties
4	July 30, 2020 (Harm)	The Army Times reported that Army National Guard Lt. Robert collapsed after completing a fitness test in preparation for the Basic Officer Leadership Course and was taken to St. Luke's Hospital in Tempe, where he died the next morning	Detect casualties

assessment, so as to reduce the injury and accident incidents during the physical fitness assessment and training.

In recent years, there are still some soldiers who get injured or pass away during physical fitness training or test in National Physical Fitness Test Center or home station. This research focus on reviewing the current procedures of physical fitness test of the ROC Armed Forces, cutting manpower and time, and building a comprehensive way to assess risk in order to monitor the soldier's body performance and status effectively. The goal is to reduce the injuries and casualties during physical fitness test and training.

There are three purposes of our research. The first one is to establish a complete assessment mechanism for the national military to assess risk factors for physical fitness in the garrison, and update the content of the national military physical fitness risk management operation manual and defense regulations to meet the actual implementation needs of the national military. According to the physical and psychological conditions of officers and soldiers during the physical fitness assessment of the National Army, the introduction of body monitoring equipment can effectively monitor the physiological monitoring data generated by officers and soldiers during assessment and training, and reduce personal injuries and casualties. The third one is to re-formulate the national military physical fitness assessment group and assessment process to reduce manpower expenditure and time consumption of the assessment process.

This research applies the Delphi method to construct framework for risk assessment. The experts selected by this research are professionals who serve as the chief officer of the National Army, the director of the training business department, the physicians and lecturers of the National Military Hospital as the selection conditions, and form an expert group to participate in the revision of the research structure and subsequent filling in order to provide professional opinions. After the literature discussion, the prototype structure was established, and 3 evaluation aspects and 13 evaluation criteria were summarized, and the prototype structure was used to interview the expert group of this study. The final research framework was determined through the results of expert interviews, including 4 evaluation aspects and 9 criteria. And then uses the decision-making laboratory method to identify key factors and results.

2   Review and Discussion of Military Fitness Risk Factors

Literature survey on the risk factors affecting the physical fitness of the national army. Jiang Yaozheng (2011) pointed out that the factors affecting the national military's physical fitness assessment and training cause sports injuries, including (1) demographic characteristics (age, place of residence, education level, parameters such as smoking habits and exercise habits), (2) physiological indicators (height, weight, BMI, body fat percentage, waist and hip circumference and other parameters), (3) sports injury prevention and cognition.

Huang (2012) pointed out the risk factors affecting the physical fitness evaluation of the 3,000 m barefoot running, the evaluation criteria include (1) BMI values, (2) cardiac and respiratory endurance indicators, and (3) human physical condition. Hong (2016) noted that the risk factors that affect physical fitness assessment of the national army, the evaluation criteria are (1) body quality index, (2) weight management, (3) ECG, pulse, systole and diastole blood pressure physical examination index.

Xu 2017) investigated the source of risk in physical fitness training and testing, according to the negligence of personnel, the characteristics of training items, the time of training and testing, the safety of training and venues, the safety of training equipment, and the physical fitness and health status of officers and soldiers. 3 risk factors and 13 risk factors were summarised.

Ye and Su (2018) suggest that understanding fatigue recovery timing and practice in athletic training are key metrics to avoid sports injuries and master effective athletic training. Most of the risk factors come from exercise-induced fatigue, the evaluation criteria are (1) exercise intensity self-perception table, (2) observation method: 1. Self-perception, 2. Perception of others, 3. EMG.

Cai et al. (2018) noted that in the future physical fitness test, wearable devices can be integrated into the physical fitness test equipment of the military to monitor the national.The real-time state of exercise during the physical fitness assessment of military personnel can reduce the risk of sports injuries and accidental injuries, as well as preventive medical reference. Through the evaluation of 4 criteria, namely (1) mental index, (2) wearable technology, (3) cardiac stress assessment, and (4) data records, it is possible to evaluate whether it is possible to incorporate wearable devices into the military's physical fitness assessment equipment to reduce the risk of sports injuries and accidental injuries.

Zhang et al. (2019) explored the major factors causing sports injuries in army football players, and summarised three dimensions, (1) "weight training", (2) "external environment", and (3) "internal environment". The "environment" criteria include (1) insufficient training time, (2) insufficient training volume, (3) excessive training volume, (4) old training equipment, (5) inadequate equipment, (6) poor site environment, (7) affected by weather, (8) inadequate psychological construction, (9) failure to wear protective gear, (10) failure to warm up before the game, (11) lack of concentration, and other 11 criteria.

Wang (2019) pointed out that risk factors should be analysed and evaluated based on three aspects, namely (1) the "basic quality" of officers and soldiers, (2) the frequency of training, and (3) the relationship between strength and physical fitness. Zhang (2019) pointed out that in order to effectively improve physical fitness and reduce sports injuries, criteria such as (1) heart rate belt, (2) heart rate value, and (3) rating of perceived exertion (RPE) should be combined.

In the above literature analysis, only Xu (2017) literature is based on the compilation rules of the Army Risk Management Operation Manual, and proposes physical fitness risk factors to distinguish personnel, environmental and mechanical factors, which is more consistent with this study's focus. Thus, this study combined some aspects and criteria with the same or similar definitions in the literature, and ultimately constructed a prototype research framework including three aspects and 13 evaluation criteria, and defined each criteria, as shown in Table 3.

Table 3. Aspects and criteria of prototype architecture.

Aspects	Criteria	Definition of Criterion	References
Human Aspects	Training time allocation	Insufficient training time, incorrect warm-up and stretching methods, inaccurate movements, lack of targeted and specificity	Xu (2017), Zhang et al. (2019), Wang 2019)
	Physiological latent risk factors of officers and soldiers	Fatigue, staying up late, injury, illness and injury training, chronic disease, old injury, family inheritance, undetected cause	Xu (2017), Ye and Su (2018)

(continued)

Table 3. (*continued*)

Aspects	Criteria	Definition of Criterion	References
	Attention during training	Concerned, play does not listen to command commands, training or inspection of the energy lax	Xu (2017), Zhang et al. (2019)
	Physical and mental state	Substandard body mass index, new recruits, underweight or overweight, lack of exercise habits, insufficient training, improper prescriptions, exhaustion, lack of rest, lack of water, staying up late and working overtime, training hard for a sense of honor	Kong (2011), Huang (2012), Hung (2016), Yip and So (2018), Zhang et al. (2019)
	Training program and organization	The progress of physical fitness training is gradual, the foundation of officers and soldiers is different, the training method is inappropriate, and the result is twice the result with half the effort. During the training, there is no distinction between good and bad, no strength and weakness, and it is impossible to form key supplementary training	Xu (2017), Zhang et al. (2019)

(*continued*)

Table 3. (*continued*)

Aspects	Criteria	Definition of Criterion	References
	Medical readiness and awareness	There are no medical staff, no emergency communication network, delayed medical treatment, improper first aid procedures, and insufficient awareness of the concept of sports injury prevention	Jiang (2011), Xu (2017)
	Exercise intensity self-perception	It is helpful for army trainers to grasp and record the heartbeat rate and self-feeling situation of each individual in the large army, so as to serve as the most immediate reference for evaluating the training effect and fatigue	Jiang (2011), Yip and So (2018), Cheung (2019)
Envirnmental Aspects	Training environment	The runway is uneven, there is no traffic control during the running test, the runway is uneven, the route is not clearly marked, and the field is slippery	Xu (2017), Zhang et al. (2019)
	Training period	Just received the leave and implemented intense training and testing, the risk factor exceeded 40, and the weather changed at that time	Xu (2017), Zhang et al. (2019)
	Weather conditions	Thunderstorms, extreme heat, extreme cold	Xu (2017), Zhang et al. (2019)

(*continued*)

Table 3. (*continued*)

Aspects	Criteria	Definition of Criterion	References
Mechanical Aspects	Protective equipment	There is a shortage of first aid equipment, ambulance stations, and tea supply stations, the planning of the site configuration and movement lines is disordered, security alerts, control and communication are not accurate, and inspections are neglected	Xu (2017), Zhang et al. (2019)
	First aid equipment	Damaged oxygen cylinders, lack of oxygen, expired or shortage of medicines, no ambulance, ambulance breakdown, etc	Xu (2017), Zhang et al. (2019)
	Body monitoring equipment	Based on the definition of relevant criteria, it is possible to measure basic physiological parameters, including heart rate, respiration rate, acceleration, maximum oxygen consumption, temperature, and monitoring of physical activity regardless of the use of heart rate belts and related technological devices	Hung (2016), Yip and So (2018), Cai et al. (2018), Zhang (2019)

3 Research Methodology and Results

3.1 Establishment of a Formal Research Framework

This study makes use of the Delphi method, which is a long-term forecasting technique that was developed by the Rand Corporation in the United States in the 1960s. Pill (1971); Shefer and Stroumsa (1981); Rowe et al.(1991). Through anonymous written discussion among experts, the Delphi method encourages experts to use their professional knowledge, experience and opinions to reach a consensus on complex issues. Based on the prototype research framework shown in Table 3, this study uses the interview method of experts' experience, intuition and value judgement (Deng 2012), and finally confirms the formal research framework. The experts selected in this study are professionals in the roles of chief officer of the national army, director of the training business department, doctor and lecturer of the national military hospital, who form an expert group to participate in the revision of the questionnaire structure and the filling of the follow-up questionnaire, and provide professional opinions.

The first expert survey on the addition and deletion of the prototype structure was conducted in March 2021. The opinions of the experts were listed. Following this, the dimensions of the prototype structure have been modified according to the modifications proposed by the expert panel, and a second expert interview has been conducted for this prototype framework in April 2021. Finally, after two rounds of expert interview revisions, a formal research framework consisting of four dimensions and nine criteria was finally determined. The formal evaluation levels and definitions were sorted as shown in Table 4.

Table 4. Aspects and criteria of Formal Architecture

Aspects	Criteria	Definition of Criterion
Human Aspects	Training time allocation	Insufficient training time, incorrect warm-up and stretching methods, inaccurate movements, lack of targeted and specificity
	Physiological latent risk factors of officers and soldiers	Fatigue, staying up late, injury, illness and injury training, chronic disease, old injury, family inheritance, undetected cause
	Attention during training	Concerned, play does not listen to command commands, training or inspection of the energy lax
	Physical and mental state	Substandard body mass index, new recruits, underweight or overweight, lack of exercise habits, insufficient training, improper prescriptions, exhaustion, lack of rest, lack of water, staying up late and working overtime, training hard for a sense of honor

(*continued*)

Table 4. (*continued*)

Aspects	Criteria	Definition of Criterion
	Training program and organization	The progress of physical fitness training is gradual, the foundation of officers and soldiers is different, the training method is inappropriate, and the result is twice the result with half the effort. During the training, there is no distinction between good and bad, no strength and weakness, and it is impossible to form key supplementary training
	Medical readiness and awareness	There are no medical staff, no emergency communication network, delayed medical treatment, improper first aid procedures, and insufficient awareness of the concept of sports injury prevention
	Exercise intensity self-perception	It is helpful for army trainers to grasp and record the heartbeat rate and self-feeling situation of each individual in the large army, so as to serve as the most immediate reference for evaluating the training effect and fatigue
Envirnmental Aspects	Training environment	The runway is uneven, there is no traffic control during the running test, the runway is uneven, the route is not clearly marked, and the field is slippery
	Training period	Just received the leave and implemented intense training and testing, the risk factor exceeded 40, and the weather changed at that time
	Weather conditions	Thunderstorms, extreme heat, extreme cold

(*continued*)

Table 4. (*continued*)

Aspects	Criteria	Definition of Criterion
Mechanical Aspects	Protective equipment	There is a shortage of first aid equipment, ambulance stations, and tea supply stations, the planning of the site configuration and movement lines is disordered, security alerts, control and communication are not accurate, and inspections are neglected
	First aid equipment	Damaged oxygen cylinders, lack of oxygen, expired or shortage of medicines, no ambulance, ambulance breakdown, etc
	Body monitoring equipment	Based on the definition of relevant criteria, it is possible to measure basic physiological parameters, including heart rate, respiration rate, acceleration, maximum oxygen consumption, temperature, and monitoring of physical activity regardless of the use of heart rate belts and related technological devices

3.2 Determination of the Causal Relationship and the Key Factors

This study uses the DEMATEL method to clarify the causal relationship between the factors and criteria that influence the risk assessment of the National Military Fitness Assessment. First, through the distribution of questionnaires, a direct influence relationship matrix X based on the results of the questionnaires is generated, and then by normalizing the direct influence matrix and bringing it into the formula $T = X(1 - X)^{-1}$. The total influence relationship matrix T (Total Influence Matrix) established according to the criteria can be obtained as shown in Table 5.

The d value can be obtained by summing each column of the total impact marix in Table 5; the r value can be obtained by summing each row. The row and column sum (d + r) of each item is called "importance", which indicates the relevance of that item to the problem. The rank difference (d-r) of each element is called the "causal degree". If the rank difference is positive, it means that this element is biased towards "active influence" and is classified as a cause; if the rank difference is negative, this element is biased towards "influence". Impact" is classified as effect. According to the overall impact relationship matrix in Table 5, calculate the importance and correlation as shown in Table 6.

Table 5. The total influence relationship matrix

	A1	A2	B1	B2	B3	C1	C2	C3	D1	Column sum (d)
A1	0.190	0.306	0.145	0.212	0.017	0.230	0.277	0.351	0.152	1.882
A2	0.212	0.104	0.052	0.125	0.010	0.151	0.191	0.245	0.068	1.159
B1	0.205	0.159	0.041	0.117	0.010	0.135	0.134	0.235	0.048	1.083
B2	0.156	0.107	0.058	0.055	0.004	0.090	0.094	0.176	0.088	0.828
B3	0.235	0.171	0.104	0.151	0.007	0.140	0.163	0.303	0.126	1.401
C1	0.186	0.183	0.104	0.118	0.010	0.075	0.172	0.210	0.119	1.179
C2	0.187	0.176	0.103	0.108	0.009	0.142	0.078	0.121	0.050	0.974
C3	0.251	0.172	0.042	0.134	0.009	0.070	0.083	0.106	0.051	0.919
D1	0.242	0.174	0.089	0.170	0.011	0.148	0.148	0.294	0.053	1.329
Row sum (r)	1.865	1.553	0.739	1.190	0.087	1.182	1.340	2.042	0.757	

Table 6. Importance and Relevance of Corresponding Criteria

Criteria	d	r	d + r	Ranking	d-r
Hazards in the physical and psychological states of officers and soldiers (A1)	1.88	1.87	3.75	1	0.02
Medical readiness and awareness (A2)	1.16	1.55	2.71	3	−0.39
Assess the site environment (B1)	1.08	0.74	1.82	8	0.34
Detection period (B2)	0.83	1.19	2.02	7	−0.36
Weather conditions (B3)	1.40	0.09	1.49	9	1.31
Hazard prevention facilities (C1)	1.18	1.18	2.36	4	−0.00
First aid equipment (C2)	0.97	1.34	2.31	5	−0.37
Body Monitoring Equipment (C3)	0.92	2.04	2.96	2	−1.12
3,000-m running test standard (D1)	1.33	0.76	2.09	6	0.57

This study adopts the DANP operation framework proposed by Hu et al. (2015), uses DEMATEL's total influence matrix as the unweighted super matrix in ANP operation, normalises the matrix, and self-normalises the results until the fourth convergence, obtaining the limit super matrix shown in Table 7. The relative weight of each criterion can be determined by the limit super matrix.

In order to determine the final overall ranking of the criteria, the importance ranking of DEMATEL and the weight ranking of DANP are used in this study. The sum of the rankings in Table 8 is also called the Borda score. The lower the score, the more important the criterion.

Table 7. Extreme Super Matrix

	A1	A2	B1	B2	B3	C1	C2	C3	D1
A1	0.1831	0.1831	0.1831	0.1831	0.1831	0.1831	0.1831	0.1831	0.1831
A2	0.1501	0.1501	0.1501	0.1501	0.1501	0.1501	0.1501	0.1501	0.1501
B1	0.0674	0.0674	0.0674	0.0674	0.0674	0.0674	0.0674	0.0674	0.0674
B2	0.1117	0.1117	0.1117	0.1117	0.1117	0.1117	0.1117	0.1117	0.1117
B3	0.0086	0.0086	0.0086	0.0086	0.0086	0.0086	0.0086	0.0086	0.0086
C1	0.1097	0.1097	0.1097	0.1097	0.1097	0.1097	0.1097	0.1097	0.1097
C2	0.1232	0.1232	0.1232	0.1232	0.1232	0.1232	0.1232	0.1232	0.1232
C3	0.1768	0.1768	0.1768	0.1768	0.1768	0.1768	0.1768	0.1768	0.1768
D1	0.0694	0.0694	0.0694	0.0694	0.0694	0.0694	0.0694	0.0694	0.0694

Table 8. Ranking of criterion weight

Criteria	DEMATEL Importance ranking	DANP sort by weight	Sort and Borda Score	Overall sort
Hazards in the physical and psychological states of officers and soldiers (A1)	1	1	2	★ 1
Medical readiness and awareness (A2)	3	3	6	★ 3
Assess the site environment (B1)	8	8	16	8
Detection period (B2)	7	5	12	6
Weather conditions (B3)	9	9	18	9
Hazard prevention facilities (C1)	4	6	10	★ 5
First aid equipment (C2)	5	4	9	★ 4
Body Monitoring Equipment (C3)	2	2	4	★ 2
3,000-m running test standard (D1)	6	7	13	7

From the results in Table 8, it can be seen that, according to the discussions with the experts, the five most important criteria for assessing the risk factors affecting the physical fitness of the national military are "Identification of potential dangers in the physical and mental state of officers and soldiers (A1)", "Physical monitoring equipment (C3)", "Medical readiness and cognition (A2)", "First aid equipment (C2)" and "Hazard prevention facilities (C1)". And causality diagram between key criteria shows in Fig. 1.

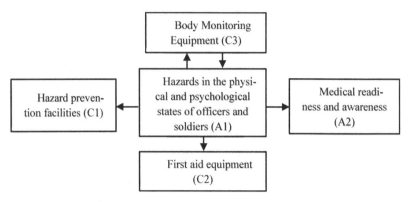

Fig. 1. Causality Diagram Between Key Criteria

4 Conclusions

The results of the research show that the key factors include: "the physiological status of soldiers", "health monitoring device", "medical care preparation and knowledge", "first aid equipment" and risk prevention facility. In addition, the results also indicate that "health monitoring device", "medical care preparation and knowledge", "first aid equipment" and "risk prevention facility" that are key factors for reducing the risk and need to be improved. The cause-effect diagram shows that "identification of potential hazards of soldier's bodies and physiological status" can be used as a source to improve other key factors. Even ROC Armed Forces applies many methods and procedures for monitoring soldiers' mental and physiological status when they take physical fitness test in order to reduce risks some means are still insufficient. The overall results are positive. However, it still needs some enhancement during assessment and evaluation. The current physical fitness test is evaluated by human at home station and no electrical device and health monitoring equipment are used during test for measure solder's basic physiological status. Therefore, many soldiers get injured or die during the test. The results of this research can help the ROC Armed Force combine advanced technologies and devices for physical fitness test and applies the risk assessment to reduce casualties, cut down extra manpower and testing time.

References

Cai, P., et al.: A wearable device that monitors the state of the heart during running training. Med. Inf. Maga. **27**(2_, 25–34 (2018)

Chen, Z.: Wearable technology. Natl. Inst. Exp. Sci. Dev. **512**, 20–22 (2015)

Cheng, S.: New thinking of big data in the era of sports industry 4.0. Sports Manag. (33), 19–44 (2016)

Deng, Z.: Multi-criteria decision analysis-methods and applications. Dingmao Publishing House, Taipei (2012)

Hong, J.: A study on factors affecting the physical fitness of volunteers in the national army. Dissertation for the master's class in the Department of Science and Technology Management, Chung Hsin University, Hsinchu, Taiwan (2016)

Hu, Y.C., Chiu, Y.J., Hsu, C.S., Chang, Y.Y.: Identifying key factors for introducing GPS-based fleet management systems to the logistics industry. Math. Probl. Eng. (2015). Article ID 413203

Huang, K.: A study on the relationship between body mass index and 3000-meter barefoot running performance in the national army. National Defense University School of Management, Department of Operations Management, Master Class, Taipei, Taiwan (2012)

Jiang, Y.: A study on the effectiveness of physical training and injury-related factors for enlisted men in the Navy. Master's Thesis, Department of Sports, Health and Recreation, National Kaohsiung University, Kaohsiung, Taiwan (2011)

Jiang, Y.: The new physical fitness system of the US Marine Corps is on the road, and the training is more diverse. Youth Daily (2020)

Martilla, J.A., James, J.C.: Importance-performance analysis. J. Mark. **41**(1), 77–79 (1977)

Ou Yang, Y.P., Shieh, H.M., Leu, J.D., Tzeng, G.H.: A novel hybrid MCDM model combined with DEMATEL and ANP with applications. Int. J. Oper. Res. **5**(3), 160–168 (2008)

Pill, J.: The Delphi method: Substance, context, a critique and an annotated bibliography. Socio-Econ. Plan. Sci. **5**, 57–71 (1971)

Rowe, G., Wright, G., Bolger, F.: Delphi: a reevaluation of research and theory. Technol. Forecast. Soc. Chang. **39**, 235–251 (1991)

Shefer, D., Stroumas, J.: The Delphi method: A planning. Socio-Economic Planning Science **15**(5), 236–276 (1981)

Wang, J.: A study on the physical fitness and related factors of officers and soldiers of a national military unit. Master's Thesis, Master of Business Administration, Department of Business Administration, Gaoyuan University of Science and Technology, Kaohsiung, Taiwan (2019)

Xu, G., Zhuang, M.: Runners should know the key data of running. City State Cultural Enterprise Co., Ltd., Taipei City (2018)

Xu, Q.: From the perspective of risk management, discuss the research on the safety and prevention of national military fitness training. Army Acad. Monthly **53**(556) (2017)

Ye, M., Su, J.: Exploration on exercise-induced fatigue of military physical fitness and combat skills. J. Cult. Sports **26**, 1–16 (2018)

Zhang, G., Lin, G., Lin, C., Lin, Y.: Assessment of sports injuries in Army rugby players: application of the Modified Deffie Method and Fuzzy AHP. Taiwan J. Sports Psychol. **19**(1), 41–57 (2019)

Zhang, K.: Feasibility study of heart rate monitoring applied to running training. College of Sports and Health Sciences, National Sports University, New Taipei, Taiwan (2019)

Chatbots in Academic Advising: Evaluating the Acceptance and Effects of Chatbots in German Student-University Communication

Annebeth Demaeght[(✉)], Natalie Walz, and Andrea Müller

Hochschule Offenburg, Badstrasse 24, 77652 Offenburg, Germany
annebeth.demaeght@hs-offenburg.de

Abstract. In order to attract new students, German universities must provide quick and easy access to relevant information. A chatbot can help increase the efficiency in academic advising for prospective students. In this study we evaluate the acceptance and effects of chatbots in German student-university communication. We conducted a qualitative UX-Study with the chatbot prototype of Offenburg University of Applied Sciences (HSO), in order to determine which features are particularly relevant and which requirements are made by the users. The results show that acceptance increases if the chatbot offers quick and adequate assistance, furthermore, our participants preferred an informal communication style and valued friendly and helpful personality traits for chatbots.

Keywords: Chatbot · User Experience · Empirical Studies · Academic Advising

1 Introduction

Digitization is advancing in leaps and bounds in almost all sectors, so it is important for universities to stay up to date. In order to attract new students, a number of measures must be taken, be it offering an attractive range of study programs or highlighting the advantages of the university location. However, this alone is usually not enough. Prospective students must be given the possibility to access relevant information about studying quickly and easily.

Today's generation is used to receiving relevant information from anywhere and at any time. Universities must keep up to date in this regard in order to secure the attention of their target group. Most universities have a homepage, but searching for information is usually time-consuming and yet sometimes questions remain unanswered. Thus, it makes sense to develop new alternatives to stand out from other institutions and at the same time advance digitalization in university communication. One solution is to introduce a chatbot for student advising in order to provide prospective students with the information they are looking for easily, quickly and at any time, and to advise them on their choice of study. A chatbot offers a number of opportunities and can help universities increase their efficiency and competitive position.

© Springer Nature Switzerland AG 2023
F. Fui-Hoon Nah and K. Siau (Eds.): HCII 2023, LNCS 14039, pp. 18–29, 2023.
https://doi.org/10.1007/978-3-031-36049-7_2

This paper examines which concrete communication measures should be taken in order to achieve the highest possible acceptance for the use of chatbots in academic advising. Furthermore, recommendations for concrete design options will be given so that a chatbot can be successfully introduced for student advising. Since it is of great importance that universities include all groups of people and provide them with unrestricted access to information, this study also identifies which elemets are necessary for barrier-free communication and interaction.

2 Related Work

2.1 Chatbots

Definition
An apt definition of chatbots is provided by Khan, who states that "a chatbot is a computer program that processes natural-language input from a user and generates smart and relative responses that are then sent back to the user. [1]" This is made possible by the Conversational Interface, which enables interaction between humans and the system in natural language [2]. A chatbot is therefore a text-based dialog system that can communicate autonomously with humans [3]. The terms "chatterbot" and "bot" or "conversational agent" can also be used synonymously. However, nowadays many chatbots are not only capable of communicating in writing, there are bots that have a speech output and can communicate with the other person using spoken language. This type of chatbot is also called a voicebot. Amazon Alexa and Apple Siri are the most prominent examples.

Historical Development
Joseph Weizenbaum's ELIZA was the first chatbot being able to communicate with humans in natural language [4]. ELIZA acted as a psychotherapist and interacted with humans using natural language to do so. This worked so well that many people were convinced that they were communicating with a human [5].

ELIZA's successor was the chatbot Parry, which was developed by Kenneth Colby in 1972 [6]. It was also developed for the medical environment and its task was to imitate a patient with schizophrenia in order to find out more about the clinical conditions [7].

After these developments by Weizenbaum and Colby, the development of chatbots became somewhat slack. The major challenges turned out to be limited computer power, high costs, and lack of functionality. As a result, there were only few new developments in this area until the 1980s [5].

Chatbots and instant messaging experienced a resurgence with the introduction of the Internet in the late 1990s [3]. Another milestone was the chatbot A.L.I.C.E (Artificial Linguistic Internet Computer Entity) developed by Richard Wallace in 1995. A.L.I.C.E has the distinction of being the first chatbot to use Natural Language Processing (NLP), the processing of natural language through machine learning [6]. It has won the Loebner Prize three times, an award given annually to a program capable of withstanding the Turing Test [8]. The Turing Test was developed by Alan Turing to evaluate the capabilities of a machine by comparing it with human interlocutors and thus to measure its credibility.

[9]. The Turing test is passed if the user cannot distinguish the human from the machine and does not know whether he is talking to a human or a machine.

Due to the rapid development of artificial intelligence, a genuine chatbot boom has taken place in recent years [6]. Examples are the Amazon Lex Framework and IBM Watson.

2.2 University Communication

Communicative Measures to Attract Students

One of the measures that universities can use to attract new students is communication policy. It is a part of the marketing mix and is used to directly address the potential target group. A precise analysis of the target group is therefore essential in order to conduct communication policy, because without knowledge of the preferences of the target group, communication policy measures can only deliver inadequate results.

The main instruments of the communication policy of universities consist of advertising, public relations and personal communication [10]. Personal communication includes, in addition to the area of student advising, reports and experiences of (former) students of the university [10]. Public relations is often the only perceived marketing activity of universities and is usually limited to the presentation of the university's services [10]. Due to changes in media use and information behavior of the target group, universities increasingly have to deal with digitization and actively engage in online marketing. Suitable measures include, for example, having their own internet presence in the form of a website or a presence in social networks [11].

Relevance of a Chatbot for Academic Advising

As explained earlier, today's society - especially the younger generation - expects the ability to access information at any time and from any place. In order to meet the "accessibility requirement" of this target group, universities must take a number of measures and encourage further development of digitization [12]. Although the provision of a dedicated homepage has become standard for universities, it cannot serve as the only means of providing information.

Due to the abundance of information, which is also directed at the university's various stakeholder groups, it can be assumed that prospective students will only find the information they are looking for - if at all - after a long search on the homepage. This leads to the fact that the personal contact to the university personnel and/or the study consultation is looked for or the search is given up.

Since the questions or information needs of first-year students and prospective students are regularly repeated at the beginning of the semester, it makes sense to develop a simple solution to relieve the student office so that they can take care of the more complex issues. A chatbot could be a suitable communication measure to provide information and advice to students and prospective students at any time and independent of opening and office hours. The chatbot can be seen as a supplement to the homepage and to personal counseling. The majority of simple questions and concerns could be handled by the chatbot, and if the case is too complex for it, it can be forwarded to the appropriate contact person or telephone contact can be established. There could also arise further advantages through its use, as this form of dialog is accessible from anywhere in the world and can thus also be used by international students, which could increase visibility on the national and international education market [13].

3   Study

3.1   Research Goal

The aim of our study was to find out what effects the use of a chatbot has on communication between students and the university and what potential for acceptance can be identified.

For universities, it is of great importance to offer suitable options for first-year students to communicate and get in touch with the university.

In order to achieve the objectives and to answer the research questions described above, a qualitative study was conducted. A chatbot prototype of the University of Applied Sciences Offenburg was tested in a qualitative UX study and its potentials were examined. Within this study, we wanted to evaluate to what extent acceptance for the use of chatbots in student-university communication exists and how this can be improved. The aim of the study is furthermore to formulate recommendations regarding design principles for accessibility, so that the chatbot can be used barrier-free by all groups of people. Furthermore, recommendations for the design of the personality of the chatbot and for the improvement of the user experience will be given.

3.2   Population and Setup

The study was conducted with nine participants, seven females and two males between 20 and 34 years old. Due to Covid 19-contact restrictions, eight of the nine testings were conducted online using the software Zoom. One testing took place in the Customer Experience Tracking Lab of Offenburg University.

3.3   Procedure

Each testing started with introducing the participant to the research project and its purpose. Then the participants were informed about the data protection policy of the study and their consent for using the data gathered in the testing was obtained.

In the next step, some pre-test questions related to the participants' previous experience with chatbots were asked, e.g.: Are you familiar with the term chatbot? If not, what do you think it means? Have you ever used a chatbot before? If yes, how was your experience so far? If you have used a chatbot before, what did you find particularly good? What did you find not so good?

After the pre-test interview, participants were asked to open the chatbot prototype of Offenburg University and to find the answers to some questions using the chatbot. The tasks were typical questions that prospective and first-year students might ask themselves. Five different problems were posed, which the test persons should solve using the chatbot. We analyzed the interaction of the participants with the chatbot to find out to what extent the users get along with the system. In the first task, the participants had to find out which courses of study in the field of computer science are offered at the university. This could be a typical question that prospective students might ask themselves. The next task concerned the application process: The participants were asked to find out the submission deadline and also how many applications can be submitted at the same time. After that, the next task was to find out when the bridge courses take place. The fourth problem was concerned with university campuses. The users were asked to find out how many locations the university has and how best to get to the campus by public transport. The last problem aimed at the element of surprise. The participants were instructed to let the chatbot tell a joke and to say goodbye afterwards. Here, a small "Easter Egg" in the form of an animal picture was included to surprise the users when they said goodbye (Fig. 1). This task was used to check the reaction of the test persons to emotional content. It was found that there was a positive reaction on the part of the test persons.

Fig. 1. Chatbot goodbye

After they had performed the tasks, participants were asked several post-test questions on their experience with the chatbot, its personality traits and the accessibility of the chatbot.

3.4 Results

Familiarity with Chatbots and Previous User Experiences
At first, we wanted to find out whether the respondents were familiar with the term "chatbot" and what experience they already had with it. The survey revealed that seven out of the nine respondents were familiar with chatbots and had already used them. Only two respondents stated that they had not yet had any contact with such a program.

As far as previous experience with a chatbot is concerned, the respondents are rather divided in their opinion. This is illustrated in Fig. 2.

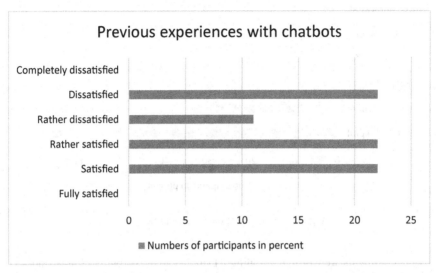

Fig. 2. Previous experiences with chatbots

The most common problems that respondents see in its use were unanimously named. These are the chatbot's lack of understanding of the specific problem, as well as the output of only generalized answers without any real reference to the question. One respondent states: " [It] Takes a long time if you don't use the targeted search terms to the chatbot until it knows exactly what [you] mean." In addition, the test subjects criticized the problem of being referred to the same topic several times if a chatbot does not understand the question. The test subjects who have not yet come into contact with a chatbot were asked in the interview to describe where they might see problems in the use of chatbots. Here, they listed that a chatbot might be "user-unfriendly", "confusing" and "not clear".

The participants largely agree on the positive features. Here, the advantage of the direct response without waiting time was mentioned most often. In addition, many participants stated that a lengthy search on the homepage can be bypassed using a chatbot. Also, a human behavior of the chatbot is described as positive, as is a direct forwarding to customer service or a forwarding by means of a link for further information.

Experience and Evaluation of the HSO Chatbot Prototype After Use
After the usability testing with the chatbot prototype, the participants were asked how helpful they perceived the chatbot to be. A clear picture emerges here. Six respondents stated that the chatbot was "very helpful" and the remaining three respondents found it "helpful" (see Fig. 3). This could be due to the fact that all participants were able to solve the tasks given to them, which led to satisfaction and acceptance.

If we look at the positive features named by the users, it becomes clear that most of them were positively impressed by the fast response time and the ease of use. The

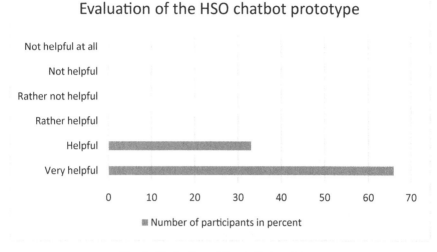

Fig. 3. Evaluation of the chatbot prototype

following comments are worth mentioning: "Detailed answers (but not long-winded) in comprehensible language with possibilities to click further to get more information" or "The chatbot responds quickly and had the right answer directly to almost all questions". Worth mentioning is that most of the participants remember the farewell picture in the form of baby animals: "Very sympathetic and purposeful. Very wide range of information. Shortcut links to obtain more information. Quick response time of answers. Ending the conversation with a picture which makes you smile."

Another aspect that was perceived as quite positive and time-saving is the use of selection options in the form of drop-down or selection fields. The users found this to be extremely practical, as they were able to click through quickly without having to type in specific questions. This also ensures that the chatbot understands the question and can provide the appropriate answer. An example of the selection options is shown in Fig. 4.

The features that could still be expanded and improved were also named very unanimously by the test subjects. One of the central shortcomings is that the chatbot sometimes outputs very long text. This was perceived as very negative by seven test persons. Figure 5 illustrates this problem. The question asked is: "Until when can I apply?" In response to this simple and short question, the participants receive a very long text with three paragraphs, an excerpt is shown in Fig. 5.

The unanimous feedback was that the users felt overwhelmed by so much text and information and were not willing to read through the entire text. Most would have liked to see the most important key points highlighted in bold or to find a link for further information. It was criticized that very precise and simple questions such as "Until when can I apply?" were not answered briefly and precisely, but instead a lot of information was provided which is not relevant at all or only to a limited extent. The answer to the question about the end of the application is presented in the second section of the text, but can quickly be drowned out by so much information. During the survey, most of

Fig. 4. An example of selection options in the chatbot prototype

Fig. 5. An example of a too long answer of the chatbot prototype

the participants took quite a long time to find this information, or to grasp it, and they formulated the question differently without reading through the entire text. Furthermore, it was asserted that by the long text, one noticed much faster, that one was communicating with a computer and not with a human being.

Communication and Areas of Use
The respondents were asked in which situations they would consult a chatbot. Six respondents immediately mentioned the search for information. Specifically, this is information

that can be found in the FAQ or cannot be found even after a long search on the home-page. Chatbots are thus understood as an information system that can clearly present the information on the website. The terms "advice", "complaints" and "support" are also mentioned. However, it is clear from the survey that a chatbot is mostly only considered as a last solution and that other communication channels are preferred beforehand.

This was addressed with the question: "Which communication channels do you usually use when you want advice?" In this question, the subjects were able to select several applicable answer options. It should be emphasized that all test persons first obtain information by searching on the homepage or by using search engines in order to gather information before a consultation. This makes sense insofar as one can then ask more specific questions during a consultation and thus already has an approximate idea. Furthermore, consulting via e-mail is very popular (eight out of nine respondents prefer this type of communication), closely followed by consulting by telephone (seven respon-dents). Consulting via a chatbot can currently only be imagined by 1/3 of the respondents. The reasons given for this were that a chatbot is often not yet available. The last place is occupied by personal consultation (two respondents). This is presumably due to the availability of other communication channels, as digital or telephone communication involves significantly less effort than face-to-face consultation. It should be noted here, however, that some respondents pointed out that the reason for the consultation plays an important role. The more important the advice is to the individual the more effort is accepted and the more likely it is that face-to-face advice is preferred.

An application of the chatbot for study counseling is seen by the participants as suitable as long as they have a rough idea of what they would like to study. Furthermore, the chatbot is considered suitable for obtaining general information about studying. Especially since many questions are often repeated at the beginning of a new semester. This type of information could easily be taken over by a chatbot, so that the student advisory office is relieved and they can take care of those cases, that are complex or more protracted. If prospective students do not yet know at all, what kind of study program suits them, the chatbot will probably only be able to contribute to the solution to a limited extent. The reason for this is that the chatbot cannot fully grasp what prior experience and interests the person has, but only decides which field of study might be suitable based on the possible areas of interest. This is too little information for comprehensive study advice. However, if prospective students already have a tendency as to which direction they might want to study, the clustering of areas of interest is quite useful.

When asked which type of communication (verbal or written) with the chatbot was is preferred, all test subjects were of the opinion that they prefer written communication. Thus, it can be deduced that a chatbot without speech output would be used more, or the input and output of speech would not be used very frequently. However, for reasons of accessibility, it is recommended to implement a speech input and output that can be used if necessary to enable all people to access the use of the chatbot.

Personality Traits and Behavior of the Chatbot
This section now deals with the personality traits and further on with questions about the user experience and the behavior of the chatbot. We asked our participants, whether they think it is important that the chatbot has a personality or not. This was by a total of

55% of the respondents answered with "rather agree" or "completely agree". Therefore, a chatbot should not serve exclusively for giving information, something more is needed to inspire the users. Above all, the test subjects noted that the chatbot should be adapted to the individual target group and that attention should be paid to what purpose it fulfills.

Since the target group consists mainly of young people who have been familiar with digital media from an early age and who are used to communicating with messenger services, it is advantageous if the chatbot adapts its language style to the communication of young people and can issue short and concise responses, as is common when communicating with messengers. The participants' comments to the statement "I prefer it when the chatbot addresses me in a formal way." made it clear that they tend to dislike formal communication, but prefer a short and clear statement.

In the further course of the survey, the respondents were asked to select which personality traits they consider particularly important in a chatbot. Here again, the result is clear. All respondents stated that the chatbot should have the characteristics "Helpful" and "Friendly". In second place, the characteristic "Authentic" was named. This includes, above all, that it should be communicated on the part of the chatbot that it is not able to understand every type of question correctly and that the chatbot is still in development. There should be no forcible attempt to humanize the chatbot, as this could have the opposite effect and the chatbot will be perceived as inauthentic. It can be clearly communicated that one is corresponding with a machine. However, care should be taken that, for example, long texts are sent in multiple messages to improve readability and give the other person more of a feeling that messages are written individually. Another attribute that was queried with regard to the personality traits is "omniscient". There is a great deal of disagreement about this among the test subjects. Four of the respondents stated this as an important attribute, the other five people were of the opinion that this was not possible and should therefore not be strived for.

Due to the fact that authenticity is very important to all subjects, it should be refrained from presenting the chatbot as omniscient or trying to implement this. This would possibly have a negative effect on authenticity, which appears to be more important than the "omniscient" feature. Another feature that was available for selection, but was not chosen by any of the test subjects, is the feature "Familiar". Accordingly, this is not a decisive personality trait for the test subjects. In addition, other characteristics were addressed that were not available for selection, namely "Speed" and "Functionality". Here it becomes clear once again that most users want to obtain information quickly without having to spend a long time using it. This confirms the question of whether it is important to the participants to also talk about other topics with the chatbot or to engage in small talk with the chatbot. Here, just under half (44%) state that it is "not at all important" for them to make small talk, and one person considers it "not important". Two respondents each rated this question as "neutral" or "rather important". None of the respondents considered it imperative to talk to chatbots beyond their topic areas. It is clear that it is much more important to the participants to have a chatbot with a personality than to talk to it about the weather or the last soccer match. The importance of emojis or smileys in texts, on the other hand, is different. Seven out of nine respondents find the use of emojis "important" or "very important." Emojis make the text and the statement livelier and give the chatbot its own personality. Thus, emojis can be used to express significantly more emotions than the written word.

4 Conclusion

In this paper we analyzed which aspects are relevant for a chatbot in student-university communication and which hurdles still have to be overcome.

Acceptance. From the user's point of view, a quick solution to a problem is the most important aspect to gain acceptance. If the chatbot does not answer the question adequately, acceptance for this medium can dwindle very quickly, so it is necessary to ensure that the most important information is included and also that the chatbot takes over the conversation by providing drop-down fields or selection options so that a satisfactory answer can be issued.

The study showed that most participants have communicated with chatbots before, which indicates a basic acceptance. Positive previous experiences can also increase the basic acceptance of such media, which is why it should be ensured at an early stage of chatbot implementation that users always have a positive experience.

Communication. The results of the study have produced clear recommendations for action regarding the communication of the chatbot. On the one hand, the test subjects prefer personal communication which is not too formal, but more like chatting with a friend. This is also supported by the fact that the majority of the test subjects state that the use of emojis in the text is perceived as positive and important. It should be ensured that the chatbot linguistically resembles a fellow student or friend rather than a computer with predefined texts. This requires a high degree of creativity on the part of the developers. It is also clearly described that short texts and highlighting of the most important aspects are preferred by the majority of respondents.

Small talk, on the other hand, is not considered important by most and is one of the things that does not necessarily have to be present. This is mainly because users are mostly just looking for information and do not want to spend too much time talking to the chatbot. It is much more important that the quality of the answers provided is high. Only when this is ensured can additional functions such as small talk or information about the weather, for example, be considered.

Personality. Giving the chatbot personal traits can bring great advantages. For example, a somewhat clumsy chatbot personality can be used to charmingly guide over the fact that it does not understand the question, for example. However, the personality is also the face to the outside world and must therefore be representative of the university. Authenticity is very important to the majority of the participants, so this character trait should be particularly emphasized. Since the personality traits "friendly" and "helpful" were mentioned by all test persons, these traits should also be considered. The chatbot should be able to react confidently and always in a friendly manner to any input, but should also be able to give quick-witted answers if necessary.

Accessibility. Accessibility is particularly important to enable everyone to participate equally in digital media. The criteria for a barrier-free usage should be considered and reviewed before the chatbot is fully implemented. The study also showed that there are still some deficits in the execution of accessibility. For one, it should be ensured that long text sections are avoided. Better is a presentation in several messages or a link for further information, if still more information is desired. Also, the texts should be presented in simple and understandable language to comply with the principles of accessibility. Secondly, care should be taken to ensure that the chat window is clearly visible and can be enlarged if necessary. This was criticized by many participants in the study and must be observed in the interests of accessibility. It should also be ensured that the chatbot has a voice output via which the written text can be read out. This is where the principle of perception comes into play in order to include people with visual impairments.

References

1. Khan, R., Das, A.: Build Better Chatbots: A Complete Guide to Getting Started With Chatbots. Apress, New York (2018)
2. Diers, T.: Akzeptanz von Chatbots im Consumer-Marketing: Erfolgsfaktoren zwischen Konsumenten und künstlicher Intelligenz. Springer Fachmedien, Wiesbaden (2020)
3. Kruse Brandão, T., Wolfram, G.: Digital Connection. Springer Fachmedien, Wiesbaden (2018)
4. Storp, M.: Chatbots. Möglichkeiten und Grenzen der maschinellen Verarbeitung natürlicher Sprache (2002). https://www.mediensprache.net/networx/networx-25.pdf. Accessed 03 Jan 2023
5. Cornelius, A.: Künstliche Intelligenz: Entwicklungen, Erfolgsfaktoren und Einsatzmöglichkeiten. Haufe, Freiburg (2019)
6. Kohne, A., Kleinmanns, P., Rolf, C., Beck, M.: Chatbots: Aufbau und Anwendungsmöglichkeiten von autonomen Sprachassistenten. Springer Fachmedien, Wiesbaden (2020)
7. Stucki, T., D'Onofrio, S., Portmann, E.: Chatbots gestalten mit Praxisbeispielen der Schweizerischen Post. Springer Fachmedien, Wiesbaden (2020)
8. Wallace, R.S.: The Anatomy of A.L.I.C.E.. In: Epstein, R., Roberts, G., Beber, G. (eds.) Parsing the Turing Test, pp 181–210. Springer, Dordrecht (2009)
9. Spierling, U.; Luderschmidt, J.: Chatbots und mediengestützte Konversation. In: Kochhan, C., Moutchnik, A. (eds.): Media Management, pp. 387–408. Springer Gabler, Wiesbaden (2018)
10. Bliemel, F., Fassott, G.: Marketing für Universitäten. In: Tscheulin, D.K., Helmig, B. (eds.) Branchenspezifisches Marketing: Grundlagen - Besonderheiten - Gemeinsamkeiten, pp. 265–286. Springer Fachmedien, Wiesbaden (2001)
11. Herdin, G., Künzel, U.: Online-Marketing im Student Recruitment - Teil 2: Instrumente und Planung (2011). https://www.wissenschaftsmanagement-online.de/sites/www.wissenschaftsmanagement-online.de/files/migrated_wimoarticle/LeseprobeOnline-MarketingimStudentRecruitment.pdf, last accessed 2023/01/03
12. Carstensen, N., Roedenbeck, M.: Chatbots in der Studienorientierung: Ein Projekt zur nachhaltigen Implementierung von digitalen Dialogsystemen in der Hochschule. In: Barton, T., Müller, C., Seel, C. (eds.) Hochschulen in Zeiten der Digitalisierung, pp. 27–40. Springer Vieweg, Wiesbaden (2019)
13. Hochschulforum Digitalisierung: 20 Theses on the Digitalisation of Higher Education. https://hochschulforumdigitalisierung.de/en/20-theses-digitalisation-higher-education. Accessed 03 Jan 2023

Study on the Impact of Service Robot Autonomy on Customer Satisfaction

Keli Li and Guoxin Li[✉]

Harbin Institute of Technology, Harbin 150001, China
21b910020@stu.hit.edu.cn, liguoxin@hit.edu.cn

Abstract. The rapid development of artificial intelligence technology has accelerated the promotion and application of service robot in the market. Although technology has provided service robot with increasingly autonomous functions, more research is needed on how service robot with different levels of autonomy affects customer satisfaction in service scenarios. Guided by the theory of affordance, this study examined whether service robot operational and decisional autonomy would have effects on customer satisfaction and explored explanatory mechanism. Adopting an experimental vignette method (EVM), the study reveals that direct effect of service robot operational autonomy and indirect effect of decisional autonomy on customer satisfaction, and functional affordance played a positive mediating role in the impact of service robot autonomy on customer satisfaction. The results extend and enrich the relevant literature on human-machine interaction and customer satisfaction research. Our results also provide marketing insights for enterprises to improve autonomous robot design and enhance customer relationships.

Keywords: Service Robot · Autonomy · Affordance Perception · Functional Affordance · Customer Satisfaction

1 Introduction

Nowadays, technological advancements make service robot become an effective way to reduce labor costs, enhance service enjoyment, and improve customer experience (Singh et al. 2017; van Doorn et al. 2017). Service robot are being increasingly introduced into multiple industries, such as hotels, catering, and tourism (Gursoy and Chi 2020; Kim et al. 2021). For example, the service robot Pepper, used at the Wynn Palace hotel in Las Vegas, can recognize gender, age, and emotions, and even take selfies with customers (Escobar 2017; Tuomi et al. 2021). The global market value of service robot is expected to grow from US \$295.5 million in 2020 to US \$3.083 billion in 2030, with a compound annual growth rate of approximately 25.5% (Allied Analytics 2021).

Although service robots are favored by businesses, there is still significant resistance to change, which may be the fundamental barrier to adoption (Chi et al. 2020). Surveys show that most customers still prefer to interact with human employees rather than AI-based robots (Leah 2022), as customers have reported issues that service robots are lack of flexibility and accuracy in understanding customers' emotional expressions. Customers resistance to service robot has a significant impact and challenge for businesses that are deploying them. Therefore, investigating the factors and potential mechanisms that influence customer satisfaction with service robot and attempting to improve customer satisfaction has become one of the most pressing issues in the industry.

Autonomy is an important benchmark for human-computer interaction (Kahn et al. 2007). Through advanced machine learning algorithms, robots can adapt their behavior to specific goals and execution environments with little (partial autonomy) or no (full autonomy) human intervention, significantly improving service efficiency (ISO 2012). However, highly autonomous robots may also have a negative impact on customer satisfaction, as the autonomy of the robot may cause concerns about privacy and security, controllability, and ethical responsibility (Landau et al. 2015; Calverley 2006). Therefore, there is currently no consensus among scholars on how the autonomy of service robot affects customer reactions, and further research is needed on how to better apply the autonomy of service robot. The relationship between the autonomy of service robot and customer reactions has become an important research topic (Złotowski et al. 2017; Mende et al. 2019; Kim et al. 2022).

Therefore, the purpose of this study is to investigate the impact of service robot autonomy on customer satisfaction, particularly from the perspective of the affordance theory, by answering the following two research questions: 1) How does service robot operational autonomy and decisional autonomy affect customer satisfaction? 2) What are the mediating mechanisms of the relationship between service robot autonomy on customer satisfaction? In this study, we examine that the service robot operational autonomy is a key factor to positively and directly influence customer satisfaction, test the mediating role of functional affordance in the process, and examine the moderating role of service robot decisional autonomy in relationship between service robot operational autonomy and customer satisfaction. This study provides a new perspective for exploring the underlying mechanism that affect customer satisfaction with service robot.

2 Literature Review and Hypotheses

2.1 Operational Autonomy

Autonomy refers to the degree to which people experience considerable freedom, discretion, and independence in their work (Kirkman et al. 2004). In the field of service robot, autonomy represents the ability of a robot to act without human control and interference and to perform some or all of the tasks initially performed by humans (Parasuraman et al. 2000; Rau et al. 2013). A robot with high operational autonomy is fully autonomous at the task level (Giralt 1995). Therefore, we consider that the degree of service robot operational autonomy reflects its ability to perform tasks independently without human intervention. Service robot with low operational autonomy can only adapt their behavior to a specific goal and execution environment and perform limited tasks under human control and supervision. Service robot with high operational autonomy can communicate with consumers, receive order information, and directly deliver drinks (perform all tasks) without human control and supervision.

Advances in technology have increased the robots autonomy, but there is currently no academic consensus on the pros and cons. For companies, service robot with high autonomy can optimize decision making, change internal operations and production processes (Xu and Xu 2020). For employees, service robot with high autonomy are able to perform complex tasks and cope with various environments with less human intervention, and are therefore considered smarter and more useful partners (Yang et al. 2017). For customers, service robot with high autonomy can significantly improve service efficiency (Tuomi et al. 2021). However, the level of intelligence and sophistication of autonomous robots is likely to be perceived as a threat to their technological power and control (Złotowski et al. 2017). And there is an increased risk of malfunctioning and regulatory issues (Yang et al. 2017). Research on robot autonomy needs to be enriched. Thus, we propose the hypothesis:

H1: Service robot operational autonomy will have a positive effect on customer satisfaction.

2.2   Functional Affordance

This study uses the affordance theory to understand the impact of service robot operational autonomy on customer satisfaction. Originating from ecological psychology, affordance refers to the action possibilities that the environment offers to an animal (Gibson 1979), it reflects the operable attributes that any object in the real world offers to users under certain conditions (Gibson 1986). Later, the affordance theory was introduced to the field of human-computer interaction (Norman 1988). According to their role in supporting users in interactions, affordances are divided into four types: cognitive affordances, physical affordances, sensory affordances, and functional affordances (Hartson 2003). Functional affordance is the relationship between a technical object and the user under specific conditions, which can be described as the action possibilities that a technical object offers to the user (Markus and Silver 2008; Grgecic et al. 2015). Functional affordance, as a function-centric concept, emphasizes the specific functions of the technology (Grgecic et al. 2015). In order to achieve functional affordance, it must be perceived and cognized by people (Leonardi 2011). In the task-oriented human-computer interaction environment, user behavior is goal-oriented and purposeful, and due to different elements such as users and interaction design, users perception of service robot functional affordance is also different (Koroleva and Kane 2017). Therefore, we believe that the design attribute of service robot (i.e., operational autonomy) will affect functional affordance and, in turn, affect customer satisfaction. We propose the hypothesis as followed:

H2: Functional affordance mediates the effect of service robot operational autonomy on customer satisfaction.

2.3   Decisional Autonomy

In addition to operational tasks, AI is gradually replacing humans in an increasing number of scenarios to make decisions. The degree of decisional autonomy reflects the extent to which AI decision systems based on big data and machine learning and deep learning algorithms operate in an independent and goal-oriented manner, free from user inter-ference (Baber 1996; Rijsdijk et al. 2007; Rijsdijk and Hultink 2009), and can exhibit proactive and spontaneous behavior (Benlian et al. 2020). The most basic and core aspect of decisional autonomy is the decisional autonomy relationship between the decision-maker and the person in decision-making, as reflected in the proportional relationship between them, or rather, the relationship between the two as a trade-off (Xiao 2022). When service robot have an absolute advantage in decision-making, their decisional autonomy will be at the highest level, and they will replace humans in decision-making (Leung et al. 2018). According to Self-Determination Theory (SDT), the decisional autonomy at different levels of the decision subject have different effects on the psycho-logical and behavioral activities of customers (Fan and Liu 2022). The fact that these decisions are made by AI rather than people may influence the perception and adop-tion of AI, regardless of the quality of the actual decision outcome (Sundar and Nass 2001), and these perceptions can influence people's attitudes and willingness to adopt AI (Castelo et al. 2019; Wu et al. 2020). Thus service robot decisional autonomy may be an important influence on functional affordance. We propose the hypothesis as followed. Figure 1 summarizes the theoretical model utilized in this study.

H3: Service robot decisional autonomy moderates the effect of operational autonomy on functional affordance. Specifically, when service robot decisional autonomy is high, it attenuates the positive effect of operational autonomy on functional affordance.

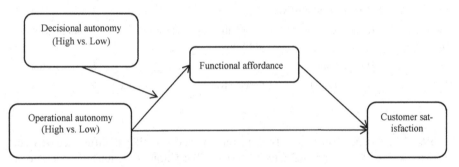

Fig. 1. Research model.

3 Method

3.1 Pretest

This study employed a one-factor (service robot operational autonomy: high vs. low) between-subjects design. 97 participants recruited through Credamo platform (30 males, $M_{age} = 29.8$, $SD_{age} = 7.33$).

In a hotel environment, robots can be used to deliver various tasks at different stages of customer service, including pre-arrival, arrival, check-in, and departure (Luo et al. 2021; Ivanov et al. 2020). In this study, the hotel service robot mainly provided four services, including room reservation, information provision, check-in, and room service, and the specific services provided varied depending on the level of operational autonomy, as shown in Table 1.

Procedure. We showed participants various pictures of service robot and informed about the latest advancements in robot technology, including robots performing tasks such as greeting guests and delivering meals. Participants were asked about their interaction experience with hotel service robot and their interest in technological development. Then, participants were randomly assigned to two groups of different scenarios: (a) high operational autonomy, where service robot can autonomously perform all tasks, or (b) low operational autonomy, where only some tasks can be executed under the supervision and control of human staff. Participants were then asked to report on service robot operational autonomy described above, we used a 7-point Likert scale (Cronbach's α = 0.791; 1 = "strongly disagree", 7 = "strongly agree"), adapted from previous research (Kim and Hinds 2006; Złotowski et al. 2017) that includes three items: "I think this service robot is independent in performing services," "I think this service robot can make judgments and decisions," and "I think this service robot can accept or refuse commands from humans.". The last part of the study collected demographic information such as participants' gender, age, education level, and field of study.

Table 1. Level of operational autonomy.

	Clear Divisions of Work			
Operational Autonomy	Room Reservation	Information Provision	Check-in	Room Service
Low	Human Staff	Human Staff	Human Staff	Service Robot
High	Service Robot	Service Robot	Service Robot	Service Robot

Results. The independent sample t-test results showed a significant difference between the two groups. Participants perceived service robot autonomy with high operational autonomy to be significantly higher than those with low operational autonomy (M_{high} = 5.12, SD_{high} = 0.83; M_{low} = 3.29, SD_{low} = 1.00; t(96) = 8.35, p < 0.001). Additionally, participants perceived the autonomy of the service robot with low operational autonomy to be below the median score of 4, while they perceived the autonomy of the service robot with high operational autonomy to be above the median score of 4, indicating the successful manipulation of service robot operational autonomy.

3.2 Study 1

The study had two objectives: first, to examine the impact of service robot operational autonomy (high vs. low) on customer satisfaction (H1); second, to investigate the mediating effect of functional affordance on the relationship between service robot operational autonomy and customer satisfaction (H2). A total of 119 participants were recruited online using the Credamo platform, with 1 participant excluded from analysis due to failure to pass screening criteria. The final sample consisted of 118 participants (47 males, $M_{age} = 31.18$, $SD_{age} = 7.04$). The study employed a one-factor (operational autonomy: high vs. low) between-subjects experimental design.

Measures

Functional Affordance. This construct was measured using previously validated scales (Grgecic et al. 2015; Lei et al. 2019), and were measured using a 7-point Likert scale (Cronbach's $\alpha = 0.801$; 1 = "strongly disagree", 7 = "strongly agree"). The measurement was used to assess participants' perception of service robot functional affordance, included four items: "I feel that this service robot is fully functional", "When interacting with this service robot, I can find products or services that meet my needs", "I feel that using this service robot can be very useful for my dining experience".

Customer Satisfaction. This construct was measured using previously validated scales (Seo 2022; Oliver 2010), and were measured using a 7-point Likert scale (Cronbach's α = 0.912; 1 = "strongly disagree", 7 = "strongly agree"). The measurement was used to assess participants' satisfaction with the service robot, included five items: "I am pleased with this robot interaction experience", "I was very satisfied with this robot experience", "I was happy with this robot experience", "Interacting with service robot did a good job for me", "Choosing interacting with this service robot was a wise choice".

Procedure. We showed participants various pictures of service robot and informed about the latest advancements in robot technology, including robots performing tasks such as greeting guests and delivering meals. Participants were asked about their interaction experience with hotel service robot and their interest in technological development. Then, participants were randomly assigned to two groups of different scenarios. Participants then reported on service robot operational autonomy, customer satisfaction, and functional affordance. The last part of the study collected demographic information such as gender, age, education level, and field of study.

Results

Manipulation Checks. The independent sample t-test results showed that participants perceived significantly higher autonomy in the service robot with high operational autonomy compared to the service robot with low operational autonomy ($M_{high} = 5.09$, SD_{high} = 0.90; $M_{low} = 2.82$, $SD_{low} = 1.37$; $t(117) = 10.56$, $p < 0.001$), indicating a significant manipulation of service robot operational autonomy.

Customer Satisfaction. An ANOVA showed that participants reported higher satisfaction with the service robot under the high operational autonomy condition than under the low operational autonomy condition, indicating that participants were more satisfied with the service robot with high operational autonomy than low operational autonomy ($M_{high} = 5.72$, $SD_{high} = 0.75$; $M_{low} = 5.27$, $SD_{low} = 1.29$; $F(1,116) = 21.16$, $p < 0.001$). H1 was hence supported.

Mediation by Functional Affordance. Firstly, an ANOVA showed that participants perceived higher functional affordance in high service robot operational autonomy compared to low operational autonomy ($M_{high} = 5.75$, $SD_{high} = 0.54$; $M_{low} = 4.53$, SD_{low}

$= 1.53$; $F(1,116) = 84.95$, $p < 0.001$). Secondly, following Hayes' (2015, Model 4) mediation analysis model, 5000 bootstrap samples were computed to test the 95% confidence interval of the mediation effect of functional affordance. The results show that controlling for the mediating variable of functional affordance, the direct effect of operational autonomy on customer satisfaction is not significant ($\beta = 0.102$, SE $= 0.072$, 95% CI $= [-0.04, 0.24]$). Functional affordance fully mediates the effect of operational autonomy on customer satisfaction ($\beta = 0.159$, SE $= 0.070$, 95% CI $= [0.027, 0.304]$). H2 was thus supported.

3.3   Study 2

Study 2 aimed to verify the moderating role of service robot decisional autonomy (H3). 156 participants were recruited through Credamo platform online, and 2 participants were excluded from the analysis due to failing the system screening, leaving a total of 154 samples. Among these participants, more than half were aged between 18–30 (57.8%), more than half were female (66.2%), two-thirds had a bachelor's degree (67.5%), 74.7% of the participants had experienced robot service in real life, and 83.8% of the participants were very interested in technology and development.

Procedure. This study featured a 2 (operational autonomy: high vs. low) × 2 (decisional autonomy: high vs. low) between-subjects experimental design. First, we showed participants various pictures of service robot and informed about the latest advancements in robot technology, including robots performing tasks such as greeting guests and delivering meals. Participants were asked about their interaction experience with service robot and their interest in technological development. Then, participants were randomly assigned to four groups of different scenarios. Then participants reported on the autonomy of the service robot, and customer satisfaction and functional affordance. The last part of the study collected demographic information such as participants' gender, age, education level, and field of study.

Results

Manipulation Check. The independent sample t-test results showed that participants perceived significantly higher autonomy in service robot with high operational autonomy than those with low operational autonomy ($M_{high} = 4.68$, $SD_{high} = 0.91$; $M_{low} = 3.34$, $SD_{low} = 1.25$; $t(153) = 7.50$, $p < 0.01$), indicating a significant manipulation of service robot operational autonomy.

Customer Satisfaction. An ANOVA revealed a significant interaction between operational autonomy and decisional autonomy on customer satisfaction ($F(1,152) = 8.236$, $p = 0.005$). Simple effect analysis showed that for the low decisional autonomy group, there was a significant difference in the effect of operational autonomy on customer satisfaction ($F(1,152) = 16.918$, $p < 0.001$), with customers in the high operational autonomy group ($M = 5.490$, $SD = 0.159$) reporting significantly higher satisfaction than those in the low operational autonomy group ($M = 4.565$, $SD = 0.159$). However, for the high decisional autonomy group, the effect of operational autonomy on customer satisfaction was not significant ($p > 0.2$). The main effect of operational autonomy was significant (F(1,152) = 7.999, p < 0.01), while the main effect of decisional autonomy was not significant (F(1,152) = 1.368, p > 0.2). H1 and H2 were thus supported.

Moderated Mediation. We performed a bootstrapping procedure with Model 7 and computed standard errors and 95% confidence intervals (CIs). We evaluated our structural model by running a bootstrap analysis with 5,000 resamples. The results showed that the moderated mediation model was significant (95% CI: [−1.417, −0.335], excluding zero). For the low decisional autonomy group, the indirect effect of functional affordance was significant (95% CI: [0.761, 1.696], excluding zero), indicating that under low decisional autonomy, functional affordance partially mediated the effect of operational autonomy on customer satisfaction. For the high decisional autonomy group, the indirect effect of functional affordance was significant (95% CI: [0.021, 0.734], excluding zero), indicating that under high decisional autonomy, functional affordance partially mediated the effect of operational autonomy on customer satisfaction. The direct effect of operational autonomy was significant (95% CI: [−0.568, −0.090], including zero). H3 was hence supported.

4 Discussion and Implications

4.1 General Discussion

This study investigates the impact of service robot operational autonomy and decisional autonomy on customer satisfaction through functional affordance. The study produces several important findings. First, service robot operational autonomy has a significant impact on customer satisfaction, suggesting that higher operational autonomy leads to higher customer satisfaction. This finding supports previous research on the impact of service robot' autonomy on customer satisfaction. Second, the study examines the mediating role of functional affordance in the above impact model. The results suggest that functional affordance fully mediates the relationship between operational autonomy and customer satisfaction, meaning that operational autonomy can promote service robot functional affordance, thereby influencing satisfaction. Third, the study examines the moderating effect of decisional autonomy in this process. It is found that when service robot decisional autonomy is low, higher operational autonomy leads to higher satisfaction, and the highest customer satisfaction is achieved when decisional autonomy is low and operational autonomy is high. When service robot decisional autonomy is high, there is no significant difference in the impact of operational autonomy on customer satisfaction.

4.2 Theoretical and Managerial Implications

First, we have revealed that service robot operational autonomy is a key factor that influences customer satisfaction. Customers appear to be more satisfied with higher service robot operational autonomy compared to those with lower autonomy. Our proposed extension of autonomy to include service robot operational autonomy broadens the research perspective on service robot autonomy.

Second, our research extends the study of affordance theory in theory. Previous research on affordance mainly focused on social commerce, social media, and product interaction design (Leonardi 2011; Treem and Leonardi 2013; Maier and Fadel 2009). We have used this theory as a mechanism for understanding how service robot affect customer satisfaction. In the traditional understanding of most people, service robots still exist as human assistants. When people perceive service robot functional affordance, they see and understand the meaning, value, and purpose of service robot, leading to higher customer satisfaction.

Last, we have identified the boundary conditions under which the effect of service robot operational autonomy on customer satisfaction is produced. Our research results indicate that people still prefer service robot with low decisional autonomy, emphasizing the importance of decisional autonomy in influencing people's perception.

Our research findings provide insights for managers that are considering the use of service robot. While adapting to the trend of digital management, companies should make decisions appropriately using artificial intelligence to serve customers. Otherwise, high decisional autonomy may backfire, leading to negative customer perception. At the same time, companies should realize that service robot autonomy can influence customer satisfaction through functional affordance, and therefore should find ways to improve service robot functional affordance to enhance customer satisfaction. At the same time, robotics research and design companies should also be aware of this, and further develop service robot operational autonomy, so that service robot can become smarter, more flexible, efficient, with more functions. Next, service robot should still be set to an appropriate level of decisional autonomy at this stage, only then can products of robot research and design companies be favored by users.

4.3   Limitations and Future Research

This study has several limitations. First, the study used pictures and scenario descriptions as stimuli instead of real service robot. The experiment effect might be better if subjects were in the real service robot service scene. Future research can conduct field experiments based on real customer service scenarios with service robot. Second, individual differences among customers may also affect the functional affordance of different service robot autonomy. Future research can incorporate customer characteristics such as their technological acceptance and readiness, cultural differences, etc., to examine their potential moderating effect on the relationship between robot autonomy and perceived affordance. Last, in addition to the interaction between customers and service provider (service robot) in hotel and catering service scenarios, external environmental factors such as spatial distance can also affect customer perception of affordance. Future research can investigate specific influencing factors.

References

Allied Analytics.: Hospitality robots market by type and end user sales channel: global opportunity analysis and industry. Forecast 2021–2030 (2021)

Baber, C.: Humans, servants and agents: human factors of intelligent domestic products. In: IEE Colloquium on Artificial Intelligence in Consumer and Domestic Products (IEE), vol. 4, pp. 1–3 (1996)

Benlian, A., Klumpe, J., Hinz, O.: Mitigating the intrusive effects of smart home assistants by using anthropomorphic design features: a multimethod investigation. Inf. Syst. J. **30**, 1010–1042 (2020)

Calverley, D.J.: Android science and animal rights, does an analogy exist. Connect. Sci. **18**(4), 403–417 (2006)

Castelo, N., Bos, M.W., Lehmann, D.R.: Task-dependent algorithm aversion. J. Mark. Res. **56**(5), 809–825 (2019)

Chi, O., Denton, G., Gursoy, D.: Artificially intelligent device use in service delivery: a systematic review, synthesis, and research agenda. J. Hosp. Market. Manag. **29**(7), 757–786 (2020)

van Doorn, J., et al.: Domo arigato mr. roboto: emergence of automated social presence in organizational frontlines and customers' service experiences. J. Serv. Res. **20**(1), 43–58 (2017)

Escobar, M.C.: Hotels put robots to work. Hospitality Technology (2017). Accessed 16 Jan 2023. https://hospitalitytech.com/november-hotel-tech-tren-d-hotels-put-robots-work

Fan, Y., Liu, X.: Exploring the role of AI algorithmic agents: the impact of algorithmic decision autonomy on consumer purchase decisions. Front. Psychol. **13**, 1009173 (2022)

Gibson, J.J.: The Ecological Approach to Visual Perception. Lawrence Erlbaum Associates, Hillsdale (1986)

Gibson, J.J.: The theory of affordances. In: The ecological approach to visual perception, Lawrence Erlbaum Associates, Inc., Hillsdale (1979)

Giralt, G.: Mobile robots: decisional and operational autonomy. In: IFAC Intelligent Components for Autonomous and Semi-Autonomous Vehicles, Toulouse, France, October (1995)

Grgecic, D., Holten, R., Rosenkranz, C.: The impact of functional affordances and symbolic expressions on the formation of beliefs. J. Assoc. Inf. Syst. **16**(7), 580–607 (2015)

Gursoy, D., Chi, C.G.: Effects of COVID-19 pandemic on hospitality industry: review of the current situations and a research agenda. J. Hosp. Market. Manag. **29**, 527–529 (2020)

Hartson, R.: Cognitive, physical, sensory, and functional affordances in interaction design. Behav. Inf. Technol. **22**(5), 315–338 (2003)

Ivanov, S., Seyitoğlu, F., Markova, M.: Hotel managers' perceptions towards the use of robots: a mixed-methods approach. Inf. Technol. Tour. **22**(4), 505–535 (2020). https://doi.org/10.1007/s40558-020-00187-x

Kahn, P.H., Jr., et al.: What is a human? toward psychological benchmarks in the field of human–robot interaction. Interact. Stud. **8**(3), 363–390 (2007)

Kim, H., So, K.K.F.: Two decades of customer experience research in hospitality and tourism: a bibliometric analysis and thematic content analysis. Int. J. Hosp. Manag. **100**, 103082 (2022)

Kim, S.S., Kim, J., Badu-Baiden, F., Giroux, M., Choi, Y.: Preference for robot service or human service in hotels? impacts of the COVID-19 pandemic. Int. J. Hosp. Manag. **93**, 102795 (2021)

Kim, T., Hinds, P.: Who should i blame? effects of autonomy and transparency on attributions in human-robot interaction. In: ROMAN 2006-The 15th IEEE International Symposium on Robot and Human Interactive Communication, Hatfield, UK, pp. 80–85 (2006)

Kirkman, B.L., Rosen, B., Tesluk, P.E., Gibson, C.B.: The impact of team empowerment on virtual team performance: the moderating role of face-to-face interaction. Acad. Manag. J. **47**(2), 175–192 (2004)

Koroleva, K., Kane, G.C.: Relational affordances of information processing on Facebook. Inf. Manag. **54**(5), 560–572 (2017)

Landau, M.J., Kay, A.C., Whitson, J.A.: Compensatory control and the appeal of a structured world. Psychol. Bull. **141**(3), 694–722 (2015)

Leah.: What do your customers actually think about chatbots? (2022). https://www.userlike.com/en/blog/consumer-chatbot-perceptions

Lei, S.I., Wang, D., Law, R.: Perceived technology affordance and value of hotel mobile apps: a comparison of hoteliers and customers. J. Hosp. Tour. Manag. **39**, 201–211 (2019)

Leonardi, P.M.: When flexible routines meet flexible technologies: affordance, constraint, and the imbrication of human and material agencies. MIS Q. **35**(1), 147–167 (2011)

Leung, E., Paolacci, G., Puntoni, S.: Man versus machine: resisting automation in identity-based consumer behavior. J. Mark. Res. **55**, 818–831 (2018)

Luo, J.M., Vu, H.Q., Li, G., Law, R.: Understanding service attributes of robot hotels: a sentiment analysis of customer online reviews. Int. J. Hosp. Manag. **98**, 103032 (2021)

Maier, J.R.A., Fadel, G.M.: Affordance based design: a relational theory for design. Res. Eng. Design **20**, 13–27 (2009)

Markus, M.L., Silver, M.S.: A foundation for the study of IT effects: a new look at DeSanctis and Poole's concepts of structural features and spirit. J. Assoc. Inf. Syst. **9**(10), 609–632 (2008)

Mende, M., Scott, M.L., van Doorn, J., Grewal, D., Shanks, I.: Service robots rising: how humanoid robots influence service experiences and elicit compensatory consumer responses. J. Mark. Res. **56**(4), 535–556 (2019)

Norman, D.A.: The Psychology of Everyday Things. Basic Books, New York (1988)

Parasuraman, A.: Technology readiness index (TRI) a multiple-item scale to measure readiness to embrace new technologies. J. Serv. Res. **2**(4), 307–320 (2000)

Rau, P.P., Li, Y., & Liu, J.: Effects of a Social Robot's Autonomy and Group Orientation on Human Decision-Making. *Advances in Human-Computer Interaction,* 263721 (2013)

Rijsdijk, S.A., Hultink, E.J.: How today's consumers perceive tomorrow's smart products. J. Prod. Innov. Manag. **26**(1), 24–42 (2009)

Rijsdijk, S.A., Hultink, E.J., Diamantopoulos, A.: Product intelligence: its conceptualization, measurement and impact on consumer satisfaction. J. Acad. Mark. Sci. **35**, 340–356 (2007)

Seo, S.: When female (male) robot is talking to me: effect of service robots' gender and anthropomorphism on customer satisfaction. Int. J. Hosp. Manag. **102**, 103166 (2022)

Singh, A., Juneja, D., Malhotra, M.: A novel agent based autonomous and service composition framework for cost optimization of resource provisioning in cloud computing. J. King Saud Univ. Comput. Inf. Sci. **29**(1), 19–28 (2017)

Sundar, S., Nass, C.: Conceptualizing sources in online news. J. Commun. **51**(1), 52–72 (2001)

Treem, J.W., Leonardi, P.M.: Social media use in organizations: exploring the affordances of visibility, editability, persistence, and association. Ann. Int. Commun. Assoc. **36**(1), 143–189 (2013)

Tuomi, A., Tussyadiah, I.P., Stienmetz, J.: Applications and implications of service robots in hospitality. Cornell Hosp. Q. **62**(2), 232–247 (2021)

Wu, J.F., Yu, H.Y., Zhu, Y.M., Zhang, X.Y.: Impact of artificial intelligence recommendation on consumers' willingness to adopt. J. Manag. Sci. **33**(5), 29–43 (2020)

Xiao, H.J.: Algorithmic responsibility: theoretical justification, panoramic portrait and governance paradigm. J. Manag. World **38**(4), 200–226 (2022)

Xu, P., Xu, X.Y.: Change logic and analysis framework of enterprise management in the era of artificial intelligence. J. Manag. World **1**, 122–129 (2020)

Yang, G. Z., et al.: Medical robotics-Regulatory, ethical, and legal considerations for increasing levels of autonomy. Sci. Rob. **2**(4), eaam8638 (2017)

Złotowski, J., Yogeeswaran, K., Bartneck, C.: Can we control it? autonomous robots threaten human identity, uniqueness, safety, and resources. Int. J. Hum. Comput. Stud. **100**, 48–54 (2017)

Interactive Robot-Aided Diagnosis System for Children with Autism Spectrum Disorder

Szu-Yin Lin[1](✉), Yi-Pei Lai[1], Hao-Chun Chiang[1], Yawei Cheng[2], and Shih-Yi Chien[3]

[1] Department of Computer Science and Information Engineering, National Ilan University, Yilan, Taiwan
szuyin@niu.edu.tw
[2] Institute of Neuroscience and Brain Research Center, National Yang Ming Chiao Tung University, Hsinchu, Taiwan
[3] Department of Management Information System, National Chengchi University, Taipei, Taiwan

Abstract. Autism spectrum disorder (ASD) is a group of complex neurodevelopmental disorders characterized by difficulties with social communication and interaction as well as restrictive interest and stereotyped behavior. Despite the behavioral symptoms of ASD often appear early in infancy, the ASD diagnosis is often cumbersome even for expert clinicians owing to characteristic heterogeneity in the symptoms and severity. Early diagnosis and intervention can help children with ASD to achieve more improvement, particularly in their social communication. Here, the study designs an interactive robotic agent and an intelligent image analysis system to assist in the ASD diagnosis of children. The children's facial expression images and body pose movement images are collected during the human-robot interaction, which three computational models are used for further data analysis. The stored database is presented as a reference for diagnosis in a visual interface. Furthermore, we incorporate multiple AI models in facial emotion recognition and eye tracking detection to automatically analyze images and visualize data, assisting clinicians in diagnostic decision making.

Keywords: autism spectrum disorder (ASD) · human-robot interaction · robot-aided diagnosis · computer vision

1 Introduction

Autism Spectrum Disorder (ASD), based on the Diagnostic and Statistical Manual of Mental Disorders (DSM-5) [1], is a complex group of neurodevelopmental disorders characterized by social communication difficulties, restrictive interests, and repetitive behaviors. In terms of social communication difficulties, patients may show (1) a lack of eye contact, (2) non-verbal communication difficulties, and (3) difficulties in developing, maintaining, and understanding social relationships [1–3]. In terms of repetitive behaviors and restrictive interests, there may be (1) repeated use of specific objects or language, as well as persistent repetition of the same actions, (2) adherence to fixed

F. Fui-Hoon Nah and K. Siau (Eds.): HCII 2023, LNCS 14039, pp. 41–52, 2023.
https://doi.org/10.1007/978-3-031-36049-7_4

rules, fixed speaking patterns, or expressing oneself through specific physical gestures, (3) highly restricted and fixed interests, and (4) high or low sensory responses or strong interests in environmental stimuli [1–3]. Early intervention as possible can help children to make up for inherent defects and fully unleash potential, and also aid in improving social difficulties in ASD. However, clinicians usually take time to diagnose ASD in children with the need of intensive clinical training, catching subtle facial emotions, eye contacts, and gestures during clinical practice. The care-givers or parents are required to fill out a series of psychological assessments. Only after the physicians conduct a comprehensive evaluation can the diagnosis of ASD be confirmed [2].

A simple definition of Human-Robot Interaction (HRI) is the dynamic relationship between humans and intelligent robots [5]. Many studies are dedicated to using robots in treating autism, where robots serve as the interacting partner with the patient to achieve therapeutic outcomes. The use of robots in [6] to treat communication obstacles in patients resulted in significant improvement in communication compared to other patients. The symptoms displayed by individuals with autism are often complex, and many patients have difficulties in communication and interaction. This increases the difficulty in diagnosis, as physicians need to spend a significant amount of time building relationships and emotions with the children to better initiate interaction and observe the child's behavior for diagnostic purposes. As in [7–8], research teams proposed using robots in four test games with children. The children's success in completing the tasks was marked by both the robots and medical professionals. The final markings were compared to each other, and the model of the robot coding the participants' behavior has adjusted accordingly, allowing the robot to automate the interaction tasks and observe the participants' behavior. The information was then provided to physicians as a reference through coding. In the study [9], a parrot-like robot was placed in a room, and the interaction between the robot and the child was recorded for about 190 s. Algorithms were used to calculate the child's position relative to the robot, and features were extracted and classified using a Gaussian support vector machine to determine if the subject was an autistic child. Other studies have also used small tabletop robots for interaction with children. The research used the children's facial features and movement features as the data set and trained the model, allowing the model to recognize children at risk of autism [10].

In the era of rapid development of hardware and artificial intelligence, this study references the groundbreaking work of Sylvia and Ricky Emanuel, pediatricians and psychotherapists at the University of Edinburgh, who used the mobile turtle-shaped robot LOGO for autism therapy in 1976 [4]. Many research labs have also begun to study the possibility of using robots as a therapy for children with autism since the late 1990s. Based on the above research, the goal of this study is to integrate the diagnostic process into the robot and use the robot as a tool to assist physicians in diagnosing children with autism through interactions with the children. The diagnostic process also involves collecting observation information from professional doctors and primary caregivers through a mobile app and collecting related image data through the robot's camera and external cameras. The collected data is analyzed and stored, and the result of the data analysis is visualized for the doctor as a reference for diagnosing and monitoring the condition. The following lists the main research objectives of this study.

1. Implement an automated human-machine interactive system for evaluating and monitoring autistic children and addressing the challenge of multiple evaluations and monitoring required for autism diagnosis.
2. Gather interaction and image data during human-robot interaction and establish a database that can aid physicians in diagnosis and image analysis.
3. Develop an AI-powered image analysis method for detecting body pose movements, facial expressions, and eye gaze during human-robot interaction to gather information for assessing autism in children and provide it to physicians for diagnosis.

2 Interactive Robot-Aided Diagnosis System

2.1 Research Framework

The framework of the research is shown in Fig. 1 which is divided into four parts: autism diagnosis process and robot interface, observer record, artificial intelligence (AI) extraction of autism children's assessment information, and assisting physicians in diagnosis. In the part of the autism diagnosis process and robot interface, data collection is the main focus. By using Zenbo and its built-in camera, we can interact with children through the robot and enable the camera to record image information in specific processes. The collected image information will be uploaded to a database for storage. The edge AI device NVIDIA Jetson Nano is utilized for detecting the body and specific actions, and the recorded frequency of these actions is uploaded to the database at the end of the process. With regard to human-computer interaction, there will also be a questionnaire for the main caregivers or nursing staff present to fill out, and the results will be stored in the database. In the part of Artificial Intelligence (AI) extraction of autism children's assessment information, facial expression recognition and eye gaze detection are performed through AI models. The data source is the image information collected in the previous part. The emotion analysis model will classify the characters' expressions in the image, calculate the number of occurrences of various emotions in the subjects during each process, and store the number in the database. The eye gaze analysis model classifies the direction of the eyes' gaze and similarly calculates the number of times the direction of the eyes' gaze is seen in each process. The final value is stored in the database. After the analysis results are completed, data visualization techniques will be used to convert the information such as emotions, eye gazes, poses, and questionnaire results from text and numbers into charts and provide them to physicians as reference materials for diagnostic evaluation.

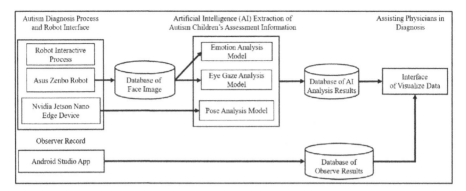

Fig. 1. The research framework of the interactive robot-aided diagnosis system

2.2 Autism Diagnosis Process and Robot Interface

The process of autism diagnosis and human-robot interaction through the robot interface is shown in Fig. 2. This research refers to Module 2 of the Autism Diagnostic Observation Schedule-2nd edition (ADOS-2) in the process. The ADOS-2 (Autism Diagnostic Observation Schedule-2nd edition, ADOS-2) [11] is a clinical assessment tool for physicians, consisting of five modules designed for different age groups. Module 2 is used for interaction and testing with patients with the lowest level of language and communication skills, consisting of 14 activities such as responding to name calls, trust games, telling stories, sharing attention, blowing bubbles, etc. The design of the interactive content is mainly used to observe the patient's communication and social interaction, as well as the presence of any restricted and repetitive behaviors. The checklist also provides key observations in each item of key observation, such as unusual eye contact, facial expressions, shared attention, unusual interests, and rigid behavior, which satisfy clinical physicians' needs to observe Autism during the interaction process [11, 12]. The study designed four human-robot interaction processes: calling names, telling stories, singing and dancing, and playing imitation games. These four processes correspond to the items observed in module 2 of ADOS-2, such as children's facial expression information, eye gaze information, and larger body pose movements during the process. The human-robot interaction process is realized through the support of the Asus Zenbo Robot. The programs needed for the robot process were completed through Android Studio and Zenbo SDK, with the assistance of Jetson Nano for simultaneous intelligent image analysis model operation.

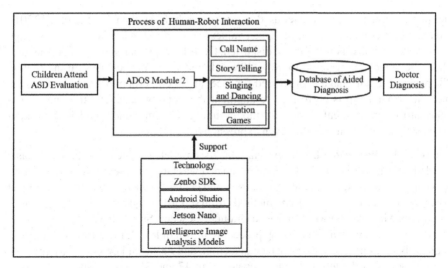

Fig. 2. The robot interface and diagnostic process for autism

2.3 Using AI to Obtain Assessment Information for Children with Autism

During the evaluation process, the system will record the participant's facial images and body pose movements. Emotions and eye gaze focus will be analyzed based on the facial image information. This experiment aims to conduct experiments with different methods and evaluate the methods that are suitable for the system.

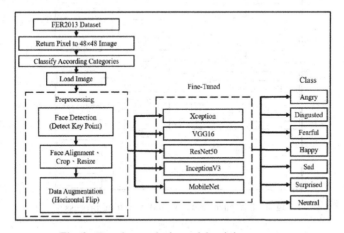

Fig. 3. Emotion analysis model training process

Emotion Analysis Model. The research uses the FER2013 dataset [13, 14] provided by the Kaggle competition website as training and validation data. FER2013 is a well-known face expression image dataset that is widely known and easy to obtain. Figure 3 shows the training process of the model used in this research. First, the CSV file of FER2013 is read, which contains the images' pixel values and expression labels. The labels range from 0 to 6 and correspond to angry, disgusted, fearful, happy, sad, surprised, and neutral. Then, the values are restored to 48 × 48 black-and-white images through the Python PIL package. Each image is processed for face alignment, cropping, and enlargement, and a horizontal flip is performed to augment the training data. The training data is input into seven different models for model tuning. The fine-tuned models can output the classification results for the seven expressions.

Eye Gaze Analysis Model. The eye gaze analysis model used the Eye-Chimera dataset [15, 16], which consisted of 1,135 image data and was labeled in seven directions: right-up, right, right-down, left-up, left, left-down, and center. In the eye gaze analysis model experiment, three machine learning methods from scikit-learn [17] were compared: Support Vector Machine (SVM), Stochastic Gradient Descent (SGD), and Nearest Centroid. The eye gaze analysis model training process is shown in Fig. 4. First, after the image is read, Dlib's Landmark is used for face detection, and 68 facial key points are marked. The key points of the left and right eyes are 42 to 47 and 36 to 41, respectively. After capturing the human eye key points, the part of the eye is cropped. The image is normalized and standardized to make all images the same size and compress the pixel values between 0 and 1. Finally, the SVM is input to train the model to classify the data into seven categories.

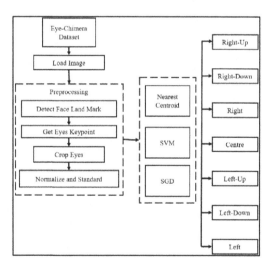

Fig. 4. Eye gaze analysis model training process

Pose Analysis Model. The process of the pose analysis model in this study is shown in Fig. 5. The pose analysis model used was implemented by Openpose [18, 19]. The training samples for the joint key points marked by Openpose were mainly from the COCO database, which defined a total of 18 joint parts. After the human body key points are found, the image is judged to have the situation of raising hands, nodding, and rough movements through the calculation of joint angles and the displacement of facial feature points. The process on the left is the hand-raising detection process. The middle is the detection of nodding and shaking, with the nose as the reference point. When the nose's horizontal or vertical coordinate displacement difference exceeds the set threshold, it is determined that there is nodding or shaking behavior. The process on the right is the detection of rough movements, which uses the elbow, knee, and ankle joints on both sides as reference points. When the joint angle change exceeds the set threshold, it is determined that rough movements have occurred.

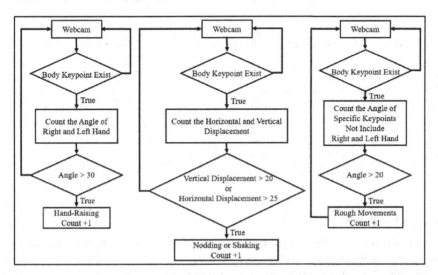

Fig. 5. Three detection processes in the pose analysis model

3 Experiments

3.1 Emotion Analysis Model

This study used the FER-2013 image dataset for model training and validation to develop an emotion recognition model suitable for the system. The following will be a discussion of the experimental results. This experiment used pre-trained models on Keras, and the weights selected were trained on ImageNet [20]. The models used are VGG16, ResNet50, InceptionV3, Xeption, and MobileNet.

Experimental Results and Discussion. The results of this experiment were evaluated using Precision, Recall, and F1-score as indicators. Table 1 shows the F1-score of five models, and it can be seen from the table that in terms of the F1-score indicator, InceptionV3 and Xception are relatively better-performing models. By comparing the performance of the two models in each emotion through Precision and Recall in Table 2, we finally decided to use Xception as the model for emotion analysis.

Table 1. F1-score of the five models in the seven emotion classification

F1-score	Angry	Disgusted	Fearful	Happy	Sad	Surprised	Neutral	Avg
VGG16	0.55	0.62	0.51	0.87	0.44	0.80	0.63	0.63
ResNet50	0.57	0.68	0.53	0.87	0.43	0.81	0.63	0.65
InceptionV3	0.55	0.63	0.53	0.87	0.48	0.82	0.65	0.65
Xception	0.56	0.69	0.53	0.87	0.46	0.82	0.65	0.65
MobileNet	0.57	0.69	0.48	0.87	0.45	0.80	0.65	0.64

Table 2. Precision/Recall of the five models in the seven emotion classification

Precision/ Recall	Angry		Disgusted		Fearful		Happy		Sad		Surprised		Neutral		Avg	
	Pr	Re	Pr	Re	Pr	Re	Pr	Re	Pr	Re	Pr	Re	Pr	Re	Pr	Re
VGG16	0.51	0.60	0.76	0.53	0.46	0.57	0.90	0.84	0.52	0.39	0.80	0.79	0.63	0.64	0.65	0.62
ResNet50	0.59	0.54	0.92	0.54	0.50	0.57	0.84	0.89	0.56	0.35	0.90	0.83	0.59	0.68	0.70	0.63
InceptionV3	0.60	0.50	0.57	0.71	0.51	0.54	0.86	0.87	0.52	0.45	0.81	0.83	0.61	0.69	0.64	0.66
Xception	0.59	0.54	0.72	0.66	0.52	0.54	0.89	0.86	0.52	0.42	0.82	0.82	0.59	0.73	0.66	0.65
MobileNet	0.53	0.61	0.87	0.56	0.57	0.41	0.88	0.86	0.47	0.44	0.76	0.85	0.61	0.69	0.67	0.63

3.2 Eye Gaze Analysis Model

In this study, the Eye-Chimera image dataset was used for model training and testing, and an attempt was made to find an eye gaze analysis model suitable for the system. The machine learning methods used in the experiment are SVM, SGD-Classifier, and NearestCentroid, all from Scikit learn. The following is a discussion of the experiment results.

Experimental Results and Discussion. Table 3 shows the results obtained from the experiment using the Eye-Chimera eye movement dataset with SVM. The table shows that the average accuracy of SVM is 0.8, and the performance of Recall is also 0.83 on average. In general, SVM performs better than SGD-Classifier and NearestCentroid. The model is more average and stable in classifying the seven different eye gaze directions.

Table 3. Result of eye gaze analysis using SVM

SVM	Centre	UpRight	UpLeft	Right	Left	DownRight	DownLeft	Avg
Precision	1	0.89	0.65	0.81	0.76	0.62	0.89	0.8
Recall	0.89	0.91	0.91	0.87	0.67	0.8	0.73	0.83
F1-score	0.94	0.9	0.76	0.84	0.71	0.7	0.8	0.81

From Table 4, we can see that SGD-Classifier performed particularly poorly in the classification of "Right," which may be due to the similarity between the images of

"Right," "Right-Up," and "Right-Down." Although the Precision of the SGD-Classifier for "Right-Up" reached 1.00 in the table, its Recall was only 0.64, meaning that some of the actual "Right-Up" images were predicted to be in other directions. Overall, SGD-Classifier only performed well in the "Centre" category, while it performed poorly in other categories, especially in the classification of "Right."

Table 4. Result of eye gaze analysis using SGD-Classifier

SGD-C	Centre	UpRight	UpLeft	Right	Left	DownRight	DownLeft	Avg
Precision	0.91	1	0.72	0.36	0.76	0.81	0.49	0.72
Recall	0.97	0.64	0.87	0.71	0.6	0.57	0.86	0.75
F1-score	0.94	0.78	0.79	0.48	0.67	0.67	0.62	0.71

Table 5 shows that NearestCentroid generally performed poorly in the seven-direction classification tasks. Only the Centre achieved around 80% Precision among the seven directions, while the results in other directions were poorer. This may be due to NearestCentroid's method, which compares the distance between the center of each category and the new data and categorizes based on the nearest. However, besides the Centre, the other directions are more similar and close, such as Right-Up, Right, and Right-Down, which mostly look right, leading to misclassification. Overall, SVM performed the best among the three experiments, and as a result, we finally chose SVM as the model for eye gaze and pose analysis.

Table 5. Result of eye gaze analysis using NearestCentroid

NC	Centre	UpRight	UpLeft	Right	Left	DownRight	DownLeft	Avg
Precision	0.81	0.74	0.65	0.52	0.62	0.66	0.7	0.67
Recall	0.86	0.69	0.83	0.65	0.5	0.62	0.57	0.67
F1-score	0.83	0.71	0.73	0.58	0.55	0.64	0.63	0.67

3.3 Pose Analysis Model

In this experiment, Openpose is used as the main tool for skeleton detection. Openpose adopts the Part Affinity Fields for Part Association (PAF) method for human skeleton detection, which is based on finding the relationship between the body parts and the individuals in the image. First, a set of detected body parts is given, and an attempt is made to assemble these points into a full-body posture. Then, calculations are performed on each body part to find the possible main body position and direction. Using the skeleton information detected by Openpose, displacement or angle calculations are performed to obtain the desired observed actions, such as nodding, shaking, rough movements, and raising hands. In the implementation, Jetson Nano is used as the edge computing device to execute Openpose and the program that calculates the changes in limb displacement or angles.

3.4 Summary

The study employed publicly available datasets to evaluate the models used in the system and determine the most appropriate models. To ensure our system diagnoses are reliable and can serve as a robust reference for doctors, a model that can handle a wide range of emotions was selected rather than only processing a small set of features. Xception was therefore chosen as the model for emotion analysis. For eye gaze analysis, SVM was selected, and Openpose was used for body pose analysis, including skeleton detection and calculation of movement or joint angles to detect specific movements.

4 System Scenario

4.1 System Scenario and Interface

The interactive robot process in this study consists of the Asus Zenbo robot, NVidia Jetson Nano, and a video camera. Zenbo is responsible for the interaction process with children, which serves as the data for subsequent emotion and eye gaze analyses. Meanwhile, the video camera connects to the Jetson Nano to detect and recognize body pose movements from video images. The overall system setup is shown in Fig. 6. During the process, the medical team will observe from the side and fill out the observation scale designed for this system and the medical-specific observation scale required during the diagnosis process.

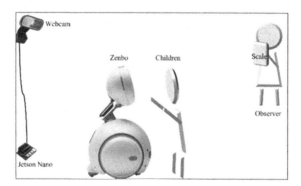

Fig. 6. System and application setup scenario

5 Conclusions

This study aims to develop a robot-interactive process and intelligent image analysis system for assisting in the diagnosis of autism in children. Through discussions with professional doctors and references to relevant assessment scales and diagnostic manuals for autism, a diagnostic process has been established on the robot, including calling names, telling stories, singing and dancing, and imitating games. The participant's image information and body pose movements are recorded during the process, and medical staff is asked to fill in the information during the diagnosis process. Finally, this information is presented to the doctor as a diagnostic reference in the form of data visualization. Experimental comparison of the models used in the system. In terms of emotion analysis models, the five models used in this experiment seem to have little difference in the average values of Precision, Recall, and F1-score. Among them, Xception performed relatively well in terms of Precision and F1-score, so Xception was ultimately adopted as the model used for emotion recognition in the system. The eye gaze analysis model uses SVM with average analysis results obtained in different eye gaze directions. The proposed assistive diagnostic robot interaction process and intelligent image analysis system in this study serve as a preliminary combination of medical and technology. In addition to improving the models used in the process, the collected data can also be used to train a classification model for better predicting the probability of autism symptoms and presenting the diagnosis to the doctor in a visualized data format.

Acknowledgement. This research was supported by the National Science and Technology Council, Taiwan, under Grant 109-2410-H-197-002-MY3 and 112-2410-H-197-002-MY2.

References

1. American Psychiatric Association and American Psychiatric Association (eds.). Diagnostic and statistical manual of mental disorders: DSM-5, 5th ed. American Psychiatric Association, Washington, D.C (2013)
2. Lord, C., et al.: Autism spectrum disorder. Nat. Rev. Dis. Primer **6**(1), 5 (2020). https://doi.org/10.1038/s41572-019-0138-4
3. Al-Dewik, N., et al.: Overview and introduction to autism spectrum disorder (ASD). In: Essa, M.M., Qoronfleh, M.W. (eds.) Personalized Food Intervention and Therapy for Autism Spectrum Disorder Management. AN, vol. 24, pp. 3–42. Springer, Cham (2020). https://doi.org/10.1007/978-3-030-30402-7_1
4. Emanuel, R., Weir, S.: Catalysing communication in an autistic child in a LOGO-like learning environment. In: Proceedings of the 2nd Summer Conference on Artificial Intelligence and Simulation of Behaviour, pp. 118–129 (1976)
5. Shamsuddin, S., Yussof, H., Ismail, L.I., Mohamed, S., Hanapiah, F.A., Zahari, N.I.: Initial response in HRI- a case study on evaluation of child with autism spectrum disorders interacting with a humanoid robot NAO. Procedia Eng. **41**, 1448–1455 (2012). https://doi.org/10.1016/j.proeng.2012.07.334
6. Silvera-Tawil, D., Bradford, D., Roberts-Yates, C.: Talk to Me: The role of human-robot interaction in improving verbal communication skills in students with autism or intellectual disability. In: 2018 27th IEEE International Symposium on Robot and Human Interactive Communication (RO-MAN), Nanjing, Aug., pp. 1–6 (2018). https://doi.org/10.1109/ROMAN.2018.8525698

7. Petric, F., et al.: Four tasks of a robot-assisted autism spectrum disorder diagnostic protocol: first clinical tests. In: IEEE Global Humanitarian Technology Conference (GHTC 2014), pp. 510–517 (2014)

8. Petric, F., Kovačić, Z.: Hierarchical POMDP framework for a robot-assisted ASD diagnostic protocol. In: 2019 14th ACM/IEEE International Conference on Human-Robot Interaction (HRI), pp. 286–293 (2019)

9. Moghadas, M., Moradi, H.: Analyzing human-robot interaction using machine vision for autism screening. In: 2018 6th RSI International Conference on Robotics and Mechatronics (IcRoM), Tehran, Iran, Oct., pp. 572–576 (2018). https://doi.org/10.1109/ICRoM.2018.865 7569

10. Javed, H., Park, C.H.: Behavior-based risk detection of autism spectrum disorder through child-robot interaction. In: Companion of the 2020 ACM/IEEE International Conference on Human-Robot Interaction, Cambridge United Kingdom, Mar., pp. 275–277 (2020). https://doi.org/10.1145/3371382.3378382

11. Pruette, J.R.: Autism diagnostic observation schedule-2 (ADOS-2). Google Sch., pp. 1–3 (2013)

12. McCrimmon, A., Rostad, K.: Test review: autism diagnostic observation schedule, second edition (ADOS-2) manual (Part II): toddler module. J. Psychoeduc. Assess., 32(1), 88–92 (2014). https://doi.org/10.1177/0734282913490916

13. Zahara, L., Musa, P., Prasetyo Wibowo, E., Karim, I., Bahri Musa, S.: The facial emotion recognition (FER-2013) dataset for prediction system of micro-expressions face using the convolutional neural network (CNN) algorithm based raspberry Pi. In: 2020 Fifth International Conference on Informatics and Computing (ICIC), pp. 1–9 (2020). https://doi.org/10.1109/ICIC50835.2020.9288560

14. FER-2013. https://www.kaggle.com/datasets/msambare/fer2013. Accessed 29 June 2022

15. Florea, L., Florea, C., Vrânceanu, R., Vertan, C.: Can Your Eyes Tell Me How You Think? A Gaze Directed Estimation of the Mental Activity (2013)

16. Vrânceanu, R., Florea, C., Florea, L., Vertan, C.: NLP EAC recognition by component separation in the eye region. In: International Conference on Computer Analysis of Images and Patterns, pp. 225–232 (2013)

17. Pedregosa, F., et al.: Scikit-learn: machine learning in python. J. Mach. Learn. Res. 12(85), 2825–2830 (2011)

18. Cao, Z., Simon, T., Wei, S.-E., Sheikh, Y.: Realtime multi-person 2d pose estimation using part affinity fields. In: Proceedings of the IEEE Conference on Computer Vision and Pattern Recognition, pp. 7291–7299 (2017)

19. Cao, Z., Hidalgo, G., Simon, T., Wei, S.-E., Sheikh, Y.: OpenPose: realtime multi-person 2D pose estimation using part affinity fields. IEEE Trans. Pattern Anal. Mach. Intell. 43(1), 172–186 (2019)

20. Deng, J., Dong, W., Socher, R., Li, L.-J., Li, K., Fei-Fei, L.: ImageNet: A large-scale hierarchical image database. In: 2009 IEEE Conference on Computer Vision and Pattern Recognition, pp. 248–255 (2009). https://doi.org/10.1109/CVPR.2009.5206848

The Role of Technology-Ethical Leadership Interaction in Minimising Unethical Acts: Implications for Research and Practice

Majd Megheirkouni[1]([✉]) [iD] and David Weir[2] [iD]

[1] School of Business, Law and Social Sciences, Abertay University, 40 Bell Street, Dundee DD1 1HG, Scotland, UK
majd.megheirkouni@gmail.com

[2] York Business School, York St John University, Lord Mayor's Walk, York YO31 7EX, UK

Abstract. Various historical events and attitudes have demonstrated that ethical leaders might intentionally or unintentionally make unethical decisions. History suggests that ethical leaders relying on strong technology alone could make unforgivable mistakes, but their interaction can limit such mistakes. In this study, we suggest that the interaction between technology and ethical leadership is proposed as a key factor in precluding or minimising unethical decisions by providing checks and balances capable of reducing the potential for unethical acts. A conceptual model is offered, along with propositions to help guide future research and practice. The degree to which technology and ethical leadership interact represents one of the key factors in understanding the potential for ethical/unethical acts. This conceptual study does not contain empirical data. This study is the first attempt that proposes the need of technology-leadership interaction to minimise unethical acts.

Keywords: Technology/ethical leadership · unethical acts · rules · code of ethics

1 Introduction

Unethical acts resulting from a reliance on human (ethical) leadership) or technology alone have significantly contributed to the topic of business ethics (Cialdini et al. 2021, Hoogervorst et al. 2010; Stylianou et al. 2013; Winter et al. 2004), and, in addition, have become an important issue in multidisciplinary and interdisciplinary literature. Some believe that any investment in the integration of both ethical leadership and technology is the primary key to understanding unethical acts (e.g., Duan et al. 2019). The existence of an association between ethical leadership practices and increased success in Business-to-Business Marketing was revealed by Lin and colleagues (2020), who found from a study of 465 IT Service companies that ethical leadership moderates the technological innovation-financial performance relationship. Accordingly, in the present paper we focus specifically on the role played by the interaction between ethical leadership and technology in the prevention or avoidance of unethical behaviour within organisations.

© Springer Nature Switzerland AG 2023
F. Fui-Hoon Nah and K. Siau (Eds.): HCII 2023, LNCS 14039, pp. 53–65, 2023.
https://doi.org/10.1007/978-3-031-36049-7_5

Well before the emergence of advanced or Artificial Intelligence technologies, a review of the literature on the unethical role of technology revealed that although there are a wide number of studies that mention this negative influence (e.g., Bush et al. 2010; Charki et al. 2017; Leonard and Cronan, 2001; Stylianou et al. 2013), it is argued that there is no need to create a separate 'ethics of every subtype of technology' or technological property because all technologies are ethically relevant (Sætra and Danaher, 2022). However, the news media frequently report errors made in every field of technology. Unfortunately, some of these errors result in long-term suffering for individuals and families as when, for example, medical technologies fail to achieve safety criteria and contribute to increased rates of injury or death. An empirical study conducted by Samaranayake et al. (2012), investigated technology-related medication errors between 2006 and 2010. The study revealed that unintended and unanticipated errors can persist even when technology that is designed to reduce error is applied. Moreover, an individual leader can play an essential role in the institutionalisation of unethical practices in an organisation by condoning or supporting unethical practices. An individual leader whose behaviour was unethical would be in position to authorise unethical practices that become pervasive within an organisation (Pearce et al. 2008).

Ill-defined rules or regulatory codes of poor quality can also be the cause of limitless differences in interpretation, resulting in the legitimation of unethical practices, especially when situational factors and/or political agendas lend authority to such practices. The evidential basis for the link between poorly devised rules and unethical acts has been shown to exist (e.g., Lindgreen, 2004; Poole-Robb and Bailey, 2002), and can be illustrated with reference to events in areas such as politics and sport. We may take as an example the controversial speech given by the American Secretary of State, Colin Powell, to the United Nations Security Council on February 5th, 2003, intended to justify war with Iraq. Intervention by the US and the waging of a war of aggression without the authority of the United Nations and the UN Security Council was not in keeping either with the American democratic tradition or with its moral laws (Zarefsky, 2007). In this example, we note that the Bush administration attempted to justify its war on Iraq by relying on the authority given to the UN Security Council under the UN Charter to authorise member states to use force to repel threats to peace. However, no such threat existed, and consequently there was no legitimate justification for war.

Another example, this time taken from sport, is the case of 'the hand of Maradona' that destroyed the England team's chances in the World Cup quarter-final of 1986. Although none of the referees spotted this unethical behaviour, TV replays of the goal clearly established that Maradona's goal was illegitimate, since it was scored with the use of his hand (Genschow et al. 2019). In this example, we can note that the rules of football at that time did not give the officials the right to disqualify a goal after the event. The amended rules of football that now exist allow referees to watch a recording of the event and then to make their decision. In Maradona's case, referees were unable to cancel the goal after the match, because football rules and laws in the 80s did not allow such action.

We present a straightforward view of the role of rules in technology and also of the part played by individual leaders in the commission of unethical acts. First, we propose that ill-defined rules and poorly devised codes of ethical practice, (i.e., those that allow hermeneutical differences within the organisation and also permit subjective variations (See also, Sims, 2002; Wasieleski and Hayibor, 2008) are negatively and directly related to unethical acts. Second, we propose that technology (e.g., Charki et al. 2017) and individual ethical leaders (see also, Weaver et al. 2005; Treviño and Nelson, 2007) are incapable of avoiding unethical acts unless they work together.

Although technology can be utilised by various kinds of leaders: responsible, authentic, transformational, etc., the focus in the present study is on the interaction of 'ethical' leadership and -technology. This interaction can be defined as "The use of information and technology to support and improve public policies and government operations, engage citizens and provide comprehensive and timely government services" (Scholl, 2008, p.23). In this paper, we assert that ethical leadership-technology interaction can be a relevant factor in deterring unethical acts. Finally, we assert that unethical acts and practice are a result of the use of technology in decision-making by an ethical leader acting alone and/or as a result of ill-defined rules and also poor-quality codes of conduct, which can also result in unlimited hermeneutical variations and subjective evaluations.

While we acknowledge that ethical leadership plays an essential role in combatting unethical behaviour in an organisation, it is unable to perform this function alone in the absence of a positive and supportive organisational climate (See also Sookdawoor and Grobler, 2022; Umphress and Bingham, 2011). Technology is one of the factors in such a climate. Conversely, technology stands in need of ethical leadership. Therefore, we posit that both ethical leadership and technology should be fully exploited in order to provide an effective buffer against all forms of unethical acts. Part of our argument concerning the role of ethical leadership and technology is that the two should be integrated in the fight against unethical activities within an organisation. Balanced use of ethical leadership and technology can be critically important in sensitive or potentially life-threatening circumstances (see Edwards et al. 2000). An example of such circumstances is the 1983 Soviet nuclear false alarm incident. During the cold war, the nuclear early warning system of the Soviet Union reported the launch of intercontinental ballistic missiles from military bases in the United States. However, it was found that these missile attack warnings were false alarms. The alarm was ignored by a Soviet officer, who thus rescued the world from a nuclear disaster, since the Soviet High Command might have decided to respond by attacking the American missile bases.

2 Theoretical Model and Research Propositions

2.1 Rules/Codes and Unethical Acts

People inside or outside organisations can be engaged in unethical acts in different ways (Brass et al. 1998; Jago and Pfeffer, 2019; Rees et al. 2021). The reasons for these behaviours are based on a variety of factors. For example, personal characteristics make individuals different in terms of their cognitive moral development, which affects their understanding of ethical behavioural standards (Giacalone et al. 2016; Reynolds, 2006). Role breadth is another reason where organisational citizenship behaviour literature justifies use of extra-role behaviour as a critical role for serving organisational purposes. However, these types of behaviour are not necessarily ethical (e.g., Turnipseed, 2002).

Additionally, culture has been used by cultural anthropologists and scholars to understand divergence and convergence among people, societies, and nations regarding values, norms, moral and ethics (e.g., Davis et al. 1998; Eisenbeiß and Brodbeck, 2014). It is known that national culture can hide within it informal forces (e.g., Wasta, Guanxi, Jeitinho) that can authorise unethical behaviour in the name of public benefit. (Dunfee and Warren, 2001; Duarte, 2006). Overall, this suggests that the public good can be perceived as a valid reason for legitimising unethical behaviour in some cultures, leading some individuals, societies, and countries to accept such behaviour (Smith et al. 2012a; Smith et al. 2012b).

Although the three reasons given above can be seen as grounds for authorising and legitimising unethical behaviour, we argue that rules and codes of ethics are the primary factor in the prohibition or minimisation of unethical acts. According to Merchant and Van der Stede (2012), rules and code of ethics enable domestic and foreign businesses to adjust and control ethical acts at personnel and culture levels.

In contrast, when there are neither strong rules nor a high-quality code of ethics inside or outside an organisation, misconduct prevails, whereas the effectiveness of rules and code of ethics lie in their capacity to deter unethical acts (See e.g., Adam and Rachman-Moore, 2004; Halter et al. 2009; Vitolla et al. 2021). No doubt, there are weaknesses that cause people to circumvent or break rules. Empirical research investigating the relationship between effective business codes of ethics and unethical behaviour revealed that the mere existence of an effective code of business ethics has the effect of minimising unethical behaviour in an organisation (Kaptein, 2011).

In the literature of business ethics there are discussions of cases in which employees engaged in unethical acts for the intended purpose of benefitting the organisations to which they belonged (Umphress and Bingham, 2011). In such cases. we should be aware of the differences between individuals in terms of their understanding of ethical standards (e.g., Duh et al. 2010; Finegan, 1994). Future research may explore how geographical location frames people's understanding of ethical behaviour. Ethical standards in one geographical location can be different from, or entirely opposite to, those of another location, particularly, when the public benefit is the primary justification for action. Empirical evidence has revealed that voluntary and non-profit leaders in the Middle East do not hesitate to practice unethical acts for a public benefit (Megheirkouni and Weir, 2019). The following proposition more formally articulates this viewpoint, the logic of which is depicted in Fig. 1

Hypothesis 1: *Ill-defined rules and low-quality codes of ethics promote unethical acts, such that those rules and codes can be a key for unlimited hermeneutics, resulting in unethical acts.*

2.2 Technology/Ethical Leader Solo and Technology- Ethical Leadership Interaction

Ethical leadership is a critical element in supporting positive functional outcomes (Eisenbeiss et al. 2015; Kalshoven et al. 2011). Brown and Treviño (2006) suggested several factors that moderate the effect of individual characteristics and situational influences in ethical leadership. The ethical context in which ethical leaders operate is one of the key factors in this regard. According to Treviño et al. (1998), the ethical context and ethical culture can determine whether the organisation supports ethical behaviour and practices. Ethical culture refers to a subset of an organisation's culture, including formal and informal behavioural control systems that can moderate the relationship between an individual's level of moral reasoning and ethical/unethical acts (Trevino, 1986; Trevino and Nelson, 2021).

Ethical climate refers to "the prevailing perceptions of typical organisational practices and procedures that have ethical content" or "those aspects of work climate that determine what constitutes ethical behaviour at work" (Victor and Cullen, 1988, p. 101). Although little is known about the ways in which an unethical climate affects an ethical leader's success and whether this kind of leadership would be accepted when the organisation reinforces an unethical culture (Morais and Randsley de Moura, 2018), we propose that an ethical leader is unable to work effectively without a supportive environment inside or outside the organisation. Technology is one powerful factor in the 21st century, where ethical leadership would be weakened by its absence. In other words, interaction between ethical leaders and technology provides an ethical leader with a capability for better control, knowledge and understanding.

Advanced technology, however, despite its rapid development over the last two decades, is not free from a reputation for being unethical and of presenting a long-term threat to human beings (Duan et al, 2019). For instance, technology failure has been discussed in its various forms, such as its financial impact on business and the market (Bharadwaj et al. 2009), manipulation and control (Cram and Wiener, 2020), which is its principal means of facilitating unethical behaviour (Chatterjee et al. 2015), and because of the practice of transferring risky technology to countries lacking in trained and skilled personnel (Velasquez, 2000). In this study, we propose that technology alone is unable to work effectively in providing satisfying results and in spreading its benefits across the whole organisational structure. It needs to be under the controlling influence of ethical leaders. This relationship can be expressed by the following proposition:

Hypothesis 2: *The dependency on technology alone or on ethical leadership alone is ineffective, and this weakness can be avoided through an integrated relationship of the two.*

2.3 Technology- Ethical Leadership Interaction and Unethical Action

The empirical evidence on human-technology interaction to date has consistently linked it to positive outcomes (Duan et al, 2019; Hudson et al. 2019; Orlikowski, 2000; Pitardi and Marriott, 2021; Thüring and Mahlke, 2007; Tay et al. 2014; Wilkens and Dewey, 2019). Most studies have examined some dimension of influence and efficacy stemming from this interaction. No study, however, has examined leadership-technology interaction in general and ethical leadership-technology interaction. The most clearly related to the present study is the work of Wilson and Daugherty (2018) who argue that collaboration between humans and artificial intelligence technologies is perceived as an effective tool to cope with today's business requirements. One of these requirements is ethical outcomes. Duan et al. (2019) found a link between advanced technologies and human decision makers. They claim that such technologies can play multiple roles in decision making by human decision makers, and thus can be perceived as a decision support and augmentation tool, particularly in environments that vary significantly in the degree of endorsement of aspects of ethical leadership (e.g., Resick et al. 2006) and need more supportive tools, such technology.

Specifically, we view ethical leadership-technology interaction as an important moderator of the relationship between ill-defined rules as well as low-quality of codes and unethical acts. The basis of our argument in the present study is that ethical leadership-technology interaction can provide a robust system of checks and balances inside the organisation, thereby enhancing the latter's creditability by acting as a moderator of the relationship between ill-defined rules as well as low-quality of codes and unethical acts. The following proposition more formally articulates this predictive relationship.

Hypothesis 3: *Technology-ethical leadership interaction will moderate the relationship between ill-defined rules as well as low-quality of codes and unethical acts. Specifically, when technology and ethical leadership are integrated for the purpose of organisational leadership and management tasks, the negative relationship between ill-defined rules as well as low-quality of codes will be weaker, and vice versa.*

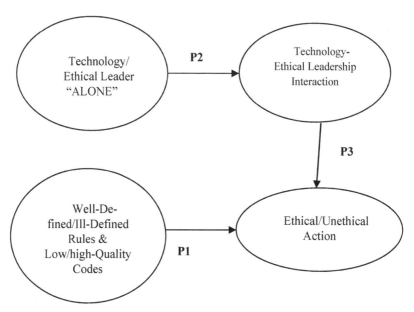

Fig. 1. Theoretical model of technology/ethical leader "ALONE" and well-defined/ill-defined rules & low/high-quality codes, technology-ethical leadership interaction, and ethical/unethical acts

3 Implications

There are several potential theoretical and practical implications of technology-ethical leadership interaction that can be derived from the model we suggest. Specifically, our model, building on the work of scholars investigating technology-human interaction (Duan et al. 2019) ethical leadership and ethical climate (Hassan et al. 2014; Kuenzi et al. 2020; Neubert et al. 2009), suggests that the orientation toward relying solely on technology or an ethical leader may need to be reconsidered if the aim is to mitigate against unethical practices or tendencies. Ethical leadership literature provides evidence that weak ethical climates (e.g., Lau and Wong, 2009) that can be seen in several forms, such as ill-defined rules as well as low-quality codes of ethical conduct, which encourage unethical behaviour, play an essential role in influencing the effectiveness of both ethical leaders and cutting-edge technology.

Technology has the power to facilitate and support the role of all kinds of leadership, including ethical leadership of an organisation. However, this power might differ based on the level of leadership and the situation (e.g., Parry et al. 2016). One perspective that can provide an explanation for this claim is offered by the analysis of expert systems adopted for business decision-making at different leadership decision levels: (strategic, tactical, and operational) (Edward, 2000). According to this analysis, technology can replace leadership decision makers at both operational and tactical levels (structure or semi structured decisions), but technology is more effective if it is used as a supportive tool for dealing with unstructured decisions at the strategic decision level. However, the analysis of expert systems adopted for business decision-making at different leadership decision levels shows the importance of technology-leadership interaction only at the strategic level. The example of the 1983 Soviet nuclear false alarm incident confirms that technology-leadership interaction is also essential at the lower levels of the military hierarchy. Despite its efficiency, advanced technology used for decision making cannot replace ethical leaders in situations where it functions as "an extended brain in making data-driven, evidence-informed decisions" that would facilitate a leader's decision making using "a blend of data-driven, evidence-informed decision-making and value-based moral decision-making" (Wang, 2021, p 256). Accordingly, to the extent that existing theories and approaches evolve toward incorporating wider conceptualizations of leadership and its interaction with cutting edge technologies in the era of artificial intelligence, more research will be necessary in order to expand our existing knowledge regarding the importance of this interaction. More specifically, further efforts directly investigating the relationships between technology-ethical leadership interaction and its role in minimising unethical practices will be necessary.

Another interesting implication of our model relates to the implications of work by researchers such a by Lin and colleagues, who studied the mediating role played by ethical leadership operating in the relationship between innovative AI and improved financial performance. Although it remains true that building ethical AI is "an enormously complex and challenging task" (Siau and Wang, 2020), the task is not merely technical or regulatory. It may also involve more emphasis on practice rather than theory, and thus practitioner-located field research is necessary for a better understanding of the behaviours and practices that may be regarded as "ethical" in real life situations, rather than further exegetic and theoretical critiques of the abstract principles or philosophical underpinnings of leadership. This may not require a reversion to the debates on "machine ethics" but an emphasis on the nature of agency that comprises both human and non-human agents. In this context the recent experience of the impact of Covid-driven lockdowns of teaching systems and programme delivery in higher education may provide useful parallels in that the outcome seems to favour a blended learning strategy for successful outcomes, rather than the either/or of traditional or online-only models.

Human systems are inevitably complex systems (Beer, 1973) and the leadership of complex systems cannot be reduced to simplistic models, whether they be Asimov's three laws of robotics or Kant's categorical imperative. Beer, in the same publication, identifies something very significant for the interpretation we are proposing in this paper when he states that "The variety attenuators to use here are not policy documents from the centre, but the managers themselves. That is what managers are for. As to the criterion of fairness, the manager—or any individual, in whatever he does—ought to be ready to take responsibility for his own decisions." (Beer, 1973, p 37). The role of the leader needs to be strengthened further in a situation of complexity, where system outputs and operations must be mediated by those who are good practical ethical decision-makers. As Boulton (2022, p. 1) points out "…while those at Number Ten were taking advantage of the lovely weather and having 'socially distanced drinks' (in apparent contradiction of the 'rule of six' in place at the time), many in the country were suffering. People weren't allowed to spend precious time with their dying relatives". That is what leaders are for. They are not there to make strategic decisions. They are there to make right decisions. If leaders break the rules that they have set for others, just because they can or because they do not care about convention, it is the whole systemic framework that is damaged (Green, 2019).

Our own generation has witnessed a recognition that AI can encompass vastly more than variety-handling and has also come to realise that the complexity of humans and machines implies a need to understand and allow for an emergent ethical competence for them both. Therefore, a fortiori, complex systems that are both technologically sophisticated and are operating in diverse and evolving environments cannot be integrated into an all-encompassing set of strategies driven by infallible procedures or unquestionable protocols. The purpose and basis of organisational leadership is to deal with the unexpected because, whatever the odds against them, unpredicted events will happen. Thus, system designers need to accept that ethical decisions are very hard to make computable and that the needs for variety, diversity and that exceptional and adaptive evolutionary change require mediation from human agencies that are varied, flexible and more likely to lead to acceptable outcomes for all system elements.

This approach may make AI systems seem more rather than less robust. As Santos-Lang (2014) argues, reliance on AI paradigms such as neural networks and genetic algorithms indicates that as scientific method changes, so it will be necessary for ethical frameworks to evolve. But at no stage can outcomes be regarded as terminal while the possibility of change exists, so we are likely always to be in a situation of never-ending learning. In such an ecology, ill-defined rules have definite value. But accepting that AI can never provide absolute organisational certainty does not imply that something has gone wrong. The wrongness lies in the expectation of certainty because this has always unreasonable. A lack of absolute definition can only be seen as a system weakness in the short term. As the French proverb has it "autres temps, autres moeurs". The role of the ethical leader does not include claiming to abolish uncertainty. The outcomes of systems, however complex, will continue to be judged on the grounds of effectiveness in relation to the achievement of humanly determined objectives rather than simplistically, on the criteria of efficiency of performance. De Rosa and Trabalzi (2016), who point out the way illegality becomes an institution, argue that legality and illegality are not universal value, but socially constructs that are the product of institutional selection. In line of this argument, we suggest that once technology is fixed and supported by heavy investment in hardware and specialist skills, then it becomes a constraining force that limited the range of possible behavioural outcomes.

4   Conclusion

Overall, it seems clear that technology-ethical leadership interaction deserves more theoretical and empirical attention. It offers a ground for the study of an important and underresearched ethical leadership perspective and a promising approach for the minimisation of unethical acts in organisations. Greater emphasis theoretically and practically on technology-ethical leadership may well offer the potential to help organisations limit the possibility of future unethical practices.

References

Adam, A.M., Rachman-Moore, D.: The methods used to implement an ethical code of conduct and employee attitudes. J. Bus. Ethics **54**(3), 225–244 (2004). https://doi.org/10.1007/s10551-004-1774-4

Albrecht, S.V., Stone, P.: Autonomous agents modelling other agents: a comprehensive survey and open problems. Artif. Intell. **258**, 66–95 (2018). https://doi.org/10.1016/j.artint.2018.01.002

Anandarajan, M.: Profiling Web usage in the workplace: a behaviours-based artificial intelligence approach. J. Manag. Inf. Syst. **19**(1), 243–266 (2002). https://doi.org/10.1080/07421222.2002.11045711

Arel, I., Rose, D.C., Karnowski, T.P.: Deep machine learning-a new frontier in artificial intelligence research [research frontier]. IEEE Comput. Intell. Mag. **5**(4), 13–18 (2010). https://doi.org/10.1109/MCI.2010.938364

Beer, S.: Designing Freedom. House of Anansi, Toronto (1973)

Bernold, T.: Aspects of technology assessment–the future role of artificial intelligence. R&D Manage. **15**(2), 179–182 (1985)

Bharadwaj, A., Keil, M., Mähring, M.: Effects of information technology failures on the market value of firms. J. Strateg. Inf. Syst. **18**(2), 66–79 (2009). https://doi.org/10.1016/j.jsis.2009.04.001

Boulton, N.: Turning a blind eye to Covid: the cover up for Boris. Psychodynamic Practice (in press), pp. 1–3 (2022). https://doi.org/10.1080/14753634.2022.2032561

Brass, D.J., Butterfield, K.D., Skaggs, B.C.: Relationships and unethical behaviours: a social network perspective. Acad. Manag. Rev. **23**(1), 14–31 (1998). https://doi.org/10.2307/259097

Brown, M.E., Treviño, L.K.: Ethical leadership: a review and future directions. Leadersh. Q. **17**(6), 595–616 (2006). https://doi.org/10.1016/j.leaqua.2006.10.004

Bush, V., Bush, A.J., Orr, L.: Monitoring the ethical use of sales technology: an exploratory field investigation. J. Bus. Ethics **95**(2), 239–257 (2010). https://doi.org/10.1007/s10551-009-0357-9

Charki, M.H., Josserand, E., Boukef, N.: The paradoxical effects of legal intervention over unethical information technology use: a rational choice theory perspective. J. Strat. Inf. Syst. **26**(1), 58–76 (2017). https://doi.org/10.1016/j.jsis.2016.07.001

Chatterjee, S., Sarker, S., Valacich, J.S.: The behavioural roots of information systems security: exploring key factors related to unethical IT use. J. Manag. Inf. Syst. **31**(4), 49–87 (2015). https://doi.org/10.1080/07421222.2014.1001257

Cialdini, R., Li, Y.J., Samper, A., Wellman, N.: How bad apples promote bad barrels: unethical leader behavior and the selective attrition effect. J. Bus. Ethics **168**(4), 861–880 (2019). https://doi.org/10.1007/s10551-019-04252-2

Daugherty, P.R., Wilson, H.J.: Human+ Machine: Reimagining Work in the Age of AI. Harvard Business Press, London (2018)

Davis, M.A., Johnson, N.B., Ohmer, D.G.: Issue-contingent effects on ethical decision making: a cross-cultural comparison. J. Bus. Ethics **17**(4), 373–389 (1998). https://doi.org/10.1023/A:1005760606745

De Rosa, M., Trabalzi, F.: Everybody does it, or how illegality is socially constructed in a southern Italian food network. J. Rural. Stud. **45**, 303–311 (2016). https://doi.org/10.1016/j.jrurstud.2016.04.009

Duan, Y., Edwards, J.S., Dwivedi, Y.K.: Artificial intelligence for decision making in the era of big data–evolution, challenges, and research agenda. Int. J. Inf. Manage. **48**, 63–71 (2019). https://doi.org/10.1016/j.ijinfomgt.2019.01.021

Duarte, F.: Exploring the interpersonal transaction of the Brazilian jeitinho in bureaucratic contexts. Organization **13**(4), 509–527 (2006). https://doi.org/10.1177/1350508406065103

Duh, M., Belak, J., Milfelner, B.: Core values, culture, and ethical climate as constitutional elements of ethical behaviour: exploring differences between family and non-family enterprises. J. Bus. Ethics **97**(3), 473–489 (2010). https://doi.org/10.1007/s10551-010-0519-9

Dunfee, T.W., Warren, D.E.: Is guanxi ethical? a normative analysis of doing business in China. J. Bus. Ethics **32**(3), 191–204 (2001). https://doi.org/10.1023/a:1010766721683

Edwards, J.S., Duan, Y., Robins, P.: An analysis of expert systems for business decision making at different levels and in different roles. Eur. J. Inf. Syst. **9**(1), 36–46 (2000). https://doi.org/10.1057/palgrave.ejis.3000344

Eisenbeiß, S.A., Brodbeck, F.: Ethical and unethical leadership: a cross-cultural and cross-sectoral analysis. J. Bus. Ethics **122**(2), 343–359 (2013). https://doi.org/10.1007/s10551-013-1740-0

Eisenbeiss, S.A., van Knippenberg, D., Fahrbach, C.M.: Doing well by doing good? analyzing the relationship between CEO ethical leadership and firm performance. J. Bus. Ethics **128**(3), 635–651 (2014). https://doi.org/10.1007/s10551-014-2124-9

Finegan, J.: The impact of personal values on judgments of ethical behaviour in the workplace. J. Bus. Ethics **13**(9), 747–755 (1994). https://doi.org/10.1007/BF00881335

Genschow, O., Rigoni, D., Brass, M.: The hand of God or the hand of Maradona? believing in free will increases perceived intentionality of others' behaviour. Conscious. Cogn. **70**, 80–87 (2019). https://doi.org/10.1016/j.concog.2019.02.004

Giacalone, R.A., Jurkiewicz, C.L., Promislo, M.: Ethics and well-being: the paradoxical implications of individual differences in ethical orientation. J. Bus. Ethics **137**(3), 491–506 (2015). https://doi.org/10.1007/s10551-015-2558-8

Green, D.A.: Boris Johnson subverts the rule of law. Financial Times (2019). https://www.ft.com/content/5f57d498-d3e0-11e9-8367-807ebd53ab77

Halter, M.V., De Arruda, M.C.C., Halter, R.B.: Transparency to reduce corruption. J. Bus. Ethics **84**(3), 373–385 (2009). https://doi.org/10.1007/s10551-009-0198-6

Hassan, S., Wright, B.E., Yukl, G.: Does ethical leadership matter in government? Effects on organizational commitment, absenteeism, and willingness to report ethical problems. Public Administration Review **74**(3), 333–343 (2014). https://doi.org/10.1111/puar.12216

Hoogervorst, N., De Cremer, D., van Dijke, M.: Why leaders not always disapprove of unethical follower behaviours: it depends on the leader's self-interest and accountability. J. Bus. Ethics **95**(1), 29–41 (2010). https://doi.org/10.1007/s10551-011-0793-1

Hudson, S., Matson-Barkat, S., Pallamin, N., Jegou, G.: With or without you? interaction and immersion in a virtual reality experience. J. Bus. Res. **100**, 459–468 (2019). https://doi.org/10. 1016/j.jbusres.2018.10.062

Jago, A.S., Pfeffer, J.: Organizations appear more unethical than individuals. J. Bus. Ethics **160**(1), 71–87 (2018). https://doi.org/10.1007/s10551-018-3811-8

Kalshoven, K., Den Hartog, D.N., De Hoogh, A.H.: Ethical leadership at work questionnaire (ELW): development and validation of a multidimensional measure. Leadersh. Q. **22**(1), 51–69 (2011). https://doi.org/10.1016/j.leaqua.2010.12.007

Kaptein, M.: Toward effective codes: testing the relationship with unethical behaviour. J. Bus. Ethics **99**(2), 233–251 (2011). https://doi.org/10.1007/s10551-010-0652-5

Kuenzi, M., Mayer, D.M., Greenbaum, R.L.: Creating an ethical organizational environment: The relationship between ethical leadership, ethical organizational climate, and unethical behavior. Personnel Psychology **73**(1), 43–71 (2020). https://doi.org/10.1111/peps.12356

Kanitz, R., Gonzalez, K.: Are we stuck in the predigital age? embracing technology-mediated change management in organizational change research. J. Applied Beha. Sci. **57**(4), 447–458 (2021). https://doi.org/10.1177/00218863211042896

Lau, V.P., Wong, Y.Y.: Direct and multiplicative effects of ethical dispositions and ethical climates on personal justice norms: a virtue ethics perspective. J. Bus. Ethics **90**(2), 279–294 (2009). https://doi.org/10.1007/s10551-009-0042-z

Leonard, L.N., Cronan, T.P.: Illegal, inappropriate, and unethical behaviour in an information technology context: a study to explain influences. J. Assoc. Inf. Syst. **1**(1), 1–31 (2001)

Lin, W.L., Yip, N., Ho, J.A., Sambasivan, M.: The adoption of technological innovations in a B2B context and its impact on firm performance: an ethical leadership perspective. Ind. Mark. Manage. **89**, 61–71 (2020). https://doi.org/10.1016/j.indmarman.2019.12.009

Lindgreen, A.: Corruption and unethical behaviour: report on a set of Danish guidelines. J. Bus. Ethics **51**(1), 31–39 (2004). https://doi.org/10.1023/B:BUSI.0000032388.68389.60

Megheirkouni, M., Weir, D.: Insights into informal practices of sport leadership in the middle east: the impact of positive and negative wasta. Int. J. Sport Policy Pol. **11**(4), 639–656 (2019). https://doi.org/10.1080/19406940.2019.1634620

Miller, T.: Explanation in artificial intelligence: insights from the social sciences. Artif. Intell. **267**, 1–38 (2019). https://doi.org/10.1016/j.artint.2018.07.007

Morais, C., Randsley de Moura, G.: The Psychology of Ethical Leadership in Organisations (2018). Springer, Cham. https://doi.org/10.1007/978-3-030-02324-9

Neubert, M.J., Carlson, D.S., Kacmar, K.M., Roberts, J.A., Chonko, L.B.: The virtuous influence of ethical leadership behaviour: evidence from the field. J. Bus. Ethics **90**(2), 157–170 (2009). https://doi.org/10.1007/s10551-009-0037-9

Nilsson, N.J.: Artificial intelligence: a modern approach. Artif. Intell. **82**(1–2), 369–380 (1996). https://doi.org/10.1016/0004-3702(96)00007-0

Orlikowski, W.J.: Using technology and constituting structures: a practice lens for studying technology in organizations. Organ. Sci. **11**(4), 404–428 (2000). https://doi.org/10.1287/orsc.11. 4.404.14600

Parry, K., Cohen, M., Bhattacharya, S.: Rise of the machines: a critical consideration of automated leadership decision making in organizations. Group Org. Manag. **41**(5), 571–594 (2016). https://doi.org/10.1177/1059601116643442

Pearce, C.L., Manz, C.C., Sims, H.P., Jr.: The roles of vertical and shared leadership in the enactment of executive corruption: implications for research and practice. Leadersh. Q. **19**(3), 353–359 (2008). https://doi.org/10.1016/j.leaqua.2008.03.007

Pitardi, V., Marriott, H.R.: Alexa, she's not human but… unveiling the drivers of consumers' trust in voice-based artificial intelligence. Psychol. Mark. **38**(4), 626–642 (2021). https://doi.org/ 10.1002/mar.21457

Poole-Robb, S., Bailey, A.: Risky Business: Corruption, Fraud, Terrorism, and Other Threats to Global Business. Kogan Page, London (2002)

Rees, M.R., Tenbrunsel, A.E., Diekmann, K.A.: 'It's just business': understanding how business frames differ from ethical frames and the effect on unethical behaviour. Journal of Business Ethics, 1-21 (2021). https://doi.org/10.1007/s10551-020-04729-5

Resick, C.J., Hanges, P.J., Dickson, M.W., Mitchelson, J.K.: A cross-cultural examination of the endorsement of ethical leadership. J. Bus. Ethics 63(4), 345–359 (2006). https://www.jstor.org/stable/25123717

Reynolds, S.J.: Moral awareness and ethical predispositions: investigating the role of individual differences in the recognition of moral issues. J. Appl. Psychol. 91(1), 233–243 (2006). https://doi.org/10.1037/0021-9010.91.1.233

Samaranayake, N.R., Cheung, S.T.D., Chui, W.C.M., Cheung, B.M.Y.: Technology-related medication errors in a tertiary hospital: a 5-year analysis of reported medication incidents. Int. J. Med. Informatics 81(12), 828–833 (2012). https://doi.org/10.1016/j.ijmedinf.2012.09.002

Santos-Lang, C.: Moral ecology approaches to machine ethics. In: van Rysewyk, S., Pontier, M. (Eds.) Machine Medical Ethics. Intelligent Systems, Control and Automation: Science and Engineering 74. Springer, Switzerland. pp. 111–127 (2014). https://doi.org/10.1007/978-3-319-08108-3_8

Scholl Hans, J.: Discipline or interdisciplinary study domain? challenges and promises in electronic government research. In: Government, D. (ed.) Chen, H, C, pp. 21–41. Springer, London (2008)

Siau, K., Wang, W.: Artificial intelligence (AI) ethics: ethics of AI and ethical AI. J. Database Manage. 31(2), 74–87 (2020). https://doi.org/10.4018/JDM.2020040105

Sims, R.L.: "Ethical rule breaking by employees: a test of social bonding theory. J. Bus. Ethics 40(2), 101–109 (2002). https://doi.org/10.1023/A:1020330801847

Sætra, H.S., Danaher, J.: To each technology its own ethics: the problem of ethical proliferation. Philos. Technol. 35, 93 (2022). https://doi.org/10.1007/s13347-022-00591-7

Smith, P.B., Huang, H.J., Harb, C., Torres, C.: How distinctive are indigenous ways of achieving influence? a comparative study of guanxi, wasta, jeitinho, and "pulling strings." J. Cross Cult. Psychol. 43(1), 135–150 (2012). https://doi.org/10.1177/0022022110381430

Smith, P.B., Torres, C., Leong, C.H., Budhwar, P., Achoui, M., Lebedeva, N.: Are indigenous approaches to achieving influence in business organizations distinctive? a comparative study of Guanxi, Wasta, Jeitinho, Svyazi and Pulling Strings. The International Journal of Human Resource Management 23(2), 333–348 (2012). https://doi.org/10.1080/09585192.2011.561232

Steiger, D.M.: Enhancing user understanding in a decision support system: a theoretical basis and framework. J. Manag. Inf. Syst. 15(2), 199–220 (1998). https://doi.org/10.1080/07421222.1998.11518214

Stylianou, A.C., Winter, S., Niu, Y., Giacalone, R.A., Campbell, M.: Understanding the behavioral intention to report unethical information technology practices: the role of machiavellianism, gender, and computer expertise. J. Bus. Ethics 117(2), 333–343 (2012). https://doi.org/10.1007/s10551-012-1521-1

Sookdawoor, O., Grobler, A.: The dynamics of ethical climate: mediating effects of ethical leadership and workplace pressures on organisational citizenship behaviour. Cogent Bus. Manage. 9(1), 2128250 (2022). https://doi.org/10.1080/23311975.2022.2128250

Tay, B., Jung, Y., Park, T.: When stereotypes meet robots: the double-edge sword of robot gender and personality in human–robot interaction. Comput. Human Beh. 38, 75–84 (2014). https://doi.org/10.1016/j.chb.2014.05.014

Thüring, M., Mahlke, S.: Usability, aesthetics, and emotions in human–technology interaction. Int. J. Psychol. 42(4), 253–264 (2007). https://doi.org/10.1080/00207590701396674

Trevino, L.K., Nelson, K.A.: Managing Business Ethics: Straight Talk About How to Do it Right, 4th edn. John Wiley & Sons, New York (2021)

Treviño, L.K., Butterfield, K.D., Mcabe, D.M.: The ethical context in organizations: Influences on employee attitudes and behaviours. Bus. Ethics Q. **8**(3), 447–476 (1998). https://doi.org/10.2307/3857431

Turnipseed, D.L.: Are good soldiers good? exploring the link between organization citizenship behaviour and personal ethics. J. Bus. Res. **55**(1), 1–15 (2002). https://doi.org/10.1016/S0148-2963(01)00217-X

Umphress, E.E., Bingham, J.B.: When employees do bad things for good reasons: examining unethical pro-organizational behaviours. Organ. Sci. **22**(3), 621–640 (2011). https://doi.org/10.1287/orsc.1100.0559

Velasquez, M.: Globalization and the failure of ethics. Bus. Ethics Quart. **10**(1), 343–352 (2000). https://www.jstor.org/stable/3857719

Verganti, R., Vendraminelli, L., Iansiti, M.: Innovation and design in the age of artificial intelligence. J. Prod. Innov. Manag. **37**(3), 212–227 (2020). https://doi.org/10.1111/jpim.12523

Victor, B., Cullen, J.B.: The organizational bases of ethical work climates. Adm. Sci. Q. **33**, 101–125 (1988). https://doi.org/10.2307/2392857

Vitolla, F., Raimo, N., Rubino, M., Garegnani, G.M.: Do cultural differences impact ethical issues? exploring the relationship between national culture and quality of code of ethics. J. Int. Manag. **27**(1), 100823 (2021). https://doi.org/10.1016/j.intman.2021.100823

Wang, Y.: Artificial intelligence in educational leadership: a symbiotic role of human-artificial intelligence decision-making. J. Educ. Admin. **59**(3), 256–270 (2021). https://doi.org/10.1108/JEA-10-2020-0216

Wasieleski, D.M., Hayibor, S.: Breaking the rules: examining the facilitation effects of moral intensity characteristics on the recognition of rule violations. J. Bus. Ethics **78**(1), 275–289 (2008). https://doi.org/10.1007/s10551-007-9376-6

Weaver, G.R., Treviño, L.K., Agle, B.: "Somebody i look up to": ethical role models in organizations. Organ. Dyn. **34**(4), 313–330 (2005). https://doi.org/10.1016/j.orgdyn.2005.08.001

Wilkens, U., Dewey, M.: The interplay of artificial and human intelligence in radiology – exploring socio-technical system dynamics. In: Ahram, T., Taiar, R., Colson, S., Choplin, A. (eds.) IHIET 2019. AISC, vol. 1018, pp. 390–395. Springer, Cham (2020). https://doi.org/10.1007/978-3-030-25629-6_60

Winter, S.J., Stylianou, A.C., Giacalone, R.A.: Individual differences in the acceptability of unethical information technology practices: the case of Machiavellianism and ethical ideology. J. Bus. Ethics **54**(3), 273–301 (2004). https://doi.org/10.1007/s10551-004-1772-6

Zarefsky, D.: Making the case for war: colin powell at the United Nations. Rhetoric & Public Affairs **10**(2), 275–302 (2007). https://doi.org/10.1353/rap.2007.0043

Systematic Literature Review on the User Evaluation of Teleoperation Interfaces for Professional Service Robots

Gaayathri Sankar[1]([⊠]), Soussan Djamasbi[1], Zhi Li[1], Jing Xiao[1], and Norbou Buchler[2]

[1] Worcester Polytechnic Institute, Worcester, MA 01609, USA
{gsankar,djamasbi,zli11,jxiao2}@wpi.edu
[2] Army DEVCOM Analysis Center (DAC), Aberdeen Proving Ground, MD 21005, USA
norbou.buchler.civ@army.mil

Abstract. Remote-controlled robots can perform common tasks, thereby minimizing the exposure to the risks faced by employees in hazardous occupations. The focus of the review in this paper is on the user evaluation of teleoperation interfaces of service robots used in organizational/professional settings. Our study is motivated by the lack of review of user evaluation methods and metrics specific to teleoperation interfaces rather than an evaluation of the robot itself. Grounded in a user-centered approach to design, the three research questions in this review were guided by the three major factors that form the design space for product and service development (Context, Technology, User). The findings of this article are categorized based on areas of application or context of use, technologies used to operate the interfaces, and tools and techniques used to assess user needs as well as user perception of developed solutions. Based on the reported findings, the article provides suggestions for future research that aims at designing interfaces for human-robot collaborations in the workplace.

Keywords: User Evaluation · Teleoperation Interfaces · Professional Service Robots · Industry 5.0 · Human-Robot Interaction (HRI) · Generative and Evaluative User Experience (UX) Research

1 Introduction

Hazardous occupations such as firefighting, search and rescue operations, etc., pose a high risk to the lives of employees. The advances in technology provide an opportunity for the use of remote-controlled robots that could perform common tasks, thereby minimizing the exposure to the risk faced by these employees. The International Organization of Standardization defines a robot as a programmable device that can move and perform tasks in its environment [1]. Robots that are controlled and manipulated remotely require an interface that can serve as a communication channel between the operator/s and the robot/s. Through this communication channel (or teleoperation interface), the operator/s can execute commands to manipulate the robot/s and receive feedback about their executed actions in real time.

© Springer Nature Switzerland AG 2023
F. Fui-Hoon Nah and K. Siau (Eds.): HCII 2023, LNCS 14039, pp. 66–85, 2023.
https://doi.org/10.1007/978-3-031-36049-7_6

Literature provides ample evidence that interaction design plays a major role in user satisfaction and that user satisfaction plays a major role in the adoption and effective usage of technology, which in turn has major implications on an organization's return on investment (ROI) [2–4]. In addition to having a positive impact on the ROI, designing user-friendly teleoperation interfaces is likely to impact a company's bottom line by reducing the time and costs that are involved in training teleoperators [5].

It goes without saying that user evaluation is a critical building block in designing and implementing user-friendly interfaces. The user evaluation of interfaces during the iterative approach to user-centered design is crucial in providing insight for improving design in each iterative cycle [6–8]. Additionally, pre-implementation user evaluations can help estimate post-implementation behavior by developing predictive models that can determine the adoption and effective usage of the interface by the operators, hence estimating the likelihood and scope of impact on an organization's ROI [6].

The focus of the review in this paper is on the user evaluation of the teleoperation interfaces of service robots used in organizational/professional settings. Our literature review is motivated by the growing trend in the development and usage of service robots with added telepresence in multiple industries [9]. This growing trend has been further heightened by the recent Covid-19 pandemic [9]. Our study is also motivated by the lack of review papers on evaluation methods and metrics specific to teleoperation interfaces. Existing review papers focus on user evaluation with respect to the robot, not the interface. That is, these reviews focus on how users evaluate the robot itself, not the interface through which the robot is controlled and/or provides feedback to the operator. This distinction is important because the operator's perception of the interface can have a major impact on how willingly the operator would work with the robot or how useful the operator perceives the robot would be.

As we move further into Industry 5.0, human-robot co-working is bound to become a major focus both in academic and industry research. Working with and alongside robots in an organization, however, is not without challenges (e.g., learning to work with robots, personal preferences for working with human or robot coworkers, negative attitudes towards robots, etc.) [10]. Since designing effective interfaces can improve employees' satisfaction [11], a systematic review of the literature helps to identify research opportunities for improving human-robot co-working in the workplace.

In the following sections, we briefly discuss the research questions that guided our literature review. Then we explain the methodology used to conduct the systematic review, report the results, and discuss our findings.

2 Materials and Methods

User experience (UX) refers to context-sensitive human-technology encounters. Hence, the design space for user-centered product development is defined by three major dimensions: Context, Technology, and User. *Context* refers to tasks and/or environment within which the human-technology encounter takes place, *Technology* refers to systems that facilitate the encounter, and *User* refers to people who use the technology within the specified context [6]. In order to provide an overview of the current state of research in designing and testing interfaces for robot teleoperation, we used the above-discussed dimensions of UX design space to develop three overarching research questions (RQs) (Fig. 1). Grounded in the *Context* dimension of UX design space, the first question in our systematic literature review (SLR) aimed at identifying the types of professional services for which teleoperated robotic interfaces were designed. The second question, grounded in the *Technology* dimension of UX design space, aimed at identifying types of interfaces that have been used to operate service robots. Inspired by the *User* dimension of UX design space, the last research question aimed at identifying tools and techniques that were used to capture users' points of view about robot teleoperation interfaces (e.g., user needs, perceptions/evaluations). Because users' subjective experience of technology has a significant impact on whether they would be willing to adopt a technology or keep using it, the last research question helps to identify the most commonly used subjective measures, including constructs and validated instruments, to capture user perceptions. Constructs refer to latent variables that are not directly observable (e.g., satisfaction, perceived usefulness, trust, etc.). They are typically measured with validated survey instruments. Constructs serve as the building block for developing behavioral theories, such as the Unified Theory of Acceptance and Use of Technology (UTAUT) [12]. Hence, they are important in advancing research that seeks to understand human perception of robotic interfaces as well as research that aims at developing predictive models for user behavior such as acceptance, adoption, and continued usage of teleoperated robots.

RQ1 (Context): For what types of applications (services) are teleoperation interfaces developed?	RQ2 (Technology): What types of technologies are used to operate teleoperation interfaces?	RQ3 (User): Which methods, metrics, and tools are used to understand and capture users' subjective reactions?

Fig. 1. Research questions (RQs) that guided the systematic literature review.

2.1 Literature Search Strategy

We used the ACM Digital Library (ACM), PsychINFO on PsychNET (PI), Springer LINK (SL), Institute of Electrical and Electronics Engineering Xplore Digital Library (IEEEX), ScienceDirect (SD), and Scopus for our review based on our analysis of the scientific databases used in prior studies focusing on human-robot interaction [13, 14].

Based on our research questions, we used 'service robot', 'teleoperation interface', 'user evaluation', and 'methodology and metrics', as keywords in our search. Using these keywords, we formulated queries (shown in Table 1) to search the scientific databases identified.

The search results were limited to articles in English published in peer-reviewed journal papers and scholarly conference papers between 2007 (when smartphones were first introduced) and the time the investigation for this literature review took place (July 2022). The choice for the timeline was grounded in prior research on robot teleoperation interfaces that shows smartphone-based interfaces are more versatile than conventional gamepad-based interfaces [15]. Our search produced a total of 985 papers, that contained 201 duplicates, leaving behind 784 unique papers for consideration.

2.2 Inclusion and Exclusion Criteria

Table 1. Online database search results by search queries (date of search: August 11, 2022)

Database	Keywords	Total
ACM Digital Library	methodology and metrics AND user evaluation AND teleoperation interface AND service robot	117
PsychINFO on PsychNET (PI)	teleoperation interface AND service robot	0
	user evaluation AND teleoperation interface	1
	methodology and metrics AND user evaluation	8
SpringerLink	methodology and metrics AND user evaluation AND teleoperation interface AND service robot	64
IEEE Xplore Digital Library	teleoperation interface AND service robot	16
	user evaluation AND teleoperation interface	26
	methodology and metrics AND user evaluation	106
Science Direct	teleoperation interface AND service robot	0
	user evaluation AND teleoperation interface	11
	methodology and metrics AND user evaluation	134
Scopus	teleoperation interface AND service robot	22
	user evaluation AND teleoperation interface	51
	methodology and metrics AND user evaluation	429
Total		**985**

In order to select papers for in-depth analysis, we first evaluated the titles and abstracts of all 784 papers retrieved from the database search against our inclusion and exclusion criteria. We only included articles that had a direct relation with methodology and metrics involved in the user evaluation of robot teleoperation interfaces in professional service. We excluded papers that (1) did not focus on teleoperation interfaces for robots as their primary goal, and (2) did not incorporate user evaluation. The review process is shown in more detail using the PRISMA flowchart of article selection in Fig. 2 [16].

2.3 Categorizing the Search Results

We used a qualitative analysis software package (NVivo 12.6) to code and categorize the 784 articles that we retrieved from our search results. We categorized these articles based on their title and abstract descriptions only. Out of the 784 retrieved articles, 210 were focused on robot teleoperation interfaces or user evaluation, whereas 574 were not relevant to user evaluation or robot teleoperation interfaces. Among these 210 articles, the total number of articles that focused on user evaluation of teleoperation interfaces as their primary objective was 100. These 100 articles were then organized based on their topics (e.g., industrial robots, interface control architecture, personal service robots, professional service robots, reviews and taxonomies, and robot learning). Since our focus was on methodology and metrics used in user evaluation of teleoperation interfaces of professional service robots, only 77 articles in this collection of 100 articles qualified for our final in-depth review. All these 77 articles were published from 2007 to July 2022.

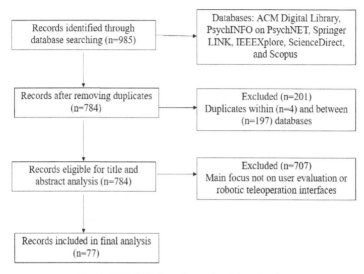

Fig. 2. PRISMA flowchart of article selection.

3   Results and Findings

In this section, we report the outcome of our in-depth review of the 77 articles on professional service robots identified through the review process discussed in previous sections. The majority of these reviewed articles (82%) were published in peer-reviewed journals. We start by reporting the areas of applications in robotics teleoperation research (RQ1) and the types of interfaces used in this line of research (RQ2). Then, we provide an in-depth analysis of methods, tools, and measures used to capture user needs and perceptions (RQ3).

3.1   RQ1 (Context): Types of Professional Services

Our first RQ focused on understanding the landscape of professional service robots for which teleoperation interfaces were designed. Professional service robots refer to robotics technologies that provide services for humans or organizations with or without necessarily involving social interaction with employees or customers [17].

Our analysis shows that almost half of the reviewed papers (47%) either did not specify any particular area of application or, they mentioned or indicated/implied that the teleoperated interfaces could be applied to service applications (named *general service* for the purpose of analysis). The rest of the reviewed articles (53%) provided a number of specific applied areas for teleoperation services such as surgery, agriculture, construction, etc. (see Fig. 3). To provide an overview of these specific applied areas, we grouped the reviewed publications into three categories: *general applications* (combining papers that either did not mention the applied area and/or those that mentioned or implied/indicated that the interface was developed for service applications), *health and wellness* (combining articles on surgery, rehabilitation services, massage therapy training, and nursing), and *other* (combining the rest of the articles). As shown in Fig. 3, almost half (47%) of papers fall under the *general applications* category. Of the remaining papers 14% fall under the *health and wellness* category, and the rest (39%) labeled as *other* in Fig. 3, cover various applications such as construction, agriculture, etc.

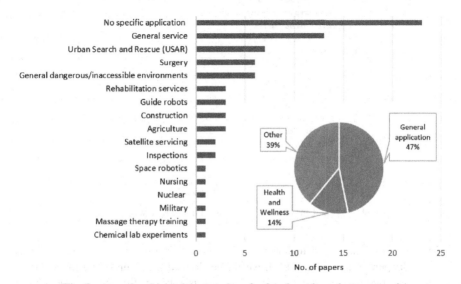

Fig. 3. Areas in which teleoperated professional service robots are used.

3.2 RQ2 (Technology): Types of Interfaces

Our second RQ focused on identifying types of teleoperation interfaces that were developed and tested for service robots at the workplace. Teleoperation systems generally comprise a complex mix of control/input and feedback interfaces. To capture the intricacies involved in designing and evaluating teleoperation interfaces, we categorized papers based on input and feedback mechanisms used in teleoperation interfaces to accommodate operators' interactions with robots.

Input Mechanisms. Our results show that all except 3 papers explicitly specified input mechanisms used in their studies. The top four explicitly specified input mechanisms were bio-signals (39%), mouse (27%), keyboard (26%), and haptic devices (18%) (see

Fig. 4). The next group of input devices that were used most frequently included touch screen (13%), stylus (13%), joystick (13%), gamepad (12%), trackpad (8%), VR controllers (6%), foot pedal (4%) and trackball (3%). The rest of the explicitly specified input devices used include knobs, buttons on a Personal Digital Assistant (PDA), joypad, handstick, fly stick, digital pen, and cutaneous device, each at 1%.

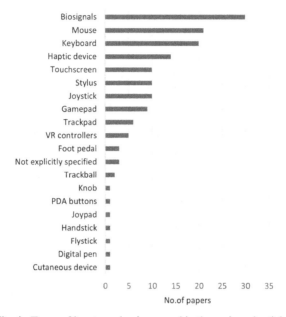

Fig. 4. Types of input mechanisms used in the reviewed articles.

Because bio-signals tend to have a major impact on improving the intuitiveness of robot teleoperation interfaces [17], we further looked into identifying mechanisms through which bio-signals were generated. Our analysis showed that the reviewed papers used hand gestures (18%), head movements (9%), body movements (8%), speech (5%), eye gaze (3%), and muscle movements (3%) to generate bio-signal inputs (see Fig. 5). Within these, hand gestures, head, and body movements formed the majority of input signals. The papers that used these bio-signal inputs reported that these inputs were collected non-invasively (13%), via Wii remote (8%), head-mounted displays (HMD) (7%), gloves (5%), rings (1%), and smartwatches (1%).

One paper out of the 77 articles reviewed used a larger mix of bio-signals for interface design (i.e., skin conductance, facial expressions, heart rate variance, eye gaze, brain activity, and muscle movements) [18]. The bio-signals used in this paper, however, were not used to give direct commands or receive feedback, rather they were used to measure the stress and workload of the user/teleoperator. The information gained from these bio-signals was then used to build an adaptive interface that could detect the operator's workload and adjust it as needed by enabling the robot to work more autonomously. Because the bio-signals were not used as direct input/feedback modalities in this case, they were not included in Fig. 5.

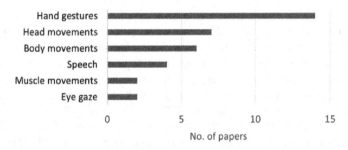

Fig. 5. Mechanisms for generating bio-signals.

Feedback Mechanisms. Our analysis shows that 3% of the papers reviewed did not explicitly specify the feedback modality that was used in their interfaces. The analysis of papers that explicitly mentioned feedback modalities in their articles showed that feedback included visual (95%), haptic (27%), auditory (9%), and cutaneous (1%) modalities (see Fig. 6). Given that explicitly specified feedback modalities included visual feedback in the majority of papers, we examined the mechanisms that were used to provide visual feedback. Our analysis shows that visual feedback was provided mostly with screen displays (74%) and Head Mounted Displays (HMD) (18%). Only two papers (3%) used other modalities: One paper used LED light to signal electric contact between a metal loop and a pipe along a 3D path to provide feedback about movement accuracy [19] while another paper used a smart interactive whiteboard as a visual feedback mechanism [20].

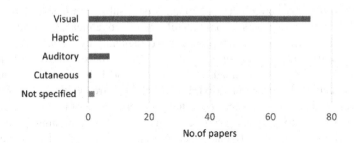

Fig. 6. Feedback modalities reported for teleoperation.

3.3 RQ3 (User): Subjective Reactions

Our last RQ focused on identifying tools and techniques that were commonly used to capture user reactions to provide insight for the design and/or evaluation of robot teleoperation interfaces. Because people's subjective reactions toward a technology play a significant role in whether they choose to use that technology or not, in this article we focused on tools and techniques that were used in the reviewed articles to capture users' perceptions of and reactions to teleoperation interfaces.

Designing positive user experiences requires both generative and evaluative UX research. Generative UX research refers to investigations that help empathize with users and understand their needs (e.g., persona development), and identify opportunities for innovation (e.g., problems for which designing a solution is needed). Evaluative UX research refers to investigations that are used to assess the quality of an existing design and/or to assess the market viability of a product (e.g., how it compares to competing products/services). Our results show that all 77 papers reviewed for this article carried out some form of evaluative research. Among the 77 papers reviewed, 4 papers have clearly reported iterative user evaluation of the teleoperation interfaces [21–24]. Methods employed by these 4 papers to capture users' subjective reactions included pilot tests, open-elicitation studies, surveys, focus groups, and interviews. These methodologies are often used in the user-centered approach to design, which emphasizes well integration of UX research in the entire product development and maintenance life cycle. For example, research prior to product development (e.g., developing personas to understand user needs), during the development cycle (e.g., iterative UX research to provide actionable insight for improving the design), and after the product is launched in the market (e.g., UX research to monitor success and competitive advantage) (see Fig. 8) [6]. Our results show that only one paper explicitly employed the user-centered approach to design including both generative and evaluative research. In this paper, personas were created, and journey maps were developed as part of generative UX research to understand user needs. Evaluative UX research was conducted to study the perceived usability of the designed interface using the system usability scale (SUS).

Next, we looked for subjective measures that were used to capture users' perceptions. Our analysis showed that not all of the reviewed papers used subjective measures in their studies. Out of the 77 reviewed papers, 51 papers used subjective measures to evaluate their interfaces. These 51 papers used a total of 19 subjective measures which are summarized in Fig. 7 and Table 2. The top three most frequently used subjective measures were workload, usability, and sense of presence: 35% of the subjective measures that occurred in the papers were workload, 21% usability, and 11% sense of presence. The rest of the subjective measures were used substantially less frequently. For example, 6% of the subjective measures that occurred in the papers had ease of use while situational awareness and simulator sickness each occurred at 3%. The remaining 13 subjective measures were each used in only one paper. In other words, 21% of the papers used one of the following subjective measures: acceptance, attitude, functionality, effectiveness, stress, sense of direction, usefulness, UX, intuitiveness, perception of robot as a teammate, self-efficacy, and speed.

The 19 subjective measures were captured with a combination of validated scales (e.g., NASA-TLX, SUS, GSE) and surveys and interview questions developed by the authors for the study (see Table 2). Of the 51 papers that used subjective measures, more than half (53%) used at least one validated scale to capture users' points of view. Among all the occurrences of subjective measures, the most frequently used validated scales were NASA-TLX (35%) and SUS (8%). Only 16 (25%) of these occurrences provided an explicit definition for these measures.

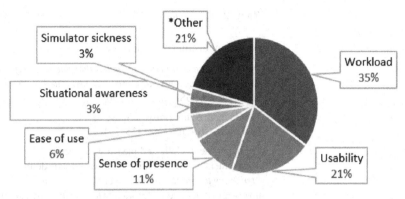

Fig. 7. Percentage of occurrences of subjective measures used in the reviewed papers: workload, usability, sense of presence, ease of use, situational awareness, and simulator sickness; *Other* refers to occurrences of the following subjective measures (acceptance, attitude, functionality, effectiveness, stress, sense of direction, usefulness, UX, intuitiveness, perception of the robot as a teammate, self-efficacy, and speed).

Table 2. List of reported subjective measures, their definitions, and the instruments used to measure them

Subjective Measure	Definition, Instrument, References
Acceptability	Not defined; Ranking based on preferences for performed tasks [25]
Acceptance	Not defined; Presence Questionnaire (PQ) [26–28]
Attitude	Not defined; own survey [29]
Ease of Use	Defined as how easy/difficult users felt in using an interface/system configuration; own survey and freeform feedback [30–33]
Effectiveness	Not defined; own survey [34]
Functionality	Defined as a measure of the number of features the user interface possesses. Used persona development to identify core functions that met the needs of target users [23]
Intuitiveness	Defined as a measure of *"observability"*, *"predictability"*, and *"directability"*; user feedback (open-ended elicitation study) [21]
Perception of robot as a teammate	Defined around the role of the robot as a teammate vs a tool, *"how much they felt that they assisted the robot, supervised the robot, and managed the robot"*; own survey [21]

(*continued*)

Table 2. (*continued*)

Subjective Measure	Definition, Instrument, References
Self-efficacy	Defined as *"an optimistic self-belief that one can perform a novel or difficult task or cope with adversity"*; General Self-Efficacy (GSE) scale [35, 36]
Sense of Direction	Not defined; Santa Barbara Sense of Direction (SBSOD) scale [35, 37]
Sense of Presence	Pre-experiment: Collected measures of focus, involvement, emotions, and games with focus defined as *"tendency to maintain focus on current activities"*, involvement defined as *"tendency to become engaged in activities"*, emotions defined as *"tendency to become involved in activities"*, and games defined as *"tendency to play video games"*; Revised version of the Immersive Tendency Questionnaire (ITQ) [28, 35] Post-experiment: Defined as *"the degree to which an individual experiences presence in a virtual environment and the influence of possible contributing factors on the intensity of this experience: Control Factors, Sensory Factors, Distraction Factors, and Realism Factors"*; MEC Spatial Presence Questionnaire (MEC-SPQ), Presence Questionnaire (PQ), Slater-Usoh-Steed (SUS) questionnaire, own survey [19, 28, 34, 35, 38–43]
Simulator Sickness	Defined as *"symptoms such as paleness, headache, nausea or vomiting caused by conflicting sensory input regarding the body's motion from the vestibular and visual system"*; Simulator Sickness Questionnaire (SSQ), own survey [34, 44, 45]
Situational Awareness (SA)	Defined as a measure of the cognitive state to represent *"how involved the operator is with the task environment"*; China Lake Situational Awareness (CLSA) scale, Post Assessment of Situation Awareness (PASA) [46–49]
*Speed	Not defined; own survey [30]
Stress	Not defined; Self-Assessment Manikin [18]
Usability	Defined as a measure of *"ease-of-use, precision, accuracy, and control"*, and by *"how the developed interface system contributes to the navigation of telepresence robots"*; System Usability Scale (SUS), Perceived Usability Rating (PUR), IBM Computer Usability Satisfaction Questionnaire, own survey and interview questions, user feedback during think-aloud protocol [19, 20, 23, 35, 50–61] and interview questions, user feedback during think-aloud protocol [19, 20, 23, 35, 50–61]

(*continued*)

Table 2. (*continued*)

Subjective Measure	Definition, Instrument, References
Usefulness	Not defined; own interview questions [33]
User Experience	Not defined; User Experience Questionnaire (UEQ) [20]
Workload	Defined as *"the nature and intensity of effort applied to the task"*, *"presenting too much, or inappropriate information"*; NASA-TLX questionnaire, interview [18, 21, 32–35, 41, 46, 48, 54, 59, 62–73]

* Perceived speed has been used as a proxy to measure user experience.

4 Discussion

In this systematic review of literature, we analyzed 77 peer-reviewed papers on the user evaluation of teleoperation interfaces of professional service robots.

Our results show that teleoperation interfaces have been mostly developed for general applications. Some that were developed for specific applications fall under the health and wellness area as shown in Fig. 3. These findings indicate that currently, research tends towards developing interfaces for general applications and or in areas specific to health and wellness. The large number of interface development focusing on general applications suggests a focus on the development of baseline interfaces that leave room for specialization and customization.

In this review, we distinguished between input and feedback mechanisms that form the complex integrated interfaces for teleoperation systems. From UX research point of view, it is important to clarify these nuances to contribute to design ideas and improvements as this area of research matures. Our review shows that bio-signals are widely used as input mechanisms. The prevalence of bio-signals used as input devices in reviewed articles is supported by the argument that using more bio-signals improves the intuitiveness of teleoperated interfaces [17]. Our results show that hand gestures, followed by head and body movements, formed the most frequently used bio-signals. Given that hand, head, and body gestures are naturally used in human communication, it is not surprising that these input mechanisms were used most frequently in the reviewed articles. Increased application of bio-signals as an input mechanism is likely to improve user experience, particularly for novice operators and/or those users who are not tech-savvy. Our review shows that traditional input mechanisms such as the mouse and keyboard are still widely used. However, the popularity of haptic devices used in the reviewed articles reflects a growing shift from traditional interfaces to the use of intuitive modalities for teleoperating professional service robots.

Only one of the reviewed papers used bio-signals to build an adaptive interface. Given the importance of personalization in user experience and task performance, adaptive interfaces are likely to become more prevalent in teleoperations. User evaluations in studies that develop adaptive interfaces, in addition to capturing user experience of the input and feedback modalities, must also include subjective measures for capturing user experience of interface adaptability. The benefits of designing intuitive input and feedback mechanisms as well as developing adaptive interfaces for teleoperating service robots could translate to reduced training times, stress, and cognitive load, thereby improving job satisfaction and other organizational outcomes.

Our results indicate that the most frequently used feedback modality in the reviewed papers was visual. This observation is consistent with previous research that indicates at least one visual display for image or sensor data visualization is used in teleoperations [14]. Our results showing that visual is the most frequently harnessed modality for feedback is also consistent with the general knowledge that vision is the dominant sense for most humans; 90% of the information transmitted to the brain is visual [74]. These results suggest that using eye tracking while operators interact with complex interfaces can be beneficial in understanding how they process information that is provided via visual feedback. Using eye tracking in user studies is shown to be effective in assessing user attributes (e.g., cognitive effort, the experience of task load, anxiety) [75, 76] as well as attention to and engagement with visual displays [7, 77]. Hence, analyzing operators' gaze when looking at visual feedback is likely to provide actionable insight for improving the design of visual feedback, to make it easier to use and/or more useful, especially for novice teleoperators.

While all reviewed papers conducted evaluative research, only one included generative research. This paper used personas and journey maps to gain a deeper understanding of user needs. Because generative UX research is a cornerstone of user-first approach to product development [78], Human-Robot Interaction (HRI) studies in general and studies focusing on designing robotics interfaces for teleoperations in particular, can benefit from including more generative UX research in their investigations.

Successful user-centered technologies are developed through a series of formative user experience studies. The objective of each formative user study is to capture users' point of view to provide timely insight for improving experience design in the next development cycle [7]. Our results showed that only a small number of papers (four out of the 77 reviewed papers) explicitly reported conducting iterative formative user studies. Increasing the number of publications that discuss iterative formative studies can contribute to advancing best practices for developing teleoperation interfaces for service robots.

What is being investigated in a user study is typically communicated formally by providing definitions for each of the measures used in the project (e.g., perceived usefulness refers to *"the degree to which a person believes that using a particular system would enhance his or her job performance"* [79]). Our results showed that only a small percentage of the papers (25%) provided explicit definitions for subjective measures used in their study. Defining subjective measures and using these definitions consistently across studies facilitates a better understanding of the scope and depth of knowledge that is gained from subjective reactions. Consequently, it can help researchers to more successfully build upon each other's work.

Using both qualitative and quantitative subjective measures is most effective in developing actionable insight for design improvements [7]. Qualitative subjective measures in the reviewed papers included user feedback collected via think-aloud protocols and interviews. Quantitative subjective measures included instruments (rankings/ratings) designed by authors specific to their study, as well as validated scales that were used in a number of other studies. Because they are well-defined and use a formal system to measure user reactions, validated scales naturally facilitate standardization across projects. Hence, using validated scales provides an invaluable tool for benchmarking progress, comparing the UX of various designs, and/or conducting meta-analysis across a large number of studies.

The most commonly used subjective measures were workload and usability. While not used as frequently, other subjective measures reported in this review (e.g., intuitiveness and user experience) can help to provide a more nuanced picture of user reactions to teleoperation interfaces. Our results show that subjective measures of usefulness and ease of use were rarely used in the reviewed papers. Ease of use and usefulness are shown to be two important predictors of technology acceptance in organizations. Other factors that are shown to impact technology acceptance at the workplace include *Social Influence* defined as *"the degree to which an individual perceives that important others believe he or she should use the new system"* and *Facilitating Conditions* defined as *"the degree to which an individual believes that an organizational and technical infrastructure exists to support the use of the system"* [12]. Because technology acceptance is shown to improve productivity [80], using subjective predictors of technology acceptance to evaluate teleoperation interfaces for the workplace is both relevant and important.

5 Conclusion and Future Directions

The objective of this systematic literature review was to provide a comprehensive picture of current methods and metrics used in subjective evaluations of teleoperation interfaces of professional service robots. Given that teleoperation interfaces typically use sensors, their design and evaluation process can benefit from the user-centered framework for designing smart and connected products and services [6]. This framework highlights the need for different types of UX research in the *design world* (i.e., the organization within which product development is initiated) and the *usage world* (i.e., where the product is used by its intended market). Grounded in design thinking, this framework relies on generative UX research to discover user needs and requires evaluative UX research to make sure that designed products meet user needs. After a product is launched, UX research is conducted to monitor and maintain market success and to identify new opportunities for value creation. For example, to stay competitive in the marketplace (*usage world*), the sensor data from teleoperation interfaces can be used by smart engines to personalize operators' experience (e.g., automatically sense operators' cognitive load and change the level of automation accordingly). The same sensor data can also be used by specialized smart engines to automatically discover contextual variables/factors that impact experience. Such automatic discoveries then translate to new opportunities for experimentation and thereby contribute to uncertainty reduction in product development (e.g., through building new behavioral models/UX theories) (see Fig. 8).

Our review showed that a great number of papers developed teleoperation interfaces for general applications. This research trend suggests a focus on developing baseline interfaces that can be assembled for specialized applications. Such an engineering-first approach to teleoperation interface design is relevant and important to solving technical problems. Developing user-centered solutions, however, requires a user-first approach to design. That is, the development process must be guided by the results of generative UX research that aims at gaining a deep understanding of user needs, challenges, and goals in respect to context of technology use. Based on the findings of the generative UX research, initial prototypes of teleoperation interfaces for workplace can be designed by either assembling existing baseline interfaces or developing new ones in a way that can best address operators' needs, concerns, challenges, and goals.

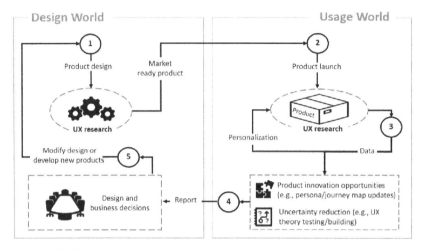

Fig. 8. User centered framework for product development, adapted from [6].

Given the impact of user experience research on minimizing costly failures and creating successful products [81, 82] conducting UX studies that are well-integrated in the development process of teleoperation interfaces are likely to have a notable impact on how readily these interfaces are adopted by their intended users and how effectively they are utilized in the workplace. Such UX studies must include quantitative (e.g., validated scales) and qualitative (user feedback) subjective measures to benchmark the progress, compare alternative designs, and create actionable insight for improving the experience of using the interfaces. By combining objective measures of experience (e.g., eye tracking) and subjective measures of experience (e.g., surveys and interviews) such UX studies can create a rich dataset for understanding users' information processing needs and preferences for visual feedback.

Acknowledgement. The research was sponsored by the DEVCOM Analysis Center and was accomplished under Cooperative Agreement Number W911NF-22-2-0001. The views and conclusions contained in this document are those of the authors and should not be interpreted as representing the official policies, either expressed or implied, of the Army Research Office or the U.S. Government. The U.S. Government is authorized to reproduce and distribute reprints for Government purposes notwithstanding any copyright notation herein.

References

1. ISO 8373:2012(en) Robots and robotic devices — Vocabulary. https://www.iso.org/obp/ui/#iso:std:iso:8373:ed-2:v1:en:term:2.10. Accessed 27 Jan 2023
2. DeLone, W.H., McLean, E.R.: Measuring e-commerce success: applying the DeLone & McLean information systems success model. International Journal of Electronic Commerce, pp. 31–47 (2004)
3. Lindgaard, G., Dudek, C.: What is this evasive beast we call user satisfaction? Interacting with Computers, pp. 429–452 (2003)
4. Menachemi, N., Burkhardt, J., Shewchuk, R., Burke, D., Brooks, R.G.: Hospital information technology and positive financial performance: a different approach to finding an ROI. J. Healthcare Manage. **51**(1), 40–59 (2006)
5. Hurtienne, J., Blessing, L.: Design for intuitive use - testing image schema theory for user interface design. In: DS 42: Proceedings of ICED 2007, the 16th International Conference on Engineering Design, pp. 829–830. Paris (2007)
6. Djamasbi, S., Strong, D.: User experience-driven innovation in smart and connected worlds. AIS Trans. Human-Comp. Interaction **11**(4), 215–231 (2019). https://doi.org/10.17705/1thci.00121
7. Alrefaei, D., et al.: Using eye tracking to measure user engagement with a shared decision aid. In: 17th edition of Augmented Cognition, 25th International Conference on Human Computer Interaction, (Forthcoming)
8. Larkin, C., et al.: ReachCare mobile apps for patients experiencing suicidality in the emergency department: development and usability testing using mixed methods. JMIR Formative Res. **7**, e41422 (2023). https://formative.jmir.org/2023/1/e41422. https://doi.org/10.2196/41422
9. Wang, X.V., Wang, L.: A literature survey of the robotic technologies during the COVID-19 pandemic. J. Manufactur. Syst. **60**, 823–836 (2021)
10. Demir, K.A., Döven, G., Sezen, B.: Industry 5.0 and human-robot co-working. Procedia Comput. Sci. **158**, 688–695 (2019)
11. Sugianto, L.-F., Tojib, D.R., Burstein, F.: A practical measure of employee satisfaction with B2E portals. ICIS 2007 Proceedings, AIS, Montreal (2007)
12. Venkatesh, V., Morris, M.G., Davis, G.B., Davis, F.D.: User acceptance of information technology: toward a unified view. MIS Quarterly, pp. 425–478 (2003)
13. Coronado, E., Kiyokawa, T., Ricardez, G.A., Ramirez-Alpizar, I.G., Venture, G., Yamayobe, N.: Evaluating quality in human-robot interaction: a systematic search and classification of performance and human-centered factors, measures and metrics towards an industry 5.0. Journal of Manufacturing Systems, pp. 392–410 (2022)
14. Ste-Croix, C., Bray-Miners, J., Morton, A.: Human-Robot Interaction Literature Review (2012)
15. Rodriguez, D., Perez, C., Jagersand, M., Figueroa, P.: A comparison of smartphone interfaces for teleoperation of robot arms. In: 2017 XLIII Latin American Computer Conference (CLEI), pp. 1–8. IEEE, Cordoba (2017)
16. Moher, D., Liberati, A., Tetzlaff, J., Altman, D.G., PRISMA Group: Preferred reporting items for systematic reviews and meta-analyses: the PRISMA statement. PLoS Medicine **6**(7), e1000097 (2009). https://doi.org/10.1371/journal.pmed.1000097
17. Leitner, J., Luciw, M., Förster, A., Schmidhuber, J.: Teleoperation of a 7 DOF humanoid robot arm using human arm accelerations and EMG signals. In: 12th International Symposium on Artificial Intelligence, vol. 20, Robotics and Automation in Space (i-SAIRAS), Montreal, Canada (2014)

18. Singh, G., Bermúdez i Badia, S., Ventura, R., Silva, J.L.: Physiologically attentive user interface for robot teleoperation: real time emotional state estimation and interface modification using physiology, facial expressions and eye movements. In: 11th International Joint Conference on Biomedical Engineering Systems and Technologies, pp. 294–302. SCITEPRESS-Science and Technology Publications (2018)

19. Almeida, L., Menezes, P., Dias, J.: Interface transparency issues in teleoperation. Appl. Sci. **10**(18), 6232 (2020)

20. Adamides, G., et al.: Design and development of a semi-autonomous agricultural vineyard sprayer: human–robot interaction aspects. J. Field Robotics **34**(8), 1407–1426 (2017). https://doi.org/10.1002/rob.21721

21. Szafir, D., Mutlu, B., Fong, T.: Designing planning and control interfaces to support user collaboration with flying robots. Int. J. Robotics Res. **36**(5–7), 514–542 (2017)

22. Michaud, F., et al.: Exploratory design and evaluation of a homecare teleassistive mobile robotic system. Mechatronics **20**(7), 751–766 (2010)

23. Nejatimoharrami, F., Faina, A., Jovanovic, A., St-Cyr, O., Chignell, M., Stoy, K.: UI Design for an engineering process: programming experiments on a liquid handling robot. In: 2017 First IEEE International Conference on Robotic Computing (IRC), pp. 196–203. IEEE (2017)

24. Latif, H.O., Sherkat, N., Lotfi, A.: TeleGaze: Teleoperation through eye gaze. In: 2008 7th IEEE International Conference on Cybernetic Intelligent Systems, pp. 1–6. IEEE (2008)

25. Awde, A., Boudaoud, M., Macioce, M., Régnier, S., Clévy, C.: A microrobotic approach for the intuitive assembly of industrial electrooptical sensors based on closed-loop light feeling. IEEE/ASME Trans. Mechatron. **27**(6), 5462–5471 (2022)

26. Lima, A.T., Rocha, F.A.S., Torre, M.P., Azpúrua, H., Freitas, G.M.: Teleoperation of an ABB IRB 120 robotic manipulator and BarrettHand BH8–282 using a Geomagic Touch X haptic device and ROS. In 2018 Latin American Robotic Symposium. In: 2018 Brazilian Symposium on Robotics (SBR) and 2018 Workshop on Robotics in Education (WRE), pp. 188–193. IEEE (2018)

27. Fidêncio, A., et al.: Metodologia para Avaliac̨ ̃ao de Interfaces de Teleoperac̨ ̃ao, XIII Simp ́osio Brasileiro de Automac̨ ̃ao Inteligente (2017)

28. Witmer, B.G., Singer, M.J., Measuring presence in virtual environments: a presence questionnaire. Presence Teleoperators virtual Environ. **7** (3), 225e240 (1998)

29. Mavridis, N., Giakoumidis, N., Machado, E.L.: A novel evaluation framework for teleoperation and a case study on natural human-arm-imitation through motion capture. Int. J. Soc. Robot. **4**, 5–18 (1998)

30. Ainasoja, A.E., Pertuz, S., Kämäräinen, J.K.: Smartphone Teleoperation for Self-Balancing Telepresence Robots (2019)

31. Antuvan, C.W., Ison, M., Artemiadis, P.: Embedded human control of robots using myoelectric interfaces. IEEE Trans. Neural Syst. Rehabil. Eng. **22**(4), 820–827 (2014)

32. Pryor, W., et al.: Experimental evaluation of teleoperation interfaces for cutting of satellite insulation. In: 2019 International Conference on Robotics and Automation (ICRA), pp. 4775–4781. IEEE (2019)

33. Macchini, M., Havy, T., Weber, A., Schiano, F., Floreano, D.: Hand-worn haptic interface for drone teleoperation. In: 2020 IEEE International Conference on Robotics and Automation (ICRA), pp. 10212–10218. IEEE (2020)

34. Schmidt, L., Hegenberg, J., Cramar, L.: User studies on teleoperation of robots for plant inspection. Industrial Robot: An Int. J. **41**(1), 6–14 (2014)

35. Adamides, G., et al.: HRI usability evaluation of interaction modes for a teleoperated agricultural robotic sprayer. Appl. Ergon. **62**, 237–246 (2017). https://doi.org/10.1016/j.apergo.2017.03.008

36. Schwarzer, R., Jerusalem, M.: Self-efficacy measurement: Generalized self-efficacy scale (GSES). In: Weinman, J., Wright, S., Johnston, M. (Eds.), Measures in health psychology: A user's portfolio, pp. 35–37. NFER-Nelson, Windsor, England (1995)
37. Hegarty, M., Richardson, A.E., Montello, D.R., Lovelace, K., Subbiah, I.: Development of a self-report measure of environmental spatial ability. Intelligence 30(5), 425447 (2002)
38. Vorderer, P., et al.: Development of the MEC spatial presence questionnaire (MEC SPQ). unpublished report to the European Community on Project Presence: MEC (IST-200137661), Hannover, Munich, Helsinki, Porto, Zurich (2004)
39. Wang, Z., Giannopoulos, E., Slater, M., Peer, A.: Handshake: realistic human-robot interaction in haptic enhanced virtual reality. Presence 20(4), 371–392 (2011)
40. Orlosky, J., Theofilis, K., Kiyokawa, K., Nagai, Y.: Effects of throughput delay on perception of robot teleoperation and head control precision in remote monitoring tasks. PRESENCE: Virtual and Augmented Reality 27(2), 226–241 (2018)
41. Nenna, F., Orso, V., Zanardi, D., Gamberini, L.: The virtualization of human–robot interactions: a user-centric workload assessment. Virtual Reality, pp. 1–19 (2022)
42. Slater, M., Usoh, M., Steed, A.: Depth of presence in virtual environments. Presence-Teleoper. Virtual Environ. 3, 130–144 (1994)
43. Usoh, M., Catena, E., Arman, S., Slater, M.: Using presence questionnaires in reality. Presence Teleoper. Virtual Environ. 9, 497–503 (2000)
44. Reason, J.T., Brand, J.J.: Motion Sickness. Academic press (1975)
45. Kennedy, R.S., Lane, N.E., Berbaum, K.S., Lilienthal, M.G.: Simulator sickness questionnaire: an enhanced method for quantifying simulator sickness. Int. J. Aviat. Psychol. 3(3), 203–220 (1993)
46. Blitch, J.G.: A neurophysiological examination of multi-robot control during NASA's extreme environment mission operations project. In: Advances in Human Factors in Robots and Unmanned Systems: Proceedings of the AHFE 2016 International Conference on Human Factors in Robots and Unmanned Systems, pp. 341–351. Walt Disney World®, Springer International Publishing, Florida, USA (2017). https://doi.org/10.1007/978-3-319-41959-6_28
47. Adams, S., Kane, R., Bates, R.: Validation of the China lake situational awareness scale with 3D SART and S-CAT. Naval Air Warfare Center Weapons Division (452330D), China Lake, CA (1998)
48. Chicaiza, F.A., Slawiński, E., Salinas, L.R., Mut, V.A.: Evaluation of path planning with force feedback for bilateral teleoperation of unmanned rotorcraft systems. J. Intell. Robotic Syst. 105(2), (2022). https://doi.org/10.1007/s10846-022-01651-y
49. Gatsoulis, Y., Virk, G.S., Dehghani-Sanij, A.A.: On the measurement of situation awareness for effective human-robot interaction in teleoperated systems. J. Cognitive Eng. Decision Making 4(1), 69–98 (2010)
50. Brooke, J.: SUS-A quick and dirty usability scale. Usability Eva. Ind. 189(194), 4–7 (1996)
51. Bangor, A., Kortum, P.T., Miller, J.T.,: An empirical evaluation of the system usability scale. Intl. J. HumaneComputer Interact. 24(6), 574e594 (2008)
52. Valero-Gomez, A., de la Puente, P.: Usability evaluation of a pda interface for exploration mobile robots. IFAC Proc. 44(1), 1120–1125 (2011)
53. Drury, J.L., Keyes, B., Yanco, H.A.: Lassoing hri: analyzing situation awareness in map-centric and videocentric interfaces. In: Proceedings of the Second ACM SIGCHI/SIGART Conference on Human-Robot Interaction, pp. 279–286 (2007)
54. Lin, T.C., Krishnan, A.U., Li, Z.: Intuitive, efficient and ergonomic tele-nursing robot interfaces: Design evaluation and evolution. ACM Trans. Human-Robot Interaction (THRI) 11(3), 1–41 (2022)

55. Jacinto-Villegas, J.M., et al.: A novel wearable haptic controller for teleoperating robotic platforms. IEEE Robotics Auto. Lett. **2**(4), 2072–2079 (2017), Article 7962162. https://doi.org/10.1109/LRA.2017.2720850

56. Lewis, J.R.: IBM computer usability satisfaction questionnaires: psychometric evaluation and instructions for use. Technical Report. IBM—Human Factors Group, Boca Raton, FL, USA (1993)

57. Cruz-Ramirez, S.R., Ishizuka, Y., Mae, Y., Takubo, T., Arai, T.: Dismantling interior facilities in buildings by human robot collaboration. In: 2008 IEEE International Conference on Robotics and Automation, pp. 2583–2590. IEEE (2008)

58. Chauhan, M., Deshpande, N., Caldwell, D.G., Mattos, L.S.: Design and modeling of a three-degree-of-freedom articulating robotic microsurgical forceps for trans-oral laser microsurgery. J. Med. Dev. **13**(2) (2019)

59. Sanguino, T.M., Márquez, J.A., Carlson, T., Millán, J.D.: Improving skills and perception in robot navigation by an augmented virtuality assistance system. J. Intell. Robotic Syst. **76**, 255–266 (2014)

60. Alonso, R., Bonini, A., Reforgiato Recupero, D., Spano, L.D.: Exploiting virtual reality and the robot operating system to remote-control a humanoid robot. Multimedia Tools Appl. **81**(11), 15565-15592 (2022)

61. Nakayama, A., Ruelas, D., Savage, J., Bribiesca, E.: Teleoperated service robot with an immersive mixed reality interface. Информатика и автоматизация **20**(6), 1187–1223 (2021)

62. Hart, S.G., Staveland, L.E.: Development of NASA-TLX (Task Load Index): results of empirical and theoretical research. In: Advances in Psychology, **52**, pp. 139–183. North-Holland (1988)

63. Kent, D., Saldanha, C., Chernova, S.: Leveraging depth data in remote robot teleoperation interfaces for general object manipulation. Int. J. Robotics Res. **39**(1), 39–53 (2022)

64. Glas, D.F., Kanda, T., Ishiguro, H., Hagita, N.: Field trial for simultaneous teleoperation of mobile social robots. In: Proceedings of the 4th ACM/IEEE International Conference on Human Robot Interaction, pp. 149–156 (2009)

65. Doisy, G., Ronen, A., Edan, Y.: Comparison of three different techniques for camera and motion control of a teleoperated robot. Appl. Ergon. **58**, 527–534 (2017)

66. Hart, S.G.: NASA-task load index (NASA-TLX); 20 years later. Paper presented at the Proceedings of the human factors and ergonomics society annual meeting (2006)

67. Okishiba, S., et al.: Tablet interface for direct vision teleoperation of an excavator for urban construction work. Automation Const. **102**, 17–26 (2019)

68. Gholami, S., Garate, V.R., De Momi, E., Ajoudani, A.: A probabilistic shared-control framework for mobile robots. In: 2020 IEEE/RSJ International Conference on Intelligent Robots and Systems (IROS), pp. 11473–11480. IEEE (2020)

69. Kikuchi, T., Takano, T., Yamaguchi, A., Ikeda, A., Abe, I.: Haptic interface with twin-driven mr fluid actuator for teleoperation endoscopic surgery system. Actuators, **10**(10), 245 (2021). https://doi.org/10.3390/act10100245

70. Bhat, R., Pandey, V., Rao, A.K., Chandra, S.: An evaluation of cognitive and neural correlates for indirect vision driving and rover teleoperation. In: 2017 2nd International Conference on Man and Machine Interfacing (MAMI), pp. 1–5. IEEE (2017)

71. Randelli, G., Venanzi, M., Nardi, D.: Evaluating tangible paradigms for ground robot teleoperation. In: 2011 RO-MAN, pp. 389–394. IEEE (2011)

72. Quintero, C.P., Dehghan, M., Ramirez, O., Ang, M.H., Jagersand, M.: Flexible virtual fixture interface for path specification in tele-manipulation. In: 2017 IEEE International Conference on Robotics and Automation (ICRA), pp. 5363–5368. IEEE (2017)

73. Jevtić, A., Colomé, A., Alenyà, G., Torras, C.: User evaluation of an interactive learning framework for single-arm and dual-arm robots. In: Agah, A., Cabibihan, J.-J., Howard, A.M., Salichs, M.A., He, H. (eds.) ICSR 2016. LNCS (LNAI), vol. 9979, pp. 52–61. Springer, Cham (2016). https://doi.org/10.1007/978-3-319-47437-3_6

74. Siddhartha, B., Chavan, A.P., Uma, B.V.: An electronic smart jacket for the navigation of visually impaired society. Materials Today: Proc. 5(4), 10665–10669 (2018)

75. Shojaeizadeh, M., Djamasbi, S., Paffenroth, R.C., Trapp, A.C.: Detecting task demand via an eye tracking machine learning system. Decis. Support Syst. 116, 91–101 (2019)

76. Alrefaei, D., et al..: Impact of anxiety on information processing among young adults: an exploratory eye-tracking study. In: Proceedings of the 56th Hawaii International Conference on System Sciences, pp. 6321–6330. Hawaii (2023)

77. Norouzi Nia, J., Varzgani, F., Djamasbi, S., Tulu, B., Lee, C., Muehlschlegel, S.: Visual hierarchy and communication effectiveness in medical decision tools for surrogate-decision-makers of critically ill traumatic brain injury patients. In: Schmorrow, D.D., Fidopiastis, C.M. (eds.) HCII 2021. LNCS (LNAI), vol. 12776, pp. 210–220. Springer, Cham (2021). https://doi.org/10.1007/978-3-030-78114-9_15

78. Jain, P., Djamasbi, S., Wyatt, J.: Creating value with proto-research persona development. In: HCI in Business, Government and Organizations. Information Systems and Analytics: 6th International Conference, HCIBGO 2019. Held as Part of the 21st HCI International Conference, HCII 2019. Proceedings, Part II 21, pp. 72–82. Springer International Publishing, Orlando, FL, USA (2019). https://doi.org/10.1007/978-3-030-22338-0_6

79. Davis, F.D.: Perceived usefulness, perceived ease of use, and user acceptance of information technology. MIS Q. 13, 318–340 (1989)

80. Igbaria, M., Tan, M.: The consequences of information technology acceptance on subsequent individual performance. Inf. Manage. 32(3), 113–121 (1997)

81. UserZoom: The State of UX 2022. UserZoom (2022)

82. Forrester Research Inc.: The Total Economic Impact of IBM's Design Thinking Practice. IBM (2018)

Safety of Human-Robot Collaboration within the Internet of Production

Minh Trinh[1]([⊠]), Hannah Dammers[2], Mohamed Behery[3], Ralph Baier[4],
Thomas Henn[5], Daniel Gossen[6], Burkhard Corves[6], Stefan Kowalewski[5],
Verena Nitsch[4], Gerhard Lakemeyer[3], Thomas Gries[2], and Christian Brecher[1]

[1] Laboratory of Machine Tools and Production Engineering, RWTH Aachen University,
Campus-Boulevard 30, 52074 Aachen, Germany
m.trinh@wzl.rwth-aachen.de

[2] Institut Für Textiltechnik, RWTH Aachen University, Otto-Blumenthal-Straße 1, 52074
Aachen, Germany

[3] Knowledge-Based Systems Group, RWTH Aachen University, Ahornstraße 55, 52074
Aachen, Germany

[4] Institute of Industrial Engineering and Ergonomics, RWTH Aachen University,
Eilfschornsteinstraße 18, 52062 Aachen, Germany

[5] Informatik 11 – Embedded Software, RWTH Aachen University, Ahornstraße 55, 52074
Aachen, Germany

[6] Institute of Mechanism Theory, Machine Dynamics and Robotics, RWTH Aachen University,
Eilfschornsteinstraße 18, 52062 Aachen, Germany

Abstract. Recent trends in globalization have led to an increased competitive
pressure, particularly affecting the manufacturing industry. The Cluster of Excellence "Internet of Production" (IoP) aims at developing innovative solutions and
reshaping production to enable local industries to thrive in a digitized world.
These developments create new possibilities for Human-Robot Interaction (HRI)
and Human-Robot Collaboration (HRC) in particular. An extended framework
for the classification, analysis, and planning of HRI use cases within the cluster
IoP was developed, whereby examples from preforming and assembly were introduced and classified. Due to their collaborative nature, these cases require high
safety standards that protect the human from the cobot. This paper describes how
the cluster IoP handles safety of human workers in HRC processes, which allows
a shift towards an increased collaboration between humans and robots. In this context, this paper proposes different methods to increase safety in HRC applications.
This includes the use and verification of Behavior Trees for process planning and
execution, the application of Computer Vision, the design of safe robot tools, and
the evaluation of human acceptance and trust.

Keywords: Human-Robot Collaboration · Safety · Behavior Trees

1 Introduction

The use of industrial robot systems that interact with humans paves the way for an
increased flexibility in production. Furthermore, new methods of data acquisition, processing and modeling enable new possibilities for human-robot interaction (HRI) and in

© Springer Nature Switzerland AG 2023
F. Fui-Hoon Nah and K. Siau (Eds.): HCII 2023, LNCS 14039, pp. 86–103, 2023.
https://doi.org/10.1007/978-3-031-36049-7_7

particular human-robot collaboration (HRC). These developments are especially driven in the Cluster of Excellence research project "Internet of Production" (IoP) [1]. A focus lies on transferable and interdisciplinary research between researchers from domains such as mechanical engineering, computer science and ergonomics. As a comparison, HRC is a research field that comprises knowledge from many different expert fields as well.

Baier et al. developed an extended framework for the classification, analysis and planning of HRI use cases, in order to support the human-oriented work system design of next generation production plants [2]. This included the analysis of use cases from HRC. Despite the continuous growth of the market for collaborative robots (cobots) [3], the industrial use of cobots is still marginal, especially for small and medium-sized businesses (SMB), which is mainly due to the strict regulations that aim to protect humans during collaboration.

Therefore, we collect current methods for safety of HRC processes within the cluster IoP in this work that are cost-efficient at the same time. This first includes the use of inherently reactive Behavior Trees (BT) and their verification. We use a verification method for BT that verifies safety properties in the program code and thus can eliminate possible errors. The BT semantics and error checks are therefore encoded in a collection of logic formulas, which in turn are verified by an existing verification algorithm. Furthermore, a Speed and Separation Monitoring (SSM) system can be integrated into a BT architecture in addition to the Power Force Limiting (PFL) function of the cobot using cameras and Computer Vision (CV) to detect the human hand. We outline how the given standards and guidelines can be integrated into the development of robot tools for new working environments as well as identifying deficits. Finally, approaches to improve the confidence range from visual indicators to anthropomorphic robots will be discussed in this paper.

This paper describes a cluster IoP-centered take on safety for HRC. Since the cluster IoP conducts research in close cooperation with the industry, this paper aims at bridging the gap between research and industry. For this purpose, two industrial use cases are presented and used for demonstrating safety regulation checks. The structure of this paper is as follows: In Sect. 2, current safety regulations regarding HRC are summarized. Section 3 shows the industrial use cases and their classification according to the extended framework. This is followed by a survey of safety concepts within the cluster IoP and a safety check for the presented use cases in Sect. 4, before finishing with a conclusion and outlook (Sect. 5).

2 Safety in Human-Robot Collaboration

According to data from the German, Austrian and U.S. labor unions, more than 55% of human injuries inflicted by robots affect the human hand, see Fig. 1. This data does not include collaborative robotic cells; however, the data shows most of the accidents between humans and robots occur due to clamping and crushing [16]. Since more than half of these accidents affect the hand, it is likely that most accidents in HRC affect the hand as well.

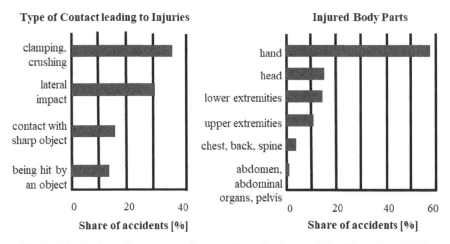

Fig. 1. Injuries depending on type of contact (l) and body parts injured by robots (r) [4].

2.1 Collaborative Safety Regulations

In 2010, the European Union developed general design principles for machine safety, risk assessment and risk reduction, called DIN EN ISO 12100:2010. It replaced its precursor DIN EN ISO 12100:2003. The main purpose of this new international standard within the European Union is to provide designers and engineers with an overview and guidance for decisions to be made during the development of machinery and to enable them to design machinery that is safe for its intended use. In DIN EN ISO 12100:2010, machine safety is defined as the ability of a machine to carry out its intended function(s) during its lifetime with the risk being sufficiently reduced. To reduce the risk sufficiently, the designer or engineer must apply a risk assessment and risk reduction scheme. This scheme consists of [5]:

1. Considering the safety of the machine in all phases of its life cycle (most important).
2. The ability of the machine to perform its function (very important).
3. The ease of use of the machine (important).
4. The manufacturing, operating and dismantling costs of the machine (less important).

In 2020 and 2021, the European Union developed DIN EN ISO 10218-2:2020 and DIN EN ISO 10218-1:2021 about safety regulations regarding industrial robots. The first part is about industrial robots in general, the second part is about their applications and industrial robot cell integration. The aim of both norms is to lay the foundation for safe construction of protective measures and to provide information for the safe use of robots in industrial environments [18, 19].

According to the first part of the norm, control systems must be designed in a way that a reasonably foreseeable human error during operation does not lead to hazardous situations. If the robot starts acting unexpectedly during a failure in the control system, there must be a protective measure to counteract harm for workers. Additionally, the robot system must provide a controller, programming pendant or external control as well as the corresponding connectivity. Furthermore, there must be a limit to the range of the robot which cannot exceed the maximal workspace that is reserved for the robot [6].

In industrial HRC, the human must be protected from the (moving) robot and the product the robot carries. If possible, sharp edges, which could harm the human must be reduced [7]. There are four different possibilities to ensure a safe HRC [8]:

- Hand-guided Control
- Power and Force Limiting (PFL)
- Speed and Separation Monitoring (SSM)
- Safety rated Monitor Stop

Hand-based control means the robot is only moving if the worker physically moves the robot with his/her hands. If the human is protected via PFL, the limits for power and force must be set beforehand. If a sensor in the robot measures a force above the defined threshold, the robot must perform an emergency stop. Another way to protect human beings is the implementation of an SSM system. The SSM system must reduce the speed of the robot if the distance between worker and robot falls below a certain threshold. Additionally, the SSM system can change the trajectory of the robot to keep a minimum distance between the worker and the robot. The safety rated monitor stop stops the robot as soon as the worker enters the workspace of the robot [8]. Therefore, the worker and the robot are not able to work together on one product in the same cell at the same time.

2.2 Risk Assessment and Reduction

Carrying out a risk assessment and risk reduction, the designer or engineer must proceed iteratively in the following order [5]:

- Defining the limits of the machine, including its intended use and reasonably foreseeable misuse.
- Identifying hazards and associated hazardous situations.
- Assess the risk for each identified hazard and hazardous situation.
- Assess the risk and make decisions about the need for risk mitigation.
- Eliminating the hazard or reducing the risk associated with the hazard through protective measures.

The risk related to the hazard under consideration is a function of the damage, which could arise due to this situation as well as the possibility that this situation will occur. The possibility that this hazardous situation will occur is itself a function of the hazard exposure, occurrence of a hazardous event as well as the possibility to avoid or limit the damage. Figure 2 visualizes this context [5]. For the long-term safe operation of a machine, it is important that the protective measures enable the machine to be used without disturbances. Additionally, the protective measures must not impair its intended use. The risk assessment must consider the possibility that protective measures can be rendered ineffective or circumvented. The designer or engineer must also consider that there may be an incentive to render protective measures ineffective or to circumvent them [5].

In 2015, the European Union published DIN EN ISO 13849-1:2015. It lays out the general principles for the design of safe control systems. There are different ways to reduce the risk of a hazard situation. The risk can be reduced by mechanical protective measures such as protective algorithms or by electrical protective measures. Before reducing the risk, the average probability of a dangerous failure is determined. If the value is standardized to one hour, the value is called performance level. There are five different performance levels ranging from more than 10^{-5} failures per hour to less than 10^{-7} failures per hour [9].

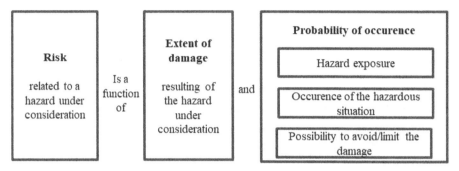

Fig. 2. Risk is a function of extent of damage and its probability of occurrence [5].

2.3   Software Safety Lifecycle

For the software safety lifecycle, a simplified V-model is used to ensure the reliability of the algorithm, see Fig. 3. Hereby, the software is designed gradually starting from the safety-related specifications of the algorithm. After each step, the system/module/code is verified to ensure the reliability and safety of the software and its integration into the system [9]. DIN EN ISO 13849-2:2012 is about the validation of safety-related control-systems. The validation of the safety functions aims to demonstrate that the machine or control system performs the safety function(s) corresponding to the specified properties. This is achieved by an analysis, which includes [10]:

- The structure or architecture of the system.
- Deterministic arguments.
- Quantifying aspects (e.g. average time until failure).
- Safety functions identified during the risk analysis, their properties and the required performance level(s).
- Qualitative aspects which influence the system behavior.

The analysis can either be performed top-down or bottom-up. [ISO12].

3   Classification of Two Use Cases Within the Cluster IoP

Two exemplary use cases of HRC within the cluster IoP are presented and classified using the extended framework as proposed in [2] to identify the kind of HRI and the characteristics of the systems used. This analysis is important in terms of safety, so that

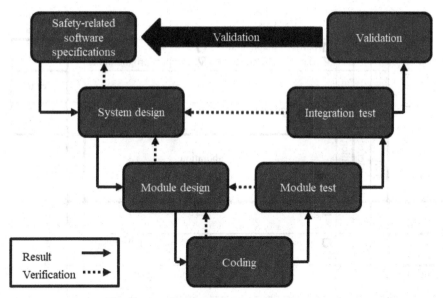

Fig. 3. Simplified V-model for safe software [9].

it is clearly shown where there are contacts between humans and robots, what risks exist and which resources (e.g. sensors) are available to counter the risks or generally increase safety.

3.1 Classification of Human-Robot Interactions

For the classification of the HRC use cases, the framework according to Baier et al. [2] is used, which was developed for the cluster IoP. Figure 4 shows the use cases described in this paper. This framework allows both the analysis and the synthesis of an HRC application, based on three dimensions: (1) On the horizontal axis, the degree of overlap of the workspaces is plotted. This ranges from completely separated to partially over-lapping and to connected to the body. (2) The dimension "precondition and implication" forms the interface of the HRC system to its subsystems. Here, the question is, which preconditions exist and which implications the HRC system possesses in the social, legal and technical domains. (3) Since the IoP project focuses on the collection, processing and use of data at distributed locations, this requirement is mapped in a separate dimen-sion. A distinction is made as to whether the data originates from sensors that are either attached to the device itself or supplied by external sensors, or whether it is available as a digital shadow. A digital shadow is the digital representation of an artifact (product or process) through a set of selected properties (see [1] for more details). Figure 4 illustrates the application of the classification scheme based on the two use cases described in the following.

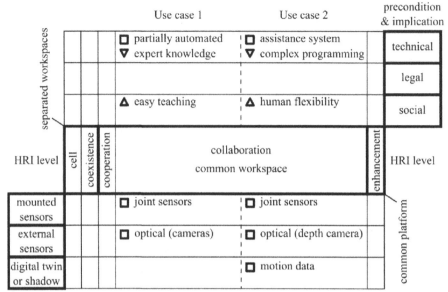

Fig. 4. Exemplary classification of use cases 1 and 2 with a selection of properties on the basis of [2] with adapted visualization to display the use cases side by side. Legend: △ positive, ▽ negative and □ not evaluated.

3.2 Use Case 1: Collaborative Assembly of Multi-variant Products

The trend towards multi-variant and customer-specific products as well as the shortening of the time to market requires an increased flexibility of the production process. Assembly is particularly affected by this due to its proximity to the market as the last production step. Collaborative assembly, which combines the advantages of manual work (flexibility and dexterity) and automated work (efficiency and quality advantages), offers potential for optimization. A disadvantage of automation is the necessary expertise, which SMB often lack.

When planning and implementing collaborative assembly processes, two aspects are particularly relevant: flexibility and safety. On the one hand, uncomplicated adaptation of the process must be possible in the event of changes, without the need for a robotics expert. The majority of robot manufacturers such as Franka Emika GmbH or Rethink Robotics Inc. Offer graphical user interfaces that replace code-based programming. Process planning was further simplified by Land et al. [11] using a simulation-based framework and, for example, a method for automated planning based on reinforcement learning was developed in [12]. Collaboration as the highest level of interaction between humans and robots (shared workspace as well as activity) requires strict safety standards. In [13], the internal safety functions of a cobot are augmented with external sensors and camera systems to prevent collisions between humans and robots while maintaining process efficiency. An algorithm has been developed in [14], for example, which can compute trajectories in real time to avoid obstacles.

The methods currently developed often represent isolated solutions that are not directly applicable to every type of robot. Furthermore, in most cases an extension by further functions requires a complex adaptation of the overall system.

In the project "CoboTrees" within the cluster IoP, BT are developed for the manufacturer-independent optimization of collaborative assembly processes while complying with safety standards at the same time [15]. The validation use case of collaborative assembly of a lamp can be seen in Fig. 5. It shows steps that can easily be done by the robot such as fastening screws but also handling of cables, which requires the dexterity of the human.

Classification. *HRI Level:* Due to the parallel manufacturing of the product by human and machine in the same workspace, this is classified as a collaboration.

Preconditions and Implications: The process can be partially automated, although individual work steps must be performed by humans. This applies in particular to the flexible cable. Expert knowledge is required to program the robot, whereas teaching is possible without prior knowledge.

Data Sources: The primary data sources used are externally mounted cameras that optically record the work process. In addition, the torque sensors installed in the robot joints are used for safety purposes by detecting unexpected forces and bringing about an emergency stop. The gripper is HRC-capable but with no dedicated sensors. A Digital Shadow is not used in this use case.

Fig. 5. Demonstrator for collaborative assembly of a lamp.

3.3   Use Case 2: Collaborative Production of FRP Parts

Due to their excellent mechanical properties combined with lightweight, Fiber-Reinforced Plastics (FRP) are used in high performances applications such as aviation, automotive, wind energy and sporting goods. At the moment, around a third of all FRP parts worldwide is produced manually [16].

On the one hand, this is due to the fact that in manual FRP production the workers' sensorimotor abilities can be fully taken advantage of. In addition, rapid product changes are possible, due to the cognitive flexibility of humans as well as low machine and tooling costs. However, waste resulting from human errors, low production speeds and high wages are disadvantageous [17, 18]. On the other hand, the automation of FRP production seems promising in reducing manual process steps and increasing reproducibility. Large corporations in the aviation and automotive industry are already using special automation machines for the production of FRP parts. Yet, these machines are expensive and their implementation is elaborate which results in a lack of flexibility [19, 20].

In most cases, the existing demands regarding FRP part quality, low part costs, productivity and ergonomics of production cannot be fulfilled by existing production scenarios. Therefore, a semi-automated production of FRP parts is a promising approach [18, 21, 22]. In this use case (see Fig. 6), the collaboration of humans and robots is investigated to exploit both, human strengths (cognitive flexibility, dexterity, sensorimotor abilities) and robot strengths (high precision and speed).

Currently, robotic tools that are able to implement the collaborative production of FRP parts are not commercially available. Therefore, these tools are developed as part of the use case. The focus lies on compatibility with the limp and partly sticky textiles to be processed. Typically, cutting devices are used for FRP part production in addition to rolling and squeegee tools. Here, special safety requirements must be met in order to protect humans from being harmed within the process.

Classification.

HRI Level: The simultaneous handling of the product in the same workspace or the assistance of the robot during human handling constitutes a collaboration according to the classification scheme.

Preconditions and Implications: Due to the complex geometry of the part and unpredictable behavior of the limp textile material, human flexibility is irreplaceable. It is combined with the power and accuracy of the machine. In combination, this results in a shift in the scope of the human task. The complex programming of the technical system is a disadvantage.

Data Sources: In this case, the torque sensors built into the robot joints are used to interact with humans. They measure the forces acting from the outside and determine the direction in which the human wants to move the arm. Furthermore, they are used to define the forces with which the textile material is brought into 3D shape. In addition to the internal sensors, external depth cameras are used to detect the human and the workpiece. Motion data from the robot controller and videos of the workers' movements for saving human expert knowledge are output as digital shadows.

4   Safe Collaboration Within the Internet of Production

In this Section, different concepts that are currently under research and enable safe collaboration for the two classified use cases are presented. These include inherently safe BT and their verification, the use and integration of CV, safe end-effector design and implementation as well as the consideration of human acceptance and trust.

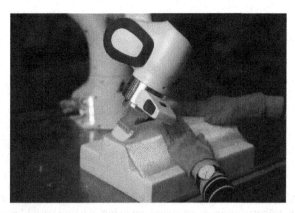

Fig. 6. Human-robot collaboration within FRP-part production.

4.1 Process Modeling and Execution Using Behavior Trees

BT are mathematical models of plan execution and widely used in the robotics community (e.g., in an unmanned aerial vehicle's control system [23]). They are reactive because of their tick-based execution [23, 24]. This allows interleaving of sensing and execution tasks. With every tick of the root node, safety branches are visited before ticking the execution node. Safety can be checked on a high level, where a "safety sub-tree" is the left most branch, i.e., ticked first.

BT synthesis approaches can create a tree in runtime from a given set of goals [24, 26, 27]. They start with a set of condition nodes, each representing a given goal and iteratively expand the tree to increase its region of attraction, i.e., the set of states from which the executing tree eventually returns success. These approaches do not focus on optimal execution but guarantee convergence. They are also able to expand the tree at runtime according to the environment, e.g., adding a branch to remove obstacles in the way before proceeding with navigation tasks.

We can also execute safety on a lower level by adding guards to the left of action nodes. Additionally, BT were combined with control barrier functions in a multi-agent system to guarantee that the tree completes a task under constraints (e.g., avoiding collisions) [25]. The tree in Fig. 7 checks whether sensors (e.g., camera) are active before executing tasks and later checks if a collision free delivery path exists before ticking the "deliver" node.

It is also possible to assure safety on the implementation level of BT nodes. Defects occur on this level due to the clear and intuitive representation of BT. The representation hides the exact implementation of actions and the control flow, which is implicitly defined. For example, in Fig. 7, it is possible, depending on the implementation, that after adding the screws the fastening step fails, which could lead to adding screws during the next tick although they are already in place. To prove the absence of these kinds of errors, different encodings of BT and their verification were presented [28–30]. These approaches often cover only a subset of nodes in a BT or need an additional input from the user, which are the semantics of actions specified in a certain logic. These restrictions prevent the simple use of verification methods by the user.

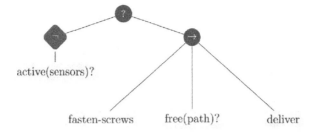

active(sensors)?

fasten-screws   free(path)?   deliver

Fig. 7. A BT to assemble and deliver a desk lamp.

The presented approach from the cluster IoP [25], depicted in Fig. 8, allows the user to write the safety property as an assert statement using existing variables in the program. The assertions are part of the program code and therefore the checked safety properties are on a lower level than those presented in Sect. 4.1. Since the assert statements are only relevant for verification, they are removed before deployment of the system and do not impact the runtime behavior. The semantics of the added assert statements and the BT are automatically encoded as Linear Constrained Horn Clauses. In [25] an existing SMT solver is used to analyze the set of clauses and either obtain proof that the assertions hold for every execution or receive a counterexample which violates at least one assert statement.

The approach works independently of the actual system and environment. In order to analyze safety properties which depend on the environment, a facility is provided to implement the semantics of the environment as a BT, which is then combined with the BT of the system. The next step encodes the BT of the system and environment together.

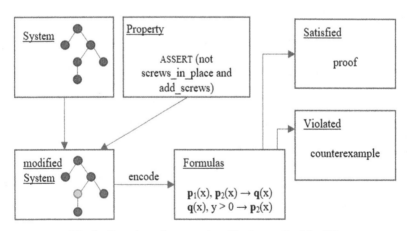

Fig. 8. Overview of proposed verification method for BT.

4.2   Application of Computer Vision

The previous section described a formal method to verify the safety of a HRC process. In order to react to dynamic changes, which mainly result from the unpredictability of the human worker, machine learning, or to be more precise, CV is proposed. CV is an interdisciplinary scientific field. It tries to process and analyze the images captured by cameras to understand their content (classification) or extract geometric information automatically (detection) [26]. In the case of HRC the latter is relevant for automatically extracting geometric information from an image to detect the position of a human being in the robot's work cell.

Since the majority of accidents occur on the human hand (see Fig. 1), we focus on CV algorithm capable of single hand detection. Many algorithms such as YOLOv5 and OpenCV. The YOLOv5 network uses bounding box regression and does not determine the exact (key) points of the hand [27]. The OpenCV hand detector on the other hand is already pretrained with more than 30.000 different images of hands [28]. Additionally, the algorithm allows for a high frame rate because as soon as a hand is detected, the algorithm searches in the local environment of the hand in the next image. A disadvantage is that the algorithm only outputs the evaluation of the hand position from an RGB image in 2D.

In use case 1, the OpenCV hand detection algorithm is integrated into the BT and tested on a real human hand. A detailed description and results can be found in [29]. The integration of the algorithm is structured according to the V-Model (see Fig. 3). After conducting the system analysis and requirements for the safety algorithm, a generic SSM system in ROS with an embedded BT is designed. The subsequent module design contains a camera module, which preprocesses and forwards the image of the camera into ROS, the detection algorithm and transformation module as well as the safety subtree. After the coding phase and testing of the individual modules, the whole system is tested. Here, the hyperparameters of the implemented algorithm are optimized, which need to be set by the engineer beforehand and depend on the robot and the environment of the robot. The stopping distance (minimum distance between hand and robot) is computed in order to meet DIN ISO 15066.

In the validation phase, the system shows high reliability for different hand gestures. The slow frame rate of 10 fps leads to a high value for the stopping distance to satisfy DIN ISO 15066, since only a CPU is used. Furthermore, the detection algorithm fails if the worker wears gloves (Fig. 9).

4.3   End-Effector Design

In addition to the use of BT for process planning and execution as well as the external application of CV, further improvement of safety can be accomplished through an appropriate design of the applied gripper. As stated in DIN 15066, different security measures are identified for distinct times of a contact phase. Here a pre-contact stage, the moment of contact and a post-contact stage are considered. Thereby, active security measures are identified to address the pre-contact phase, while passive security measures can improve safety within the stage of contact and the post-contact stage (see Table 1).

Fig. 9. Hand detection algorithm using OpenCV.

Table 1. Active and passive safety measures for end-effector design

Active safety measures (pre-contact phase)
Limitation of forces and torques by the Body Atlas
Limitation of cobot speed
External sensors to predict collisions
Passive safety measures (moment of contact and post-contact phase)
Increased contact area (rounded edges, smooth surfaces)
Compliant surfaces (padding, deformable components)
Reduction of kinetic energy through lightweight construction
Limitation of drive energy

Active security measures include limiting the maximum forces and pressures for cobots, which are derived from the Body Atlas found in DIN 15066 [8]. The Body Atlas divides the human body into 29 areas. For each area, biomechanical limits are described, each of which specifies the maximum pressure and force values that a body area can withstand without sustaining irreversible or major damage.

In addition, a distinction is made between two types of contact. Quasi-static contact (static clamping situation) describes a contact in which a part of the operator's body is clamped between the robot and another fixed part. In the case of transient contact, the body part that is hit can rebound from the robot (dynamic impact). The permissible contact forces are higher for transient contact because the person can escape the movement. Figure 10 shows an overview of the maximum forces and pressure loads for the quasi-static contact case.

Overall, the maximum cobot speed at which no damage occurs in the event of a collision can be determined from the prescribed pressure and force values in each case. Furthermore, additional sensors can be used to prevent collisions [8, 30, 31].

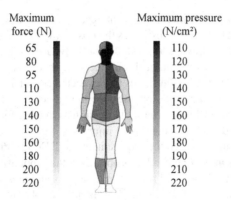

Maximum force (N)		Maximum pressure (N/cm²)
65		110
80		120
95		130
110		140
130		150
140		160
150		170
160		180
180		190
200		210
220		220

Fig. 10. Body Atlas showing the maximum forces and pressure prescribed in DIN 15066 [8].

Besides the discussed methods for safe collaboration through measuring, monitoring and control, passive approaches exist to reduce the risk of injuries solely through the mechanical design of the gripper. Thereby, the common injuries, as given in Fig. 1, can be addressed through different approaches.

First, contact with sharp objects and bruises can be avoided through an appropriate housing design. Sharp edges should be avoided, and functional parts of the mechanism must be protected in a way to prevent the possibility of bruises in general.

The risk of being hit by the gripper, as well as lateral impacts, clamping and crushing is mainly reduced by the presented approaches that try to eliminate the occurrence of dangerous situations at all. Passive security measures are applied for further support. In the case of occurring impacts, despite precautionary measures, resulting forces and thus the risk of injuries shall be minimal.

One possibility is the general limitation of forces. This can be realized actively, e.g. by controlling the applied tendon force in a tendon-actuated gripper, where the force should be dependent on the detected body part that is at risk of collision. However, passive realization is more reliable and robust in execution. It provides internal mechanisms that prevent the tendon forces from exceeding a defined level. This can be realized, by way of example, with any form of mechanical clutch. Although this approach does not allow to adjust the forces to different levels, considering the requirements of the Body Atlas, a standard value of 40 N for the whole gripper is accepted as the operation force comfort limit when operating with humans [32].

In contrast to generally limiting the applicable forces, other approaches solely focus on reducing occurring impact forces without necessarily limiting the grippers' overall performance. On the one hand, a reduction of the colliding mass by using light weight structures can be used to reduce the impact energy. On the other hand, compliant materials often are implemented.

The application of light weight structures is plausible, considering an imaginary scenario with being struck by a metal pipe versus being struck by a polymer pipe. The application of elastic materials, on the other side, does not effectively limit the occurring forces but allows to lengthen the contact phase and thus distributes the impact energy over a longer period, which consequently lowers the force amplitude [33].

A third approach becomes increasingly interesting for more complex grippers such as multi-fingered robot hands. It investigates the possibility of designing structures with orientation-dependent maximum forces. In the work by Grebenstein [33], for example, the fingers are connected to the palm through a joint that is specifically designed to dislocate, when experiencing lateral forces that exceed a certain limit, while withstanding greater forces in grasping direction. Besides the advantage of limiting the risk of damaging the expensive hardware in case of collision, occurring forces in case of lateral impacts can be adjusted to reduce the risk of injuries.

To enable a safe collaboration, the requirements of DIN 15066 are implemented in use case 2 (Sect. 3.3). On the one hand, this applies to the robot used from the company Franka Emika, which has integrated force torque sensors and thus stops moving in case of contact with the human worker. On the other hand, the robot tools used for handling and forming the textile materials are developed and designed in accordance with VDI standard 2221, with special regard to DIN 15066.

For example, handling tools are being developed which are made of carbon fiber reinforced plastic tubes and are therefore very lightweight. For both handling and forming tools, all edges are rounded and the contact area is maximized. Furthermore, soft materials, e.g., silicones, are used to ensure that humans cannot be harmed when working together with the robot.

To complete the safety concept, external camera systems are used, which can detect and predict human movements using CV (see Sect. 4.2). In this way, collisions can be largely avoided. During the implementation and regularly during usage of the collaborative workstation, the TOP model is being used which includes Technical, Organizational and Personal measures. In addition, a risk analysis is being performed so that all risks can be identified and eliminated at an early stage.

4.4 Human Acceptance and Trust

In addition to the soft- and hardware-based components presented in the previous sections, trust and acceptance are two important human factors for safe human-machine interaction:

In this context, trust is a multidimensional construct [34], for which there are many models [35]. Based on Hancock et al., Khavas et al. identify as trust-influencing factors those that are (1) robot-related, (2) human-related, and (3) task- and environment-related. As for the robot, a further distinction can be made between performance-related, behavior-related, and appearance-related. [36, 37] Distrust is expressed by the fact that users either do not use the machines or robots at all or intervene prematurely in processes because they believe that the machine cannot perform the task correctly [34]. Overtrust, on the other hand, may lead to a situation where the human, relying on the machine, does not intervene even though it would be necessary [35]. Both distrust and overtrust are subsumed under the term mistrust.

Acceptance is also a multidimensional construct and in the context of work primarily affects productivity. It is strongly influenced by trust [8, 36]. Furthermore, other factors - such as culture [37] - play a role. A connection with personality is also suspected [38]. Lack of acceptance may lead to a reluctance to use collaborative robots.

A study is currently being prepared on this topic. It will examine acceptance and trust in handling dangerous tools such as knives in a use case from textile production. The HRC will be evaluated depending on the speed of the dummy tool and its distance to the human as well as the possibility to influence the process.

5 Conclusion and Outlook

In this paper we provided an overview about currently researched concepts on safety for HRC within the cluster IoP. These range from the use and verification of BT the integration of CV algorithms, to a safe end-effector design and the consideration of human acceptance and trust. Furthermore, two use cases for HRC in the assembly process as well as production of FRP part were presented, classified using the extended framework developed by Baier et al. and validated regarding their compliance with current safety regulations [2]. Since the scope of this paper emphasized the context of the cluster IoP, we do not claim completeness, but are rather focused on transferability and interdisciplinarity.

For safe HRC, future robots embedded in an Internet of Production should be able to plan their task for highly dynamic environments. This should be done while not only maximizing the safety of the human coworker, but also the overall performance of the production tasks. Additionally, the robot should use the digital shadows of the processes, materials, and products to modify the behavior models designed by the human operator if needed.

Complex environments make it difficult to analyze the system, since it must be modeled as well. A possible solution would be to analyze the system during runtime. This approach makes the modelling of the environment unnecessary but would no longer verify the complete system. Regarding the CV algorithm, a GPU will be used to repeat the tests while detecting the whole human body as well.

For a safe collaboration, it is inevitable to not only use a safe robot, but also design safe robot tools with regard to the active and passive safety measures described in DIN 15066. Furthermore, the installation of external sensors (e.g., camera equipment) is inevitable to ensure safety.

Finally, the system should incorporate human digital shadows to model and take into account human behavior and human factors. For that, it is necessary to collect data from the human workers to predict their movements and fatigue as well as save their knowledge. This requires external sensors and intelligent software.

In conclusion, these developments pave the way for more flexibility and digitization in production through the use of HRC as part of the vision of the cluster IoP.

Acknowledgements. Funded by the Deutsche Forschungs gemeinschaft (DFG, German Research Foundation) under Germany's Excellence Strategy – EXC-2023 Internet of Production – 390621612.

References

1. Pennekamp, J., et al.: Towards an infrastructure enabling the internet of production. In: 2019 IEEE International Conference on Industrial Cyber Physical Systems (ICPS), Taipei, Taiwan, pp. 31–37 (2019)
2. Baier, R., Dammers, H., Mertens, A., et al.: A framework for the classification of human-robot interactions within the internet of production. In: Kurosu, M. (ed.) Human-Computer Interaction. Technological Innovation: Thematic Area, HCI 2022, Held as Part of the 24th HCI International Conference, HCII 2022, Virtual Event, June 26 – July 1, 2022, Proceedings, Part II, pp. 427–454. Springer International Publishing, Cham (2022). https://doi.org/10.1007/978-3-031-05409-9_33
3. International Federation of Robotics (IFR): World Robotics 2022 Industrial Robots. Statistics, Market Analysis, Forecasts and Case Studies. https://ifr.org/downloads/press2018/2022_WR_extended_version.pdf
4. Haddadin, S.: Towards Safe Robots: Approaching Asimov's 1st Law. Springer, Berlin, Heidelberg (2014). https://doi.org/10.1007/978-3-642-40308-8
5. ISO: DIN EN ISO 12100:2010. Sicherheit von Maschinen (2010)
6. ISO: DIN EN ISO 10218-1:2021-09. Robotik – Sicherheitsanforderungen – Teil 1: Industrieroboter (ISO/DIS 10218-1.2:2021) (2021)
7. ISO: DIN EN ISO 10218-2:2020. Robotik – Sicherheitsanforderungen für Robotersysteme in industrieller Umgebung – Teil 2: Robotersysteme, Roboteranwendungen und Integration von Roboterzellen (2020)
8. ISO: ISO/TS 15066:2016-02. Robots and robotic devices - Collaborative robots. ISO (2017)
9. ISO: DIN EN ISO 13849-1:2015. Sicherheit von Maschinen – Sicherheitsbe-zogene Teile von Steuerungen – Teil 1: Allgemeine Gestaltungsleitsätze (2015)
10. ISO: DIN EN ISO 10218-1:2011 Industrieroboter – Sicherheitsanforderungen, Teil 1: Roboter (ISO 10218-1:2011)
11. Land, N., Syberfeldt, A., Almgren, T., Vallhagen, J.: A framework for realizing industrial human-robot collaboration through virtual simulation. Procedia CIRP **93**, 1194–1199 (2020). https://doi.org/10.1016/j.procir.2020.03.019
12. Yu, T., Huang, J., Chang, Q.: Mastering the working sequence in human-robot collaborative assembly based on reinforcement learning. IEEE Access **8**, 163868–163877 (2020). https://doi.org/10.1109/ACCESS.2020.3021904
13. Liu, H., Wang, L.: Collision-free human-robot collaboration based on context awareness. Robot. Comput. Integr. Manuf. **67**, 101997 (2021). https://doi.org/10.1016/j.rcim.2020.101997
14. Scimmi, L.S., Melchiorre, M., Troise, M., Mauro, S., Pastorelli, S.: A practical and effective layout for a safe human-robot collaborative assembly task. Appl. Sci. **11**(4), 1763 (2021). https://doi.org/10.3390/app11041763
15. Trinh, M., Petrovic, O., Brecher, C., Behery, M., Lakemeyer, G.: Kollaborative Montageprozesse mit Behavior Trees/Collaborative Assembly Processes using Behavior Trees. wt Werkstattstechnik online **112**(09), 565–568 (2022). https://doi.org/10.37544/1436-4980-2022-09-37
16. Estin & Co.: JEC Observer - Current trends in the global composites industry 2021–2026. JEC Group, Paris (2022)
17. Elkington, M., Bloom, D., Ward, C., Chatzimichali, A., Potter, K.: Hand layup: understanding the manual process. Adv. Manuf. Polym. Compos. Sci. **1**(3), 138–151 (2015). https://doi.org/10.1080/20550340.2015.1114801
18. Eitzinger, C., Frommel, C., Ghidoni, S., Villagrossi, E.: System concept for human-robot collaborative draping. In: SAMPE Europe Conference. Baden/Zürich, Schweiz

19. Frketic, J., Dickens, T., Ramakrishnan, S.: Automated manufacturing and processing of fiber-reinforced polymer (FRP) composites: an additive review of contemporary and modern techniques for advanced materials manufacturing. Addit. Manuf. **14**, 69–86 (2017). https://doi.org/10.1016/j.addma.2017.01.003
20. Fleischer, J., Teti, R., Lanza, G., Mativenga, P., Möhring, H.-C., Caggiano, A.: Composite materials parts manufacturing. CIRP Ann. **67**(2), 603–626 (2018). https://doi.org/10.1016/j.cirp.2018.05.005
21. Dammers, H., Vervier, L., Mittelviefhaus, L., Brauner, P., Ziefle, M., Gries, T.: Usability of human-robot interaction within textile production: insights into the acceptance of different collaboration types. Usability and User Experience (2022)
22. Dammers, H., Lennartz, M., Gries, T., Greb, C.: Human-robot collaboration in composite preforming: chances and challenges
23. Iovino, M., Scukins, E., Styrud, J., Ögren, P., Smith, C.: A survey of behavior trees in robotics and AI. Robot. Auton. Syst. **154**, 104096 (2022). https://doi.org/10.1016/j.robot.2022.104096
24. Colledanchise, M., Ögren, P.: Behavior Trees in Robotics and AI. CRC Press (2018). https://doi.org/10.1201/9780429489105
25. Henn, T., Völker, M., Kowalewski, S., Trinh, M., Petrovic, O., Brecher, C.: Verification of behavior trees using linear constrained horn clauses. In: Groote, J.F., Huisman, M. (eds.) Lecture Notes in Computer Science, Formal Methods for Industrial Critical Systems, pp. 211–225. Springer International Publishing, Cham (2022)
26. Spencer, B.F., Hoskere, V., Narazaki, Y.: Advances in computer vision-based civil infrastructure inspection and monitoring. Engineering **5**(2), 199–222 (2019). https://doi.org/10.1016/j.eng.2018.11.030
27. Jocher, G., et al.: ultralytics/yolov5: v7.0 - YOLOv5 SOTA Realtime Instance Segmentation: Zenodo (2022)
28. Bradski, G.: The OpenCV library. Dr. Dobb's Journal of Software Tools (2000)
29. Trinh, M., et al.: Safe and flexible planning of collaborative assembly processes using behavior trees and computer vision. Intell. Hum. Syst. Integr. (IHSI) **2023**(69)
30. Hofbaur, M., Rathmair, M.: Physische sicherheit in der mensch-roboter kollaboration. Elektrotech. Inftech. **136**(7), 301–306 (2019). https://doi.org/10.1007/s00502-019-00743-2
31. Kossmann, M.-R.: Sicherheit in der Mensch-Roboter-Interaktion durch einen biofidelen Bewertungsansatz: Dissertation
32. Zhang, L., Wang, Z., Yang, Q., Bao, G., Qian, S.: Development and simulation of ZJUT hand based on flexible pneumatic actuator FPA. In: 2009 International Conference on Mechatronics and Automation, Changchun, China, pp. 1634–1639 (2009)
33. Grebenstein, M.: Approaching Human Performance. Springer International Publishing, Cham (2014). https://doi.org/10.1007/978-3-319-03593-2
34. Parasuraman, R., Riley, V.: Humans and automation: use, misuse, disuse, abuse. Hum. Factors **39**(2), 230–253 (1997). https://doi.org/10.1518/001872097778543886
35. Aroyo, A.M., et al.: Overtrusting robots: Setting a research agenda to mitigate overtrust in automation. Paladyn, J. Behav. Robot. **12**(1), 423–436 (2021). https://doi.org/10.1515/pjbr-2021-0029
36. Yagoda, R.E., Gillan, D.J.: You want me to trust a ROBOT? The development of a human-robot interaction trust scale. Int J of Soc Robotics **4**(3), 235–248 (2012). https://doi.org/10.1007/s12369-012-0144-0
37. Bröhl, C., Nelles, J., Brandl, C., Mertens, A., Nitsch, V.: Human–robot collaboration acceptance model: development and comparison for Germany, Japan, China and the USA. Int. J. Soc. Robot. **11**(5), 709–726 (2019). https://doi.org/10.1007/s12369-019-00593-0
38. Esterwood, C., Essenmacher, K., Yang, H., Zeng, F., Robert, L.P.: A meta-analysis of human personality and robot acceptance in human-robot interaction. In: Proceedings of the 2021 CHI Conference on Human Factors in Computing Systems, Yokohama Japan, pp. 1–18 (2021)

Fuzzy-Set Qualitative Comparative Analysis (fsQCA) in Xiamen's High-Quality Industrial Development

Yin Zheng, Hang Jiang[ID], and Yuyang Lin[✉]

School of Business Administration, Jimei University, Xiamen 361021, China
2361104941@qq.com

Abstract. The industrial economy serves as the "ballast stone" for the national economy's persistent and steady growth, acting as a potent stimulus for macroeconomic expansion and assuring societal stability. This paper employed fuzzy set qualitative comparative analysis (fsQCA) method to explore how different combinations of factors link to the high-quality industrial development. Based on the results of the configuration path analysis, this paper recognized three configuration paths suitable for Xiamen's industrial high-quality development, including high-quality industrial development mainly based on the domestic general cycle path, domestic and international dual cycle of industrial high-quality development path, and outward-oriented path of high-quality industrial development path. In the context of the economic new normal and high-quality development, this paper focuses on the influence mechanism of industrial high-quality development in Xiamen provides new empirical evidence for the government to formulate the relevant policy.

Keywords: fsQCA · configuration path · mechanism of industrial high-quality development

1 Introduction

The industrial economy serves as the "ballast stone" for the national economy's persistent and steady growth, acting as a potent stimulus for macroeconomic expansion and assuring societal stability. The industrial economy has been severely impacted due to the different dangers and difficulties both domestically and internationally. In such a situation, the government should give policy coordination more focus in order to support the industrial economy's sustained recovery and ensure that industry "walk steady to achieve the distance" in high-quality development.

Based on the BRICS Partnership on New Industrial Revolution, Xiamen, a city with the Special Economic Zone, Free Trade Zone, Economic Zone on the West Side of the Strait, and the core area on the 21st-Century Maritime Silk Road, pushes forward the reform and opening up in depth while fostering the high-quality development of the industrial economy in all aspects. The total value of Xiamen's industrial output has been

© Springer Nature Switzerland AG 2023
F. Fui-Hoon Nah and K. Siau (Eds.): HCII 2023, LNCS 14039, pp. 104–117, 2023.
https://doi.org/10.1007/978-3-031-36049-7_8

continuously increasing over the years, but due to the complex international environment and the current epidemic, the industrial economy is still struggling to thrive. Therefore, to boost the high-quality industrial development, it is important to investigate the key variables impacting industrial development and the influencing mechanisms.

In order to accelerate high-quality development beyond the new pattern and promote high-quality industrial development, it is particularly important to clarify the key factors affecting industrial development and their influence mechanisms, which can support government decision-making, improve the quality of operation, boost market confidence, stabilize industrial growth expectations. Previous studies focused on the impact of industrial development. The impact factors spanned from labor force, capital, and technology in traditional economic theory to environmental control and energy use in ecological civilization. Therefore, in addition to the traditional factors impacting industrial output such as R&D costs, labor force, and foreign direct investment used in prior studies, the inclusion of factors specific to Xiamen's economic and social growth, such as the digital economy and business environment, provides a platform for designation of countermeasures. However, the majority of the research currently available on the effects of the industrial development uses multiple regression analysis, which has significant drawbacks when handling triple and above relationships [1, 2]. In order to overcome the constraints of econometrics approaches, fuzzy set qualitative comparative analysis (fsQCA), based on set theory, employed multiple conditions as sets rather than more widely used variables to convey causality [3]. Therefore, the fsQCA will be applied to discuss the configuration path of the industrial high-quality development in Xiamen.

2 Literature Review

2.1 Industrial High-Quality Development

The industrial economy is the foundation of the modern economic system and the key to the transformation of the national industrial economy. High-quality development of the industrial economy is an inevitable requirement to adapt to the transformation of the economy from the stage of high-speed growth to high-quality development. In recent years, scholars have explored the high-quality development of industrial economy from multiple latitudes. Shi and Li [4] reviewed and analyzed the characteristics and connotations of industrial economic development quality in the planned economy period since the founding of New China, the early reform and opening-up period, after joining the WTO and the new period, and the analysis results showed that the overall industrial development quality in China showed a fluctuating upward trend, but the industrial development quality of each province was differentiated. Using 248 municipalities' intelligence indices, Shao and Wu [5] analyzed that the intelligent development achieved by three mechanisms of increasing R&D investment, enhancing technology diffusion, and improving human capital accumulation significantly contributed to the quality development of China's industrial economy. Zhou and Huang [6] analyzed the mediating effect of high-quality development of industrial economy from the perspectives of efficiency and effectiveness of industrial economy, and pointed out that environmental regulation could promote the efficiency of industrial economy, but did not significantly improve the effectiveness of industrial economy, and technological innovation did not play a mediating role in this path. Lin [7] empirically investigated that environmental regulation promotes the quality development of industrial economy in China, taking "haze" control as an example, and concluded that "haze" control has a negative impact on the quality of industrial economy development. By constructing a comprehensive evaluation index system for industrial enterprise development, Zhang and He [8] found an M-shaped nonlinear and oscillating relationship between industrial enterprise development and economic quality from a microscopic perspective. Lin and Meng [9] studied the impact of biased technological progress and factor allocation on the high-quality development of China's industrial economy in 35 industrial sectors from 1999 to 2017 from an industrial perspective, and found that a higher factor substitution elasticity could better explain the reality of China's rapid industrial economic growth based on a variable framework of technological progress rate under the assumption of a non-perfectly competitive market. Chen et al. [10] found that the single implementation of environmental "two control zones" (acid rain control zone and sulfur dioxide pollution control zone) policy at the city level is the most effective environmental policy at this stage, which can significantly improve the "quality" and "quantity" of the industrial economy through innovation compensation. However, the pilot policy of sulfur dioxide emissions trading failed to promote high-quality industrial development.

2.2 Influencing Factors of Industrial Output

As an important component of GDP, the analysis of the factors influencing industrial output mainly follows the framework of the analysis of the factors influencing GDP. As a result, most of the prior studies focused on the Cobb Douglas production function and suggested that the influencing factors affecting the growth of industrial output were labor force, capital, and technological development [11, 12], which showed significant positive correlation with industrial output. From the perspective of direct investment, bilateral FDI would have spillover effect on technology, which in turn increased industrial output [13, 14]; from the perspective of economic development, the development of digital economy [15, 16] and the optimization of business environment [17, 18] will also have an impact on industrial output. Numerous scholars have analyzed the influencing factors of industrial output from the perspectives of labor force, capital investment, and technological innovation, but few studies have included the business environment and digital economy into the scope of examination. In order to better understand the factors influencing industrial output value, this paper uses the economic situation of Xiamen as its starting point and the goal of increasing industrial output value as its end goal. It then thoroughly examines the roles that various factors played in the growth of industrial output value and offers workable suggestions for its transformation and uplift.

Fixed Asset Investment. The fixed asset investment directly affects the creation capacity of industrial economy. The more investment in fixed assets, the larger the scale of production, the stronger the economic creation capacity. As an important source of capital for infrastructure construction, the perfect infrastructure construction can not only improve the development of industrial enterprises, but also promote high-quality economic development[12].

Human Capital Investment. The development of industrial enterprises is inseparable from the participation of people, and the quality and quantity of personnel directly affect the production scale of enterprises, which in turn affects their economic output capacity. Among the factors of production analyzed for industrial development, human capital, as one of the factors of production, can promote the rapid development of the industry and the transformation of the economic growth mode through accumulation and investment. At the same time, human capital can promote technological progress and the transformation of the economic growth mode [19]. An essential requirement and the basis for the diffusion and application of technology, human capital is a significant source of technological innovation and advancement. Technological advancement can increase labor productivity, optimize industrial structure, and change the economic growth mode. Additionally, its diffusion and dissemination effects can hasten knowledge accumulation to create a positive feedback loop, which can result in a steady and continuous rise in the value of industrial output [20].

Technological Innovation. Technological innovation activity refers to the sum of a series of related activities from technology development to the final commercialization of products, and the technological innovation capability of enterprises includes the technological innovation input capability, activity capability and output capability of enterprises. The primary engine of industrial economic growth at the moment is technical innovation, which has steadily supplanted the old investment drive. Compared to the traditional investment drive market environment, industrial economic growth has entered a more stable state under the innovation drive [21]. For industrial enterprises, the degree of technological innovation specialization and resource pooling is high, but the degree of technological innovation is not active enough, and the strengthening of technological innovation investment capacity is beneficial to the technological upgrading of industrial enterprises [22].

Foreign Direct Investment. Foreign investment can promote the growth of industrial output value scale, which has a positive contribution to improve the economic creation capacity of industrial enterprises. Chen [23] analyzed the impact of foreign investment on the factor distribution pattern and manufacturing output value growth in china and concluded that foreign investment has a significant effect on manufacturing output value growth. The circular cumulative effect of foreign direct investment on industrial agglomeration is reflected in the scale effect and knowledge spillover effect [24]. The technology spillover effect has a prominent contribution to industrial technological progress, prompting technological progress and industrial structure adjustment and upgrading, and promoting the steady growth of industrial economy. in addition, foreign direct investment will also improve the labor force level and technology level, thus reducing the production cost of enterprises and improving production efficiency, and ultimately achieving synergistic socio-economic development.

Digital Economy. In the context of the digital economy, while continuously breaking through key technologies and key equipment, securing the support of the talent team and promoting the digital transformation of the manufacturing industry have become the key to high-quality industrial development [15]. In addition, the digital economy has an impact on the industrial green production efficiency as well. Xiao and Jiang [16] found that the digital economy effectively improves CHina's Industrial green production efficiency by accelerating the marketization process, upgrading the industrial structure and enhancing human capital. The digital economy represents the future development direction and is accelerating its integration with the industrial economy, with a broad prospect and unlimited potential. We should promote the accelerated integration of digital technology and industrial development in a wider range and deeper degree, and vigorously exert the amplification, superposition and multiplication of digital technology on the industrial economy to help invigorate the operation of the industrial economy and promote high-quality industrial development.

Business Environment. A good business environment is not only necessary for enterprises and development, but also can safeguard the legitimate rights and interests of market players and create a stable, fair, transparent and predictable development space for all kinds of market players to invest and develop their business. A good business environment can closely follow the needs and expectations of enterprises, stimulate market vitality, effectively promote stable industrial growth and cost reduction, and achieve steady growth of industrial economy. By optimizing the business environment, we can build a communication platform between government, enterprises and banks, and solve the problems of factor security and financing needs for industrial enterprises [18]. The Improvement of Industrial Output Value Must Be Based on the Expansion of Domestic Demand, Reasonable Expansion of the Market as the Goal, and Continuous Improvement of Product Influence and Market Share [17].

3 Methodology and Data Collection

3.1 Fuzzy-Set Qualitative Comparative Analysis

Fuzzy set qualitative comparative analysis is abbreviated as fsQCA, which has emerged as a highly objective technique for deriving predictive conclusions as the technique is based on a statistically-informed configurational approach [25–27]. More specifically, fsQCA is backed by Boolean algebra and set theory principles to analyze the configurational relationships rather than the current prevalent research paradigm of combining matrix algebra with additive-based statistical methods, exemplifies the asymmetrical thinking of complexity theory [28–31]. Therefore, fsQCA approach has a relative advantage in depicting the combinations of conditions that lead to the absence of an outcome, a positive outcome, and a negative outcome, which distinguishes it from the rest of the analytical pack [32–34].

Based on complexity theory, a qualitative comparison analysis is considered to be a configurational model approach [35]. By using fsQCA, the researcher is able to analyze the combined effect of the casual factors. Configuration modelling with fsQCA software involved serval phases. The initial stage was to calibrate all of the data, resulting in the fuzzy values that were used moving forward. The study's suggested conclusion was determined by a series of complicated causal relationships, which were revealed in the study's second stage using truth table analysis,. FsQCA also did essential conditions analysis to help the researchers to identify the circumstances that are necessary to produce the suggested outcome.

Step 1. Calibration.

In fsQCA, each variable is considered as a fuzzy set with fluctuating membership values. As consequence, the very first procedure of fsQCA is to calibrate all the data into fuzzy set value systems. Calibration can be achieved by identifying three unique positions of membership in a set: full non-membership, crossover, and full membership. This approach entails determining the scale's three points of membership. As a result of the scale applied in this research, a direct calibration strategy with three anchors was used: 5 percentiles of the data were classified as complete non-members; 95 percentiles of the data were used as the threshold for full membership; and 50 percentiles of the data were used as the crossover point. Following the establishment of the threshold limit for each construct, a non-linear stepwise logistics function integrated into the fsQCA was used to transform all potential values for the fuzzy set values.

Step 2. Necessary condition analysis (NCA).

For a fsQCA analysis, the second step is the necessary condition analysis, which is establishing what conditions are required. As a scientific phrase, NCA serves to investigate whether each antecedent condition's distinct influence is essential for the outcome. If the outcome is virtually always predictable, then a condition is necessary, according to previous research [36]. A necessary component's consistency may be validated if its score is greater than or equal to 0.90 [37].

Step 3. Sufficiency analysis.

fsQCA evaluation entails investigating the relationships among configurations, or pairings of cause elements (factors), and the outcome (dependent variable), also known as assessment of the truth table or sufficiency analysis. Sufficiency analysis produces a series of adequate conditions – sometimes known as configurations, causal models, solutions or recipes [28]. To generate a truth table of 2^k rows, the fuzzy set algorithm must be first executed in fsQCA. This method entails every conceivable combination of the k predictors. The truth table is used to establish the importance of the data using two parameters: consistency and coverage levels of 0.8 and 0.2, accordingly. Coverage is analogous to the coefficient of determination (R^2), which measures the degree to which every causal solution affects an outcome. Coverage values range from 0 to 1, with a higher value indicating more comprehensive coverage. Consistency is equivalent to the path coefficient (β), which assesses the extent to which a causal condition leads to the observed outcome [38]. When truth tables are evaluated using the fsQCA software, three solutions are generated: complicated, moderate, and parsimonious. The authors conducted their research using intermediate solutions, which are a subset of complicated solutions that also contain parsimonious answers [28].

Step 4. Robustness Check.

The robustness check can determine whether the results are stable. Therefore, a robustness check is necessary for fsQCA analysis. The robustness tests for fsQCA mainly include appropriately increasing or decreasing the number of cases, adjusting the number of antecedent conditions, adjusting the threshold selected during calibration, changing the consistency threshold, and changing the frequency of samples [39]. There are two criteria frequently used for determining whether the original analysis is robust. The first criterion is the difference of fitting parameters. The specific method is whether significant differences exist in consistency and coverage obtained by using different robustness test methods. If the difference is not large, then the result can be considered robust [32]. The second criterion is the configuration results [39]. If the configuration solutions produced after different robustness methods have similar subset relations, then the results can be considered robust.

3.2 Data Collection

Based on the literature review, total economic volume, fixed asset investment, labor force, research and development, foreign direct investment, digital economy, and the business environment are considered to be the major factors affecting industrial output. Table 1 provides measurement indicators for each of these influencing factors.

Table 1. Variable and measurement

Variables	Measurement
Industrial development (IND)	Total industrial output above the scale
Economic volume (ECO)	Gross domestic product
Fixed investment (FIX)	Total fixed asset investment
Labor force (LAB)	Human capital
Technology (TEC)	R&D investment
Foreign direct investment (FDI)	Foreign direct investment
Digital economy (DIG)	Score of digital economy
Business environment (BUS)	Score of business environment

Since there is no data for the digital economy and business environment in Xiamen, this paper, referring to previous studies, constructed a digital economy evaluation index system and a business environment evaluation index system, and then measure the comprehensive scores of the digital economy and business environment in Xiamen through principal component analysis.

All the data used are collected from the Xiamen Statistic Yearbook (2011–2021) and Economic and Social Development Statistics Bulletin (2010–2020), and also Statistic Yearbook (2011–2021) and Economic and Social Development Statistics Bulletin (2010–2020) for Fujian Province. The descriptive statistics are shown in Table 2.

Table 2. Descriptive statistics

Variables	Unit	Mean	S.D	Min	Max
IND	10 thousand Yuan	52947562.27	10418266.98	36796605.00	68612119.00
ECO	10 thousand Yuan	40617958.36	14075079.46	21490950.00	63840233.00
FIX	10 thousand Yuan	19472309.64	7280930.48	10099850.00	31103266.00
LAB	10 thousand people	987798.91	101584.83	799156.00	1124615.00
TEC	10 thousand Yuan	1159745.36	466363.34	509794.00	1968400.00
FDI	10 thousand USD	197116.82	31253.37	159453.00	253805.00
DIG	-	842.97	271.65	448.16	1337.56
BUS	-	91.15	2.45	87.39	94.87

4 Empirical Study

4.1 Calibration and Measurement

In order to employ fsQCA to analyze the configuration path between the influencing factors and industrial output, each influencing factor was used as a conditional variable and industrial output was set as an outcome variable. Following the previous studies, this research sets the full membership points of five antecedents and one results variable as three thresholds of 0.95, 0.5, and 0.05 for complete membership, intermediate, and complete non-membership points of the case data by using direct calibration [32, 40, 41]. The anchor point calibration was shown in Table 3.

Table 3. Calibration of variables

Variable classification	Variables	Fully in	Crossover	Fully out
Result variables	IND	68255265.50	50286788.00	40395890.50
Antecedent variables	ECO	61995327.50	38069438.00	23856266.00
	FIX	29842786.50	18965209.00	10690361.00
	LAB	1092533.50	1021462.00	809159.50
	TEC	1872494.00	1034180.00	599180.00
	FDI	245817.50	197101.00	160976.50
	DIG	1285.62	794.87	501.50
	BUS	94.68	91.00	87.76

4.2 Necessary Conditions Analysis

A necessary condition is "a condition that must be present for the outcome to occur, but its presence does not guarantee that occurrence" [42]. The consistency threshold of the necessity analysis is chosen at 0.9 in accordance with earlier studies [43, 44]. The fsQCA 3.0 software is used to run the necessary conditions analysis and the results are displayed on Table 4.

Table 4. Necessary conditions for industrial high-quality development

Condition	Consistency	Coverage
ECO	0.945	0.993
~ ECO	0.419	0.422
FIX	0.913	0.972
~ FIX	0.398	0.395
LAB	0.664	0.689
~ LAB	0.706	0.718
TEC	0.970	0.982
~ TEC	0.409	0.426

Notes: Following the nomenclature, the symbol " ~" represent the negation of the characteristics

In accordance with the necessary conditions results, the consistency of economic volume, fixed investment, technology level, digital economy and business environment are all higher than 0.9, and the coverage degree is greater than 0.5, which can be considered as the necessary conditions. All of these conditions are crucial for the high-quality growth of the industrial economy in Xiamen. A single antecedent condition has insufficient explanatory power for corporate social responsibility. Hence, a further analysis needs to be done to test the combined effect of antecedent conditions.

4.3 Sufficiency Analysis

Configuration analysis mainly reveals the sufficiency of different configurations of various antecedent conditions that result in industrial high-quality development. Based on the set theory, sufficiency explores whether the configuration set composed of multiple antecedent conditions is a subset of the result set. This paper generates a truth table to show all configuration states of a given data set using the fsQCA algorithm, then condenses the truth table into useful configuration [39]. Following Du and Jia's study, the consistency threshold in this paper is set to 0.90, and proportional reduction in inconsistency (PRI) consistency threshold is set to 0.65 [45].

The fsQCA reports three different types of solution for the sufficiency analysis: the parsimonious, the intermediate, and the complex solution. The difference among these solutions is the extent to which they include logical remainders in the analysis [46]. Following the mainstream research of fsQCA, this research reports intermediate solutions, assists in reporting reduced solutions, and distinguishes the core conditions and edge conditions of the configuration according to the intermediate solutions and reduced solutions. The core condition, which significantly affects the outcome, is present if the antecedent condition is shared by both the intermediate and reduced solutions. An antecedent condition is regarded as an edge condition, which is a prerequisite for auxiliary contributions, if it only emerges in the intermediate solution. However, if an antecedent condition only appears in the intermediate solution, it is considered as an edge condition, which is a condition for auxiliary contributions. Three configuration paths for industrial high-quality development are calculated by fsQCA which indicate the multiple concurrencies of industrial high-quality development. See Table 5 below for the sufficient configuration of antecedent conditions for industrial high-quality development.

Table 5. Sufficient configuration of antecedent conditions

Antecedent	Path 1	Path 2	Path 3
ECO	●	●	⊗
FIX	●	●	⊗
LAB	●		●
TEC	●	●	⊗
FDI		●	●
DIG	●	●	⊗
BUS	●	●	⊗
Consistency	0.997	0.989	0.922
Unique coverage	0.540	0.648	0.357
Raw coverage	0.150	0.258	0.067
Solution coverage	0.961		
Solution consistency	0.866		

Notes: ● indicates the core condition; ● indicates the edge condition; ⊗ indicates the absence of edge condition; and blank cells indicate that the condition is not relevant within that configuration

The consistency of the overall solution exceeds 0.9, indicating that these 3 configurations are all sufficient conditions for the occurrence of industrial high-quality development. In addition, the coverage of the single solution and the overall solution shows that each configuration can explain a considerable proportion of the industrial high-quality development, and the configurations together explain a large degree of industrial high-quality development.

4.4 Configuration Paths for Industrial High-Quality Development

Configuration path 1 can be summarized as the path of high-quality industrial development mainly based on the domestic general cycle, with the specific combination of conditions: ECO*FIX*LAB*TEC*DIG*BUS. The configuration shows that utilizing the digital economy and optimizing the business environment while simultaneously promoting traditional economic factors, relying on R&D investment and improving technological level, boosts industrial growth. The combination of factors mainly depends on the domestic general circulation-based, raising domestic demand, deepening supply-side reform (labor force as a marginal condition), focusing on conventional industrial innovation industry, focusing on the digital economy and business environment, and empowering the high-quality growth of industry.

Configuration 2 can be summarized as a path of domestic and international dual cycle of industrial high-quality development path, with a specific combination of conditions: ECO*FIX*TEC*FDI*DIG*BUS. This path suggests that, based on the traditional economy and relying on the scale market advantage, the domestic circulation attracts global resource factors to enhance the endogenous strength and reliability of the domestic circulation. The fact that this path has the biggest unique coverage among these three configuration paths indicates that it best suits the outward-oriented economy development strategy for Xiamen.

Configuration 3 can be summarized as an outward-oriented path of high-quality industrial development with the following specific combination of conditions: ~ ECO* ~ FIX*LAB* ~ TEC*FDI* ~ DIG* ~ BUS. Only foreign direct investment and labor force (edge condition) are included in this path, and all other influencing factors are absence conditions, which can be attributed to the traditional export-oriented economic development model. Xiamen can selectively "bring in", focus on the BRICS industry, promote the construction of the base to go deeper and deeper, and achieve high-quality industrial development in Xiamen.

5   Conclusion and Discussion

In the context of the economic new normal and high-quality development, industrial development plays a crucial role in economic high-quality development. This paper uses the economic situation in Xiamen as its starting point and the improvement of industrial high-quality development as its end goal. It takes into account the current state of the business environment and the digital economy as influencing factors, thoroughly analyzes the roles played by various factors in industrial high-quality development. The paper puts forward the following policy suggestions.

First and foremost, a variety of measures and targeted aid to guarantee the industrial economy's steady development. Second, boost independent innovation's capacity and the vigor of technological advancement. Third, take advantage of the strategic opportunity presented by the digital economy's explosive growth. Deeply integrate manufacturing, the industrial internet, and the digital transformation of the manufacturing sector. Lastly, to create new benefits in the company environment and drive market vitality with exact measures Improve the system for communication between the government and businesses while fostering the development of such relationships.

References

1. Abbasi, G.A., Sandran, T., Ganesan, Y., Iranmanesh, M.: Go cashless! determinants of continuance intention to use E-wallet apps: a hybrid approach using PLS-SEM and fsQCA. Technol. Soc. **68** (2022). https://doi.org/10.1016/j.techsoc.2022.101937

2. Amara, N., Rhaiem, M., Halilem, N.: Assessing the research efficiency of Canadian scholars in the management field: evidence from the DEA and fsQCA. J. Bus. Res. **115**, 296–306 (2020)
3. Beynon, M.J., Jones, P., IPickernell, D.: Country-level entrepreneurial attitudes and activity through the years: a panel data analysis using fsQCA. J. Bus. Res. **115**, 443–455 (2020)
4. Cao, Z.Y.: Research on the new manufacturing model to promote high-quality development of China's industry under the background of digital economy. Theor. Invest. (2), 99–104 (2018)
5. Chen, B.L., Qi, Y.W.: The impact of reverse technology spillover effect of OFDI on enterprise innovation: empirical analysis based on provincial panel data. Jiangxi Soc. Sci. **41**(12), 58–65 (2021)
6. Chen, L.: An investigation of FDI effect on factor income distribution and the development of the manufacturing industry. J. Suzhou Univ. Sci. Techonol. (Soc. Sci. Edn.) **29**(2), 26–31 (2012)
7. Chen, R.J., Huang, S.Q., Ji, X.C., Tao, W.J.: Selection of urban environmental policy to promote high-quality industrial development. Asia-Pacific Econ. Rev. (3), 113–124 (2019)
8. Chi, J.M.: Research on factors affecting the provincial and municipal annual total industrial output. Think Tank Era (39), 36–37 (2019)
9. Di Paola, N.: Pathways to academic entrepreneurship: the determinants of female scholars' entrepreneurial intentions. J. Technol. Transf. **46**, 1417–1441 (2021)
10. Du, Y.Z., Liu, Q.C., Cheng, J.Q.: What kind of ecosystem for doing business will contribute to city-level high entrepreneurial activity? A research based on institutional configurations. J. Manag. World **36**(9), 141–155 (2020)
11. Dul, J.: Identifying single necessary conditions with NCA and fsQCA. J. Bus. Res. **69**(4), 1516–1523 (2016)
12. Fang, L., Guo, X.C.: The impact of technological innovation on the development of the intelligent industry system: evidence from Henan, China. J. Intell. Fuzzy Syst. **38**(6), 6905–6909 (2020)
13. Fiss, P.C.: Building better casual theories: a fuzzy set approach to typologies in organizational research. Acad. Manag. J. **54**(2), 393–420 (2011)
14. Gupta, K., Crilly, D., Greckhamer, T.: Stakeholder engagement strategies, national institutions, and firm performance: a configurational perspective. Strateg. Manag. J. **41**(10), 1869–1900 (2020)
15. Harms, R., Kraus, S., Schwarz, E.: The suitability of the configuration approach in entrepreneurship research. Entrep. Reg. Dev. **21**(1), 25–49 (2009)
16. Hu, J.P., Ma, S., Zou, Q., Zeng, Y.X.: An empirical analysis of the factors affecting the increase of output value of enterprises in Tibet. Hubei Agric. Sci. **59**(S1), 452–454+458 (2020)
17. Huang, Y.J., Li, S.H., Xiang, X.Y., Bu, Y.J., Guo, Y.: How can the combination of entrepreneurship policies activate regional innovation capability? A comparative study of Chinese provinces based on fsQCA. J. Innov. Knowl. **7**(3) (2022). https://doi.org/10.1016/j.jik.2022.100227
18. Hui, S.P., Wang, X.H., Shan, J.R.: Research on the driving path and driving effect of China's industrial high-quality development. Shanghai J. Econ. (10), 53–61+76 (2021)
19. Jovanovic, J., Morschett, D.: Under which conditions do manufacturing companies choose FDI for service provision in foreign markets? An investigation using fsQCA. Ind. Mark. Manag. **104**, 38–50 (2022). https://doi.org/10.1016/j.indmarman.2022.03.018
20. Kraus, S., Ribeiro-Soriano, D., Schüssler, M.: Fuzzy-set qualitative comparative analysis (fsQCA) in entrepreneurship and innovation research - the rise of a method. Int. Entrepreneurship Manag. J. **14**, 15–33 (2018)
21. Lee, C.K.H.: How guest-host interactions affect consumer experiences in the sharing economy: new evidence from a configurational analysis based on consumer reviews. Decis. Supp. Syst. **152** (2022). https://doi.org/10.1002/smj.3204

22. Lin, P., Meng, N.N.: Biased technological progress, factor allocation distortion and high-quality development of China's industrial economy: from the perspective of technological congruence. Shanghai J. Econ. (8), 72–91 (2021)
23. Lin, Y.Q.: An empirical study on "Haze" governance promoting the high-quality development of industrial economy: empirical analysis based on PSM-DID model. J. Ind. Technol. Econ. **41**(7), 130–137 (2022)
24. Llopis-Albert, C., Palacios-Marqués, D., & Simón-Moya, V.: Fuzzy set qualitative comparative analysis (fsQCA) applied to the adaptation of the automobile industry to meet the emission standards of climate change policies via the deployment of electric vehicles (EVs). Technol. Forecast. Soc. Change **169**, 120843 (2021). https://doi.org/10.1016/j.techfore.2021.120843
25. Mehran, J., Olya, H.G.: Canal boat tourism: application of complexity theory. J. Retail. Consum. Serv. **53**(1) (2020). https://doi.org/10.1016/j.jretconser.2019.101954
26. Misangyi, V.F., Greckhamer, T., Furnari, S., Fiss, P.C., Crilly, D., Aguilera, R.: Embracing causal complexity. J. Manag. **43**, 255–282 (2017)
27. Pappas, I.O., Woodside, A.G.: Fuzzy-set qualitative comparative analysis (fsQCA): guidelines for research practice in information systems and marketing. Int. J. Inf. Manag. **58** (2021)
28. Prentice, C.: Testing complexity theory in service research. J. Serv. Mark. **34**, 149–162 (2019)
29. Ragin, C.C.: Measurement versus calibration: a set-theoretic approach: the Oxford Handbook of Political Methodology (2008)
30. Ragin, C.C.: Qualitative Comparative Analysis Using Fuzzy Sets (fsQCA). SAGE Publications, London (2009)
31. Ragin, C.C., Fiss, P.C.: Net effects analysis versus configurational analysis: an empirical demonstration. Redes. Soc. Inq.: Fuzzy Sets Beyond **240**, 190–212 (2008)
32. Schneider, C.Q., Wagenmann, C.: Qualitative comparative analysis (fsQCA) and fuzzy-sets: agenda for a research approach and a data analysis technique. Comp. Sociol. **9**(3), 376–396 (2010)
33. Shao, W., Wu, T.L.: Intelligence, factor market and high-quality development of industrial economy. Inquiry into Economic Issues (2), 112–127 (2022)
34. Shi, D., Li, P.: Quality evolution and assessment of China's industry over the past seven decades. China Ind. Econ. (9), 5–23 (2019)
35. Song, X.N., Xue, H.F.: Environmental regulation, FDI spillover and green technology innovation in manufacturing industry. Stat. Decis. (4), 81–85 (2022)
36. Sun, J.S., Dong, Y.J.: Human capital, material capital and economic growth: Based on empirical research on Chinese Data. J. Shanxi Univ. Finan. Econ. (4), 37–43 (2007)
37. Wu, F.: Econometric analysis of factors affecting China's total industrial output. Guide to Business (4), 16–17 (2015)
38. Xiao, Y.F., Jiang, Y.: Research on the influence of digital economy on industrial green production efficiency. Mod. Manag. Sci. (8), 100–109 (2021)
39. Xie, A.M., Luo, Z.H., Xie, W.J.: To study on the relationship between S & T input of large & medium industrial enterprises and gross industrial output value in Guangxi province. J. Guangxi Minzu Univ. (Nat. Sci. Edn.), **16**(1), 81–84+100 (2010)
40. Yu, W.C., Liang, P.H.: Unvertainty, business environment and private enterprises' vitality. China Ind. Econ. (11), 136–154 (2019)
41. Yuan, L.J., Du, X.P.: Business environment and industrial total factor productivity: An empirical analysis based on panel data of provincial industrial sectors in 1994–2014. J. Harbin Univ. Comm. (Soc. Sci. Edn.) (5), 55–67 (2018)
42. Zhang, H., Long, S.: How business environment shapes urban tourism industry development? Configuration effects based on NCA and fsQCA. Front. Psychol. **13**, 947794 (2022). https://doi.org/10.3389/fpsyg.2022.947794

43. Zhang, L., He, L.Y.: Have industrial enterprises contributed to the high-quality development of China's economy? An empirical study based on 952 listed enterprises. J. China Univ. Min. Technol. (Soc. Sci.) **23**(2), 102–117 (2021)
44. Zhang, Y.W., Liu, L.B.: Does the improvement of business environment promote foreign direct investment? Int. Bus. (1), 59–70 (2020)
45. Zhang, Y.W., Lu, X.H., Zhang, M.M., Ren, B., Zou, Y.C., Lv, T.G.: Understanding farmers' willingness in arable land protection cooperation by using fsQCA: roles of perceived benefits and policy incentives. J. Nat. Conserv. **68** (2022). https://doi.org/10.1016/j.jnc.2022.126234
46. Zhou, Y.X., Huang, J.: Environmental regulation, technological innovation and high quality development of industrial economy: analysis of dynamic spatial SAR model based on prefecture-level cities in China. Soft Sci. 1–14 (2022)

Applications of AI in Business and Society

Effect of AI Generated Content Advertising on Consumer Engagement

Duo Du, Yanling Zhang, and Jiao Ge[✉]

Harbin Institute of Technology Shenzhen, Shenzhen, China
jiaoge@hit.edu.cn

Abstract. With a series of major breakthroughs in AI technology, the use of AI generated content (AIGC) has become one of the most popular research topics. Social media marketing has become also a mainstream and irreplaceable way for marketing subject. Many companies are starting to use AI technology to optimize existing marketing activities to reduce costs and improve efficiency. However, how companies use the latest AIGC to generate ads that affects consumers engagement needed to be further discussed. Thus, the purpose of this paper is to investigate how firm used AIGC advertising influence customer psychological engagement and behavioral engagement with the consideration of ad emotion level and whether the firm used AIGC is explicitly labeled. Online experiments are conducted and data is analyzed through MLR models. Results show that AIGC could positively influence customer both psychological and behavior engagements, with psychological engagement being mediating factor between AIGC and behavioral engagement. Emotion level plays a negative moderating role. In addition, we find that labeling AI-generated advertisements does significantly affect customer behavior engagement.

Keywords: Social media advertising · AI generated content · Consumer engagement · Emotion · Advertising Labeling

1 Introduction

Social media are 'a group of Internet-based applications that build on the ideological and technological foundations of Web 2.0, and allow the creation and exchange of User Generated Content' [1]. Users and firms, today even Artificial Intelligence (AI), could generate different contents in different social media platforms. Among different social media platforms, there are generally three type of platforms, social network site (SNS), microblogging application, and content community are empirically used by firms and individuals.

Social media has been used as platforms for marketing and advertising increasingly. Many firms and organizations have spent a lot of resources on social media advertising [2]. Although collectively referred to as social media, due to different content forms, environments and experience methods of different social media, the level of customer engagement will also be different [3]. Therefore, different standards need to be formulated based on different platforms and different marketing models. There are many

© Springer Nature Switzerland AG 2023
F. Fui-Hoon Nah and K. Siau (Eds.): HCII 2023, LNCS 14039, pp. 121–129, 2023.
https://doi.org/10.1007/978-3-031-36049-7_9

criteria, but many companies still regard the level of customer engagement as the criteria to measure the marketing effect [4]. Many factors will affect consumers' emotions and behaviors, including consumer engagement. Some studies have shown that consumers with high perception of collusion will feel more aggressive in advertising [5]. Emotions embedded in marketing stimuli influence decision-making through processes driven by cognitive assessment. Decision-independent emotions influence decision making through ongoing appraisal tendencies [6]. The interactivity, perceived relevance, information and entertainment of social media advertising are positively correlated with brand engagement [7]. The form and method of content generation will also affect consumers. Some scholars figured out that FGC has a positive impact on brand awareness, brand loyalty, electronic word of mouth (eWOM) and purchase intention [8].

Traditionally, the person in charge of enterprise generated content is responsible for editing text, selecting images, and shooting videos for their advertising. However, the emergence of AIGC has changed this situation. AI generated content (AIGC) is defined as content generated by AI, including text, pictures and videos. This technology has been rapidly advancing in recent years, and has the potential to revolutionize various industries, including media, marketing, and education [6].

AI generated content can be used to generate product descriptions, ad copy, and social media posts in marketing. This can help companies to save time and resources by automating the content creation process. Additionally, AI generated content can be used to personalize the messaging and targeting of advertising campaigns, by analyzing large amounts of data and identifying patterns and trends.

The information and persuasive content generated by enterprises directly and indirectly affect enterprise performance by influencing consumer participation [9]. All these firm generated content could influence the customer purchasing behaviors [10]. With the increasing availability of AI generated content, in the foreseeable future, more and more enterprises and marketing personnel will adopt AI for generating these content. This is believed to greatly improve marketing efficiency and reduce marketing costs.

However, how firm use of AIGC advertising affect consumer engagement? Will the use of AIGC make consumers feel alienated, thereby reducing consumer engagement? These questions are investigated in this research.

We constructed a model to investigate whether AIGC can directly affect consumer engagement. Referring to previous research findings, we divide consumer participation into psychological consumer engagement and behavioral consumer engagement. We also added a set of moderating variables to the model to explore whether the emotional level of advertising can regulate the relationship between AIGC and consumer participation.

Results show that AIGC could positively influence customer both psychological and behavior engagements, with psychological engagement being mediating factor between AIGC and behavioral engagement. Emotion level plays a negative moderating role. In addition, we find that labeling AI-generated advertisements does significantly affect customer behavior engagement.

2 Literature Review and Hypotheses

2.1 Effect of AIGC on Customer Engagement

Customer engagement is defined as a multidimensional concept that reflects the psychological state generated through the customer experience in the service relationship [11]. Customer engagement with online communities is the degree to which clients are motivated to actively participate in activities related to social media groups [12]. It is observed that customer participation includes three dimensions, namely: cognition, emotion, and behavior [13]. Consumer engagement consists of two parts, including the visible part and the invisible part [14]. In the context of social media, we define consumer participation as unobservable (psychological) and observable (behavior) participation attitudes towards generated contents. Psychological consumer engagement includes two dimensions: cognition and emotional. Behavioral consumer engagement includes a dimension of behavior.

Firm generated contents, for instance, brand advertising, and user generated contents, such as eWOM, have effects on the consumer engagement even though the effects between these two may be different [15]. Firm generated content (FGC) is defined as the messages posted by firms on their official social media page [16]. Past studies showed that FGC could help set up a mutually beneficial relationship between customers and firms [17]. With the development of AI technology, FGC creation by AI instead of human has become a reality. AIGC for firm and FGC by human could be similar in terms of other contents except for the difference in generation form. Therefore, we have reason to infer that AIGC for firm, like FGC by human, will have positive impact on consumer engagement. Therefore, we propose that.

H1: AIGC positively influences both consumer psychological and behavioral engagement.

2.2 Moderating Effect of AIGC'S Emotion Level

Academic circles have long noticed that ad content influences ad effectiveness at generating online search and sales, two important stages of the purchase funnel. The prior studies often divided ad content into informational and emotional content. This classification is often used by practitioners and researchers [18]. Literature reviews have identified emotion as a major component of web content and word of mouth [19]. Some scholars believed that each specific emotion is associated with a set of cognitive assessments that drive the influence of emotion on decision-making through subtle psychological mechanisms. Beth L. Fossen's research suggested that the effectiveness of advertising varies with changes in advertising emotions, with more emotional advertisements, especially interesting and emotional ones, experiencing the greatest growth in online shopping activities [20]. Some scholars also found that emotional advertising and information advertising are different in application scenarios and effects [21]. Rocklage's research presented results that were different from previous studies. Previous studies have generally believed that a higher emotional level implies better advertising effectiveness and higher consumer engagement. However, the scholar conducted research on when content with higher emotional levels may have negative impacts [22]. However, in the future, under the context of AIGC advertising, the situation will be different. Due to technology constraints, the existing AI technology cannot truly understand human emotions [23]. Therefore, when a consumer browses an advertisement with a high emotional level and discovers that the advertisement is generated by AI, consumers may have alienation or discomfort due to the inconsistency between the content trait and the characteristics of the generation method. Thus, we assume that the advertising by AIGC is expected to have less emotion. Therefore, we propose that.

H2: In the relationship between AIGC and consumer psychological engagement, emotion plays a negative moderation effect.

2.3 Moderating Effect of AIGC Advertising Labeling

One another interesting question is whether the information that one advertising was generated by AI should be disclosed, especially with the latest use of ChatGPT. Information disclosure is one of the most popular fields of research in the past decade. The original researchers focused on the financial services area. A 2013 study showed that when consumers see financial disclosure in mutual fund advertisements, consumers respond more positively to advertising responsibility, recall and cognitive responses, as well as higher risk perception [24]. When teenagers learn advertising through information video blogs, advertising disclosure will also have a positive impact on the influencers, and thus have an advertising impact [25]. In some cases, governments and other regulatory authorities may force disclosure of some information on advertisements, and researchers have also conducted research on this issue. For instance, a 2014 research focused on the influence of the forced disclosure of nutrition disclosure and initial product category healthfulness perceptions, finding that such disclosure may help consumers make more informed choices [26]. Another research investigated the influence of the requirement to disclose the existence of sponsorship in advertisements. And found that the sponsored content posted by influencer leads to higher brand liking, while the brand one leads to higher brand awareness [27]. From previous research, it can be found that different types and properties of advertising disclosure can have an impact on consumers' psychology, thereby affecting their behavior. Therefore, it is reasonable for us to assume that if marketers choose to show AIGC ads while letting consumers clearly know that the ads are created by AI, the marketing effect may be different. Especially from the perspective of consumer psychology, AIGC advertising can have a very different impact on consumers compared to traditional advertising generation methods. For example, AIGC advertising may provide avantgarde, fashion, technology and other labels to improve the level of consumer engagement. These labels are likely to have a positive impact on consumer psychological participation. Therefore, we hypothesize that.

H3: The explicitly labeled AIGC ads increase consumer engagement.

3   Method and Data

3.1 Measurement of Variables

For the measure of variables, AIGC is measured by scale suggested by Thomas and others in 2020 [28]. Psychological engagement is measured from the research results of Habib et al. in 2022 [29]. The scale of behavior engagement comes from the research of brodie et al. in 2013 [30]. Besides, we also revised the original questionnaire by referring to the questionnaires of Dwivedi [31] and Li Gao [32].

3.2 Data Collection

Online experiments were conducted in the beginning of 2023 for data collection. A specific brand, Nestle, is identified to be the object brand. AI is used to generate text and picture adverting for the identified brand. One experiment was conducted with AIGC ads with one at higher emotion-including level, the other is at the lower level. The other experiment was conducted with AIGC ads with one group is explicitly labeled as the advertising of AIGC. ALL variables are asked during the experiments.

A total of 472 experimental results were received, among which 428 valid questionnaires were obtained after screening out the questionnaires with response time less than 2.5 min. In the valid questionnaire, there are 213 male subjects. There are 215 female subjects. The age of the subjects is also different, and they are distributed in all age groups. Among them, the number of people aged between 20 and 40 is the largest. As for the attitude questions, although most of our subjects are willing to support the use of AIGC, some of whom have reservations and refuse to support. The quantitative MLR models are used to test the hypothesis.

4 Results and Discussion

We first analyzed the reliability and validity of the questionnaire, and the results are shown in the Table 1. The Cronbach's Alpha coefficient of emotional-include, psychological engagement, and behavioral engagement are 0.855, 0.863 and 0.838 respectively. Therefore, the reliability of the questionnaire is good. According to the validity results, the KMO index is 0.924 which proves that the validity structure of the questionnaire is good.

Table 1. Reliability analysis

Variables	Cronbach's alpha
Emotional-include	0.855
AIGC	0.863
Psychological engagement	0.838
Behavioral engagement	0.877

The regression results are shown in Table 2. It suggests that AIGC has significant positive effect on consumer behavior engagement ($\beta = 0.5253{***}$) in model 1. In model 2, it shows that AIGC has significant positive effect on consumer psychological engagement ($\beta = 0.6004{***}$). Thus, H1 is supported. In addition, psychological engagement has significant positive effect on behavior engagement ($\beta = 0.1324{***}$) in model 1, which means the psychological engagement plays a mediation role between AIGC and consumer behavior engagement.

Model 2 shows that emotion negatively moderates the relationship between AIGC and consumer psychological engagement ($\beta = -0.0793^{***}$) while AIGC still plays significant positive effect on consumer psychological engagement ($\beta = 0.6004^{***}$). Thus, H2 is supported.

Table 2. Model results

model	Model 1: BE	Model 2: PE	Model 3: BE	
AIGC	0.5253***	0.6004***	0.6048***	
	(0.1099)	(0.1109)	(0.1072)	
Emotion	0.5257***	0.7398***	0.6237***	
	(0.1263)	(0.1268)	(0.1225)	
PE	0.1324***			
	(0.0467)			
Emotion*AIGC	−0.0607**	−0.0793***	−0.7131***	
	(0.0242)	(0.0250)	(0.0241)	
Age	0.0694	0.1103**	0.0840*	
	(0.0460)	(0.0477)	(0.0460)	
Gender	−0.0907	0.0298	−0.0867	
	(0.0928)	(0.0968)	(0.0935)	
Attention	0.0507	0.0813	0.0616	
	(0.0403)	(0.0419)	(0.0405)	
Attitude	−0.0142	0.0106	−0.0128	
Observations	(0.0349)	(0.0364)	(0.0352)	
R-squared	428	428	428	
F	0.4031	0.4303	0.3917	
	35.38	45.33	38.64	

BE is behavior engagement, PE is psychological engagement, Standard errors in parentheses *** represents $p < 0.01$, ** represents $p < 0.05$, * represents $p < 0.1$

In addition, to test hypothesis H3 for the possible different results of AIGC labeled vs. not labeled advertising. we conducted paired t test on the labeled group and the unlabeled group, respectively. Results are shown in Table 3 . We found that consumer Psychological Engagement (PE) had no significant difference between the AIGC labeled one and the non-labeled one. However, it does have significant difference for consumer Behavior Engagement (BE) between the AIGC labeled one and the non-labeled one. This suggests that if firm choose to label their AIGC advertising explicitly on social media, consumer behavior engagement, for instance, the likes, comments, shares, and re-shares could be significantly increased.

Table 3. t test results.

t test	mean	Lower 95%	Upper 95%	t	df	Sig
PE1 -PE2	.0218	−.1181	.1617	.306	427	.76
BE1 -BE2	.1215	.0117	.2313	2.175	427	.03***

5 Conclusion

We investigate whether firms' AI generated content (AIGC) advertising could influence customer psychological engagement and customer behavioral engagement trough online experiments. The results show that AIGC could positively influence customer psychological and behavioral engagement. Psychological engagement shows significant mediating effect between AIGC and behavioral engagement. Emotion plays a negative regulatory role between the relationship of AIGC and consumer behavioral engagement. In addition, labeling AI-generated advertisements does significantly affect customer behavior engagement.

There are three main contributions of this paper. First, this paper has made contributions to the expansion of FGC research. In this paper, AI is introduced as a possible generation method of FGC. It breaks through the limitation of the original research mainly limited to human FGC. It further expands the direction of FGC research. Second, this paper studies the whole process of AIGC affecting consumer psychological engagement and ultimately behavior engagement which fills the research gap. Last but not the least, it enriches the research on the mechanism that AIGC affects consumer psychology and ultimately affects consumer behavior by studying the emotional participation and labeling of AI advertisements.

However, due to the limitation of the experimental form, the experimental background is set relatively simple. Other products or social media platforms are not considered. We should choose more brands and kinds of products to make the results universal. The display form is also relatively simple. Future research of different marketing scenarios could be analyzed. In addition, the experiment lacks immersion. Due to the combination of graphics and text used in this article, compared with the combination of video and text and other forms of advertising, there is a lack of immersion. It is difficult for subjects to fully understand or understand the relevant content in a short time. Therefore, in the follow-up study, the subjects can choose the main content of the advertisement, and then display it after AI generation. Another unreasonable aspect is that we chose Nestle Coffee as the experimental background during the experiment. As a well-known brand, participants are likely to have a pre existing attitude towards it - like or dislike it. This attitude may affect the final experimental results. Therefore, we should choose less famous brands as our research background in the future. Besides, this paper emphasizes the use of AIGC in social media advertising. This is only a small part of AI marketing application. How will other application directions, such as AI content real-time generation, and real-time push, affect variables at the company level or consumer level? Further research is needed.

AIGC advertising is a newly emerging thing. Therefore, research on it is very limited. There are many problems that need to be solved urgently. The problem mainly focuses on the following aspects: the acceptance level of AIGC advertising towards consumers; the impact of AIGC advertising content and style; how to achieve personalized customization in AIGC advertising; how to measure the ethical and legal issues of AIGC advertising, as well as the ROI of AIGC advertising. We hope this article can serve as a catalyst for further research in this field.

Funding. The research was financially supported by the National Natural Science Foundation of China under grant [number 71831005], Natural Science Foundation of Shenzhen under grant [number JCYJ20220531095216037], and Natural Science Foundation of Guangdong Province [number 2023A1515012520].

References

1. Kaplan, A.M., Haenlein, M.: Users of the world, unite! the challenges and opportunities of Social Media. Bus. Horiz. **53**(1), 59–68 (2010)
2. Alalwan, A.A.: Investigating the impact of social media advertising features on customer purchase intention. Int. J. Inf. Manag. **42**, 65–77 (2018)
3. Voorveld, H.A.M., Van N, G., Muntinga, D.G.: Fred Bronner Engagement with social media and social media advertising: the differentiating role of platform type. J. Advert. **47**(1), 38–54 (2018)
4. McCarthy, J.: Adidas Chief Casts Doubt on TV Ads: 'Digital Engagement Is Key for Us,'" The Drum, 17 March (2017). http://www.thedrum.com/news/2017/03/17/adidas-chief-casts-doubt-tv-ads-digital-engagement-key-us
5. Tohid, G., Easa, S., Anders, G.: Consumer response to online behavioral advertising in a social media context: the role of perceived ad complicity. Psychol. Mark. **39**(10) (2022)
6. Du, H., et al.: AI-Generated Content (AIGC) services in wireless edge networks. arXiv preprint arXiv:2301.03220 (2023)
7. Hanaysha, J.R.: An examination of social media advertising features, brand engagement and purchase intention in the fast food industry. Br. Food J. **124**(11) (2022)
8. Nisar, T.M., Prabhakar, G., Ilavarasan, P.V., Baabdullah, A.M.: Up the ante: electronic word of mouth and its effects on firm reputation and performance. J. Retail. Consum. Serv. **53**, 101726 (2020)
9. Bai, L., Yan, X.: Impact of firm-generated content on firm performance and consumer engagement: evidence from social media in China. J. Electron. Commer. Res. **21**(1), 56–74 (2020)
10. De Costa, F., Abd Aziz, N.: The effects of user generated content and firm generated content on millennials' purchase intention of Shariah-compliant stocks. Jurnal Pengurusan **62** (2021)
11. Brodie, R.J., Hollebeek, L.D., Jurić, B., Ana, I.: Customer engagement: Conceptual domain, fundamental propositions, and implications for research. J. Serv. Res. **14**(3), 252–271 (2011)
12. Algesheimer, R., Dholakia, U.M., Herrmann, A.: The social influence of brand community: evidence from European car clubs. J. Mark. **69**(3), 19–34 (2005)
13. Patterson, P., Yu, T., De Ruyter, K.: Understanding customer engagement in services. In: Proceedings of ANZMAC 2006 Conference Advancing Theory, Maintaining Relevance, Brisbane, vol. 4, no, 6 (2006)
14. Asante, I.O., Jiang, Y., Luo, X.: The organic marketing Nexus: the effect of unpaid marketing practices on consumer engagement. Sustainability **15**(1), 148 (2022)

15. Viswanathan, V., Malthouse, E.C., Maslowska, E., Hoornaert, S., Van den Poel, D.: Dynamics between social media engagement, firm-generated content, and live and time-shifted TV viewing. J. Serv. Manag. (2018)
16. Kumar, A., Bezawada, R., Rishika, R., Ramkumar, J., Kannan, P.K.: From social to sale: the effects of firm-generated content in social media on customer behavior. J. Mark. 80(1), 7–25 (2016)
17. Lea, W.: The New Rules of Customer Engagement. Inc.com. Accessed 30 Oct 2015(2012). http://www.inc.com/wendy-lea/new-rules-of-customer-engagement.html
18. Chandy, R.K., Tellis, G.J., MacInnis, D.J., et al.: What to say when: advertising appeals in evolving markets. J. Mark. Res. 38(4), 399–414 (2001)
19. Berger, J.: Word of mouth and interpersonal communication: a review and directions for future research. J. Consum. Psychol. 24(4), 586–607 (2014)
20. Fossen, B.L., Schweidel, D.A.: Social TV, advertising, and sales: are social shows good for advertisers? Mark. Sci. 38(2), 274–295 (2019)
21. Guitart, I.A., Stremersch, S.: The impact of informational and emotional television ad content on online search and sales. J. Mark. Res. 58(2), 299–320 (2021)
22. Rocklage, M.D., Fazio, R.H.: The enhancing versus backfiring effects of positive emotion in consumer reviews. J. Mark. Res. 57(2), 332–352 (2020)
23. De Bruyn, A., et al.: Artificial intelligence and marketing: pitfalls and opportunities. J. Interact. Mark. 51(1), 91–105 (2020)
24. Lee, T., Yun, T.W., Haley, E.: Effects of mutual fund advertising disclosures on investor information processing and decision-making. J. Serv. Mark. 27(2), 104–117 (2013)
25. De Jans, S., Cauberghe, V., Hudders, L.: How an advertising disclosure alerts young adolescents to sponsored vlogs: the moderating role of a peer-based advertising literacy intervention through an informational vlog. J. Advert. 47(4), 309–325 (2018)
26. Burton, S., Cook, L.A., Howlett, E., et al.: Broken halos and shattered horns: overcoming the biasing effects of prior expectations through objective information disclosure. J. Acad. Mark. Sci. 43, 240–256 (2015)
27. De Jans, S., Van de Sompel, D., De Veirman, M., et al.: # Sponsored! how the recognition of sponsoring on Instagram posts affects adolescents' brand evaluations through source evaluations. Comput. Hum. Behav. 109, 106342 (2020)
28. Thomas, T.G.: How user generated content impacts consumer engagement. In: 2020 8th International Conference on Reliability, Infocom Technologies and Optimization (Trends and Future Directions) (ICRITO). IEEE, pp. 562–568 (2020)
29. Habib, S., Hamadneh, N.N., Hassan, A.: The relationship between digital marketing, customer engagement, and purchase intention via OTT platforms. J. Math. 2022 (2022)
30. Brodie, R.J., Ilic. A., Juric, B., Hollebeek, L.: Consumer engagement in a virtual brand community: an exploratory analysis. J. Bus. Res. 66(1), 105–114 (2013)
31. Dwivedi, A.: A higher-order model of consumer brand engagement and its impact on loyalty intentions. J. Retail. Consum. Serv. 24, 100–109 (2015)
32. Gao, L., Li, G., Tsai, F., et al.: The impact of artificial intelligence stimuli on customer engagement and value co-creation: the moderating role of customer ability readiness. J. Res. Interact. Mark. 17(2), 317–333 (2023)

AI-Enabled Smart Healthcare Ecosystem Model and Its Empirical Research

Qianrui Du[1], Changlin Cao[2], Qichen Liao[3], and Qiongwei Ye[4(✉)]

[1] Experimental Teaching and Network Technology Management Center, Guangzhou Huashang College, Guangzhou, China
[2] Guangzhou Institute for Industrial Development in Greater Bay Area, China Business School, Guangzhou, China
[3] International Business School, Beijing Foreign Studies University, Beijing, China
[4] Business School, Yunnan University of Finance and Economics, Kunming, China
yeqiongwei@163.com

Abstract. This paper combines the ecosystem theory to study the AI-enabled healthcare ecology and build a smart healthcare ecosystem. The ecosystem growth model is employed to discuss the growth and change patterns of innovative companies adopting AI technology in the smart healthcare ecosystem from a micro perspective. Combining the Lotka-Volterra competition model, this paper delineates that the larger the competition coefficient of AI companies cluster, the more stable their development in the smart healthcare ecosystem will be. The findings are supported by two case studies.

Keywords: Artificial Intelligence (AI) · Smart Healthcare · Business Ecosystem · Ecosystem Growth Model · Lotka-Volterra Competition Model

1 Theoretical Background

Due to the highly specialized nature of the healthcare industry, hospitals have always occupied an absolute core position in the traditional healthcare ecosystem. When hospitals operate at overloaded capacity, the efficiency of the entire ecosystem will decrease, and poor communication among stakeholders will lead to problems such as uneven distribution of healthcare resources. With the development of artificial intelligence (AI) in healthcare, connections in communication and information exchange among stakeholders in the healthcare ecosystem gradually improve, forming a dynamically balanced digital information network. This paper endeavors to build a business model of the future healthcare ecosystem and explore the development patterns of the AI-enabled healthcare ecosystem as well as the competition mechanism of healthcare enterprises based on the new technological revolution.

1.1 Business Ecosystem

The concept of "business ecosystem" has remained popular among researchers and practitioners since James F. Moore proposed it in his article published in Harvard Business

© Springer Nature Switzerland AG 2023
F. Fui-Hoon Nah and K. Siau (Eds.): HCII 2023, LNCS 14039, pp. 130–139, 2023.
https://doi.org/10.1007/978-3-031-36049-7_10

Review in 1993 for the first time [1]. Inspired by ecology, Moore's concept of business ecosystem provides a metaphor for understanding the intertwined quality of the industries [2]. In this paper, we introduce the business ecosystem into digital economic activities and use ecological analysis to study the dynamic evolutions of different industries [3], offering a new perspective to explain the interactions among stakeholders.

The business ecosystem and the biological ecosystem have various unique features. Similar to its counterpart, the business ecosystem emphasizes the symbiotic and competitive relationship between stakeholders [4, 5]. Viewing business and innovative network management from the perspective of ecosystems, this paper analyzes the key players of the business ecosystem [6], and explains the dynamic evolution of the increasing business and innovation activities in the contemporary world [7–9].

1.2 Ecosystem Growth Model

The growth of new species' populations in the natural world follows a certain pattern. At the primary stage when a new species enters the ecosystem, there is a continuous increase in its number, which will reach the peak at a particular time and then slow down.

The diffusion mechanism of new products in the market is similar to the growth pattern of ecosystems. Reference [10] proposed a product growth model, arguing that the growth model for the timing of the first purchase of a new product was based on the assumption that the probability of purchasing at any given time was linearly related to the number of prior buyers. After the exponential growth of initial purchases reaches a peak in a given period, the exponential decay occurs [11].

1.3 Lotka-Volterra Competition Model

In nature, competitors are everywhere. Populations often do not exist alone, and when competition occurs, the growth of different ecological populations may be inhibited. In ecology, species do not exist alone, but compete with other multiple species to achieve a state of coordinated symbiosis.

We assume that there are only two populations in the ecosystem, and use the Lotka-Volterra equations to describe the growth trends of these two populations under the constraint of competition [12, 13].

The Lotka-Volterra competition equations [14]:

$$\begin{cases} \frac{dN_1}{dt} = r_1 N_1 (1 - \frac{N_1 + \alpha N_2}{k_1}) \\ \frac{dN_2}{dt} = r_2 N_2 (1 - \frac{N_2 + \beta N_1}{k_2}) \end{cases} \tag{1}$$

where N_1 and N_2 denote the number of the two populations, k_1 and k_2 denote the environment capacity of N_1 and N_2. α is the competition coefficient for the effect that species N_1 has on species N_2, β is the competition coefficient of the effect of species N_2 on species N_1. r_1 and r_2 represent the growth rates of populations N_1 and N_2, respectively.

Currently, AI technology has presented new opportunities for business ecosystems, particularly in terms of applications in the field of healthcare [15–17]. Despite the increasing theoretical and practical relevance of AI-enabled business models to healthcare, there has been a lack of research to date that examines AI in healthcare ecosystem from a micro perspective [18]. Because of the similarities between the nature of the study of organisms in ecology and that of economic and social enterprises, application of ecological models will provide a lesson for the study of healthcare ecosystems. This paper primarily focuses on introducing AI technology into the healthcare ecosystem and constructing a network model of the ecosystem. The ecological growth model is used to uncover the growth patterns of new AI technology in the healthcare ecosystem, and the Lotka-Volterra competition model is employed to reflect the changing trend of the number of clusters of two types of AI companies in a competitive environment.

2   Constructing the Smart Healthcare Ecosystem

2.1   Smart Healthcare Ecosystem

The entrance of AI technology to the healthcare ecosystem has brought about changes in consumer behavior and habit. What with the rapid development of AI medical devices and new ideas about health, people's purchasing behavior has also been changed, and medical device products are no longer limited to hospitals, but are adopted by households as well as professional testing services [19, 20]. Medical device companies have also seized new opportunities for transformation through technological innovation and service improvement. AI image recognition in medical diagnostics can accurately and quickly identify a patient's lesion through deep learning algorithms [21] incubating numerous professional medical imaging and testing agencies who, with their state-of-the-art medical devices, issue professional, accurate and authoritative medical reports for patients in a timely manner.

The implementation of regulations on the multiple-site practice by physicians and the Internet sale of prescription drugs has propelled the rapid development of AI-assisted teleconsultation, which has attracted the Internet giants to the field of healthcare ecology, allowing doctors and patients to enjoy the convenient teleconsultation services with AI-assisted consultation systems. The medical AI can guide the patients to retail pharmacies using the information from the medical consultation. This behavioral change has rejuvenated the traditional healthcare market. Through "AI plus" medical logistics, pharmacies are now able to perfectly match the scale requirements of the increasing needs of pharmaceutical supplies, and pharmaceutical companies can accurately record the corresponding batch number, date of manufacturing and other key information to ensure medication safety (Fig. 1).

2.2   Constructing the Smart Healthcare Ecosystem Model

The fast-growing AI technology has facilitated the formation of an emerging ecosystem of the healthcare industry. Ecologically speaking, the smart healthcare ecosystem shares remarkable similarities with the ecosystem in nature. The stakeholders of the smart healthcare ecosystem include doctors, patients, hospitals, mobile devices, pharmaceuticals, and public healthcare institutions, which form an interdependent and mutually promoting organism that achieves a state of dynamic balance in competition and adds value to the whole system (Fig. 2).

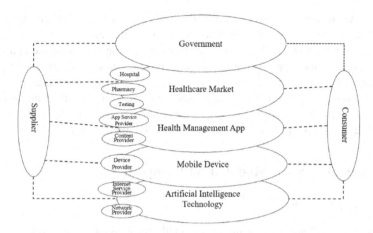

Fig. 1. Smart healthcare ecosystem

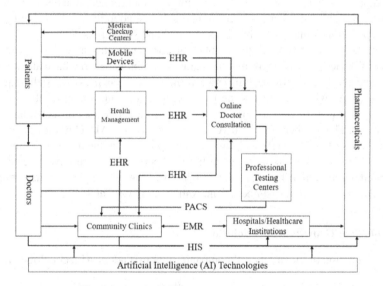

Fig. 2. Model of the smart healthcare ecosystem

2.3   Dynamic Analysis of AI-Enabled Smart Healthcare Ecosystem

The entire smart healthcare ecosystem is supported by AI technology. The activities of stakeholders, such as doctors, patients, health management, professional testing centers, hospitals/healthcare institutions, and pharmaceuticals, close the loop of the smart healthcare ecosystem.

When notified by the health management of illness, the patient can go to the medical checkup center for health screening. The results will be uploaded to the AI online consultation system for online doctor consultation. Depending on the severity of the condition, the patient can choose to directly buy medicines, do a further specialized examination, or go to a community clinic or hospital for treatment. The professional testing center transmits the AI-assisted diagnostic imaging results to the community clinic, which treats the patient according to his or her condition or, if surgery is required, the patient will be referred to an appropriate hospital or medical institution. As freelancers, doctors can communicate with patients directly, or answer their queries at any time through remote consultation. Some doctors provide medical services in multiple sites, such as community clinics, hospitals and medical institutions.

3 Ecological Growth Model of AI Companies

3.1 Model Specification

The entry of AI technology into the smart healthcare ecosystem has led to a surge in innovative companies adopting AI technology. Based on the ecosystem growth model, we can divide the suppliers in the smart healthcare ecosystem into two categories: one is the companies that first use AI technology for product production and sale, which we call innovators; and the other is called imitators whose adoption of AI technology for product innovation is influenced by environmental and social pressures.

We assume that, at time T, the number of companies that adopt AI for product innovation for the first time is M, and the number of companies that have already done so is N, and the linear relationship between them can be expressed by the equation $M = p + (\frac{q}{k})N$, where p denotes the innovation coefficient, q denotes the imitation coefficient, k stands for the environment capacity. p, q and k are constants, and when $T = 0, N = 0$. The number of innovative companies adopting AI technology in the smart healthcare ecosystem increases with time, and we predict that the number of innovative companies will reach the maximum at a certain time period. Next, we solve the ecological growth model to find time T* when the number of innovative companies is at its highest, and calculate the maximum number of innovative companies at T*.

3.2 Ecological Growth Model Solution

The linear function of the number of companies that first adopt AI technology is:

$$M = p + (\frac{q}{k})N \tag{2}$$

$F(T)$ is the probability of all companies in the ecosystem that are likely to adopt AI technology for innovation:

$$F(T) = \int_0^T f(t)dt \tag{3}$$

The number of companies that adopt AI technology for innovation is:

$$N = k * F(T) \tag{4}$$

For companies that have not adopted AI, the probability of their initial use at time T is:

$$\frac{f(T)}{[1-F(T)]} = M(T) = P + \left(\frac{q}{k}\right)N = p + q * F(T) \tag{5}$$

$$\frac{dN}{dt} = k * f(T) = M(T)(k-N) \tag{6}$$

$$\frac{dN}{dt} = \left[p + \left(\frac{q}{k}\right)N\right] * (k-N) \tag{7}$$

$$f(T) = \left[p + q * F(T)\right] * [1 - F(T)] \tag{8}$$

solving the above equations, we have:

$$F = \frac{q - p * e^{-(T+C)(p+q)}}{q * [1 + e^{-(T+C)(p+q)}]} \tag{9}$$

Because $F(0) = 0$, we have:

$$C = -\left[\frac{1}{(p+q)}\right] * \ln\left(\frac{q}{p}\right) \tag{10}$$

Substituted into the above equation, then

$$F(T) = \frac{1 - e^{-(p+q)T}}{\frac{q}{p}e^{-(p+q)T} + 1} \tag{11}$$

$$N(T) = kF(T) = \frac{k * [1 - e^{-(p+q)T}]}{\frac{q}{p}e^{-(p+q)*T} + 1} \tag{12}$$

Solving the above differential equations, we get:

$$N(T) = \frac{k * (1 - e^{-(p+q)T})}{\frac{q}{p}e^{-(p+q)*T} + 1} \tag{13}$$

When the system reaches the state of equilibrium, the growth of companies using innovative technologies is 0. At this point, let $\frac{dN}{dt} = o$,

$$T^* = \frac{1}{(p+q)}\ln\left(\frac{q}{p}\right) \tag{14}$$

$$N(T^*) = \frac{k(q-p)}{2q} \tag{15}$$

3.3 Model Analysis

After obtaining the innovation coefficient p, imitation coefficient q and environment capacity k of companies that use AI technology in the healthcare ecosystem, we can calculate that the number of companies adopting AI technological innovations in the smart healthcare ecosystem reaches its highest $N(T^*) = \frac{k(q-p)}{2q}$ at time $T^* = \frac{1}{(p+q)}\ln(\frac{q}{p})$. In addition, we also find that only when innovation coefficient p is greater than imitation coefficient q, can the innovative technological changes in the healthcare ecosystem be successfully achieved. Because the understanding of imitators and innovators is closer to the actual meaning represented by the model, the obtained calculation results can accurately reflect the growth trends of innovative companies in the smart healthcare ecosystem.

4 Lotka-Volterra Competition Model

4.1 Model Specification

It's extremely rare to see a single population develops on its own in an ecosystem. Generally, multiple populations coexist in a coordinated manner in a competitive environment. In the smart healthcare ecosystem, there are multiple populations of companies adopting AI technology for innovation, between which business cooperation and competition coexist. For example, in China's smart healthcare market, it is often the Internet conglomerates that adopt innovative technologies, and each of them hold different healthcare ecological resources. Competition among them remains fierce in some healthcare service areas.

Based on the Lotka-Volterra competition model, this paper explores the evolution of two clusters of innovative companies adopting AI technology in the healthcare ecosystem under competitive pressure.

4.2 Model Solution and Analysis

According to the Lotka-Volterra competition model, when the healthcare ecosystem reaches equilibrium, the growth rate of the two types of innovative companies is 0, i.e., when $\frac{dN_1}{dt} = \frac{dN_2}{dt} = 0$, a dynamic equilibrium is reached. Solving the equation, we have:

$$\begin{cases} N_1 = \frac{r_2}{\beta} \\ N_2 = \frac{r_1}{\alpha} \end{cases} \tag{16}$$

where α and β represent the competition coefficients of the two companies respectively, and their values are directly related to the intensity of competition between the two companies. If α is greater than β, then the growth rate of the innovative companies cluster N_1 is greater than that of the innovative companies cluster N_2, and N_1 has a more obvious competitive advantage in the ecosystem, and vice versa when β is greater than α .

4.3 Theoretical Contribution

This paper integrates the ecological growth model and the Lotka-Volterra competition model, and combines the computational derivation process of the ecosystem growth model. The same parameters of the two models are set consistent. The inferences of the fine-tuned Lotka-Volterra competition model are fully consistent with the real situation.

5 Practical Implications of the Smart Healthcare Ecosystem – The Cases of Tencent and Alibaba

The gradual application of AI technology in the field of smart healthcare has resolved the hospital treatment-centered deadlock in the traditional healthcare ecosystem, attracting an overwhelming number of Internet companies to the smart healthcare ecosystem, the most typical being the layout of the three Chinese Internet giants, BAT, the acronym for Baidu, Alibaba and Tencent, in the healthcare business ecosystem. Tencent's Internet healthcare layout has been built around WeChat, leveraging its advantage in the field of communication and socializing to connect doctors and patients and complete the resources layout from doctors to patients. Alibaba's venture into the Internet healthcare industry tends to layout through the existing Tmall platform, targeting primary level community healthcare organizations to attract Internet traffic, and has partly achieved the strategic deployment to close the loop of Internet healthcare ecology. Tencent and Alibaba represent not just individual companies, but the entire ecological chain of companies behind them. We use the two enterprise clusters of Tencent and Alibaba as cases to introduce the practical significance of the smart medical ecosystem model.

Based on the growth model of innovative companies adopting AI technology, we obtain the innovation coefficients, imitation coefficients and environmental capacity of Tencent and Alibaba respectively, which enables us to predict that, the number of companies in the two clusters of Tencent and Alibaba will reach the highest at a certain point in time in the future. This will provide reasonable support for the business and economic layout later on.

For two enterprise clusters in the same smart healthcare ecosystem, Tencent and Alibaba are bound to compete in overlapping businesses. The analysis of the Lotka-Volterra competition model revels that the larger the competition coefficient, the higher the number of enterprise clusters. As an AI company, research and innovation abilities should be the first priority and placed at the core for estimating the competition coefficient. The more advanced the company's technology is, the more evident its advantage in the ecosystem.

6 Conclusion

To study the interactions and changing trends of the stakeholders in the AI-enabled healthcare ecosystem, this paper draws on the business ecosystem as well as the ecosystem model, and builds a smart healthcare ecosystem model. It is found that after introducing AI technology to the healthcare ecosystem, hospitals no longer occupy the core position, and the said system will become more like an interconnected ecological network. After decentralization, the stakeholders in the healthcare ecosystem also undergo some new changes, most notably the gradual increase in the number of AI technology companies. In a competitive environment, companies with a greater competition coefficient will be more likely to gain a survival advantage. Thus, it is of vital importance to keep promoting technological innovations in the field of healthcare technologies so that companies can be invincible in the business ecosystem.

References

1. Moore, J.F.: Predators and Prey - a new ecology of competition. Harv. Bus. Rev. **71**(3), 75–86 (1993)
2. Koenig, G.: Business ecosystems revisited. Management **15**(2), 209–224 (2012)
3. Gomez-Uranga, M., Miguel, J.C., Zabala-Iturriagagoitia, J.M.: Epigenetic economic dynamics: the evolution of big Internet business ecosystems, evidence for patents. Technovation **34**(3), 177–189 (2014)
4. Yoon, C., Moon, S., Lee, H.: Symbiotic relationships in business ecosystem: a systematic literature review. Sustainability **14**(4) (2022)
5. Bengtsson, M., Kock, S.: "Coopetition" in business networks - to cooperate and compete simultaneously. Ind. Mark. Manage. **29**(5), 411–426 (2000)
6. Tian, C.H., et al.: BEAM: a framework for business ecosystem analysis and modeling. IBM Syst. J. **47**(1), 101–114 (2008)
7. Aarikka-Stenroos, L., Ritala, P.: Network management in the era of ecosystems: systematic review and management framework. Ind. Mark. Manag. **67**, 23–36 (2017)
8. Gao, R.Z., et al.: Modelling the emergence and evolution of e-business ecosystems from a network perspective. Stud. Inform. Control **22**(4), 339–348 (2013)
9. Aarikka-Stenroos, L., Lehtimaki, T.: Commercializing a radical innovation: probing the way to the market. Ind. Mark. Manag. **43**(8), 1372–1384 (2014)
10. Bass, F.M.: New product growth for model consumer durables. Manag. Sci. Ser. A-Theory **15**(5), 215–227 (1969)
11. Mahajan, V., Muller, E.: Innovation diffusion and new product growth-models in marketing. J. Mark. **43**(4), 55–68 (1979)
12. Bhargava, S.C.: Generalized Lotka-Volterra equations and the mechanism of technological substitution. Technol. Forecast. Soc. Change **35**(4), 319–326 (1989)
13. Fisher, J.C., Pry, R.H.: Simple substitution model of technological change. Technol. Forecast. Soc. Change **3**(1), 75–88 (1971)
14. Andrewartha, H.G., Birch, L.C.: The Lotka-Volterra theory of interspecific competition. Aust. J. Zool. **1**(2), 174–177 (1953)
15. Ismail, A., et al.: Artificial intelligence in healthcare business ecosystem: a bibliometric study. Int. J. Online Biomed. Eng. **18**(9), 100–114 (2022)
16. Kim, H., Lee, J.N., Han, J.: The role of IT in business ecosystems. Commun. ACM **53**(5), 151–156 (2010)
17. Rajpurkar, P., et al.: AI in health and medicine. Nat. Med. **28**(1), 31–38 (2022)
18. Schiavone, F., et al.: Digital business models and ridesharing for value co-creation in healthcare: A multi-stakeholder ecosystem analysis. Technol. Forecast. Soc. Change **166** (2021)
19. Nasr, M., et al.: Smart healthcare in the age of AI: recent advances, challenges, and future prospects. IEEE Access **9**, 145248–145270 (2021)

20. Ghazal, T.M., et al.: IoT for smart cities: machine learning approaches in smart healthcare-a review. Future Internet **13**(8) (2021)
21. Bi, W.L., et al.: Artificial intelligence in cancer imaging: clinical challenges and applications. CA-A Cancer J. Clin. **69**(2), 127–157 (2019)

Using Co-word Network Community Detection and LDA Topic Modeling to Extract Topics in TED Talks

Li-Ting Hung[1], Muh-Chyun Tang[1]($\boxtimes$) ⓘD, and Sung-Chien Lin[2]

[1] National Taiwan University, Taipei, Taiwan
mctang@ntu.edu.tw
[2] Shih Hsin University, Taipei, Taiwan

Abstract. Two topic detection techniques—co-word network analysis and topic modeling—were applied to extract topics in the Ted Talks. Ted Talks was chosen for its enormous impact worldwide and the rich descriptive data accompanying each talk that allow us to compare the topics resulting from different methods. The co-word network was built based on the "related_tags" field so that modularity analysis can be performed to classify the tags according to their co-occurrence patterns. Topic modeling was applied to the description field and the full-text transcript separately to detect the topics present in the free-text. The results of network modularity analysis revealed 13 interpretable topics consisting of closely knitted tags. Topic modeling generated 25 topics for the description and 40 for the transcript, respectively. Our results showed that both topic extraction methods were able to successfully identify the range of topics in the TED Talks. While the co-word network gave a broad overview and afforded visualization, the topic model revealed topics with greater granularity. We compared the semantics of the topics produced by different methods and discussed the methodological implications of our research.

Keywords: LDA topic modeling · Co-word analysis · TED Talk

1 Introduction

One of the main tasks of text analysis is to detect underlying topics from a large collection of documents. In scientometrics, the co-word network has long been used to identify major topics and map the intellectual structure of a knowledge domain [1, 2]. Co-word networks can be constructed based on the pair-wise co-occurrence patterns of keywords or other bibliographic elements within documents in a domain. Multivariate techniques (e.g., [3, 4]) and, more recently, network community detection algorithms (e.g., [5–7]) can then be applied to the co-word network to discover the topics in the domain. One of the limitations of a co-word network is its reliance on human-assigned subjects, whether in the form of index terms or author-assigned keywords in the structured metadata. Recently, topic modeling, because of its ability to efficiently process large volumes of

© Springer Nature Switzerland AG 2023
F. Fui-Hoon Nah and K. Siau (Eds.): HCII 2023, LNCS 14039, pp. 140–154, 2023.
https://doi.org/10.1007/978-3-031-36049-7_11

unstructured texts, has also been adopted in scientometrics for the purpose of discovering the underlying semantic structure of the collections of documents [8]. Blei and Lafferty (2006) demonstrated that topic modeling could identify interpretable themes in science and the word usage within them over time using a large corpus of full-text articles published by Science [9]. The basic idea behind topic modeling is to characterize each document as a mixture of topics, which, in turn, is represented by a statistical distribution of words. By examining the distribution of the terms within a topic, one can interpret its semantics. Previous studies have compared co-word network analysis and topic modeling to uncover the main topics in a knowledge domain [3, 10, 11]. Suominen & Toivanen (2016) argued that human-assigned metadata, such as subject headings, is ill-suited to reflect the ever-expanding nature of science [11]. Using pre-defined subject categories, they pointed out, risks fitting new knowledge into the historical categories. The unsupervised technique, such as topic modeling, on the other hand, draws hidden themes directly from the free-texts and therefore is more current than the human-created subject category.

The metadata of TED Talks presents an excellent opportunity to explore different topic extraction techniques as it contains both human-created subject categories in the form of tags and free-text descriptions and transcripts. We applied co-word network analysis on the tags assigned by TED and LDA topic modeling on the free-texts contained in the description field and the transcripts separately. The result will allow us to explore and compare the merits of each method and how they complement each other.

2   Data and Methodology

A total of 5050 TED Talks were available as of April 2021, when the metadata and transcripts of the talks were retrieved. TED has used a rich set of metadata elements to describe each talk, which includes title, description, tag, transcript, speaker, and view counts (See Table 1 for the available metadata fields). "related_tags" is a set of subject terms TED uses to index the talks, which is akin to subject headings in the context of human indexing. In contrast, the description and transcript provide a free-text representation of the video content.

Table 1. Metadata for TED Talk videos

	Fields
Basic information	ID, talk_id, talk_name, video_type_name, event, view_count, duration, recording_date, published_timestamp, language
About speaker	number_of_speakers, speaker_id, speaker_name, speaker_description, speaker_who_he_is
Content	related_tags[*], talk_description[*], speaker_why_listen[*], transcript[*]

[*] fields used for this research

There is a total of 464 tags representing various subjects covered in the talks. Each talk is indexed by, on average, 7.48 tags. Figure 1 shows the ten most frequently used tags in different periods, which gives an indication of how the subjects of the talks evolved. In the early years, the talks about technology, design, and science dominated; in the later years, with the rapid growth of talks published, topics related to global issues, society, and social change also gradually raised to prominence.

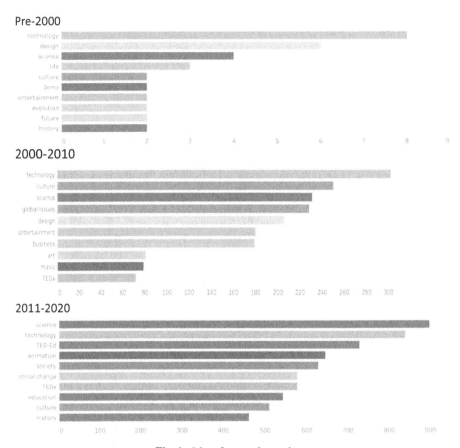

Fig. 1. Most frequently used tags

2.1 Construction and Filtering of Co-word Network

A co-word network was constructed based on the co-occurrence patterns of the tags in the metadata. An edge is created whenever a pair of tags co-occurs in the "related_tags" field of a talk. The edge weight was determined by the inclusion index [12].

$$I_{ij} = \frac{C_{ij}}{\min(C_i, C_j)}$$

where C_{ij} represents the co-occurrence of t_i and t_j, and C_i and C_j represents the occurrence of t_i and t_j, respectively.

The Leuven modularity-maximization algorithm [13] performed on the initial network yielded a low modularity value of 0.251 and six classes. A low modularity score indicates the network is not well-delineated. To enhance the resolution and interpretability of the resulting classes, filtering procedures were performed on the initial network. Following [7], we experimented with different filtering parameters with the goal of achieving both high modularity and class interpretability (See Table 2 for different filtering criteria and resulting modularity values). The highest modularity value of 0.712 was researched with the term frequency threshold at one and edge weight at 0.7, resulting in thirteen classes of closely connected tags. The most prominent tags within each class were then used to interpret its topic. We will detail the interpretation of each modularity class in the result section.

Table 2. Different filtering parameters and the resulting networks

Occurence $\geq$	Inclusion index	Nodes remain	Edges remain	Modularity	Number of modularity class
1	**0.7**	**201 (43.32)**	**297 (1.04%)**	**0.712**	**13**
2	0.5	369 (79.52)	853 (3.00%)	0.690	12
3	0.5	364 (78.45)	828 (2.91%)	0.690	12
4	0.5	358 (77.16)	811 (2.85%)	0.692	13

2.2 LDA Topic Modeling

Two corpora, one combined the free-texts in the "talk_description" and speaker_why_listen" fields (henceforth "description" corpus), and the other, the free-texts in the transcript, were used for our topic modeling analysis (See Table 3 for the descriptive statistics).

Table 3. Descriptive statistics of the corpora analyzed

	Medium	Average	SD	(word types)
Description	215.00	214.91	110.34	1509
Transcript	1967.00	2071.69	1226.62	3764

After tokenization and lemmatization, LDA topic modeling was performed on the two corpora. Based on coherence scores and interpretability of the topics, 25 topics were chosen for the description and 40 for the transcript.

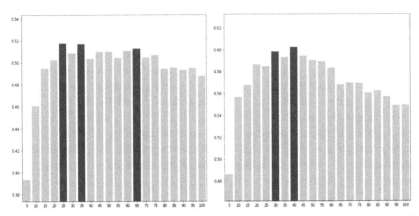

Fig. 2. Coherence scores at different numbers of topics for description (left) and transcript (right)

The labeling of the topics was based on two pieces of information, the ten most probable terms within each topic and the five talks it has the highest presence. Detailed procedures for topic interpretation and labeling will be given in the results section.

3 Results

3.1 Co-word Network Modularity Analysis

We labeled each of the thirteen modularity classes by its nodes with the highest weighted node degrees. Seven of the modularity classes contain just one most prominent tag, including M0 "Design," M2 "Science," M3 "technology," M4 "Ted-ED," M8 "Music," M10 "Ocean," M11 "Education." The rest of the classes were named considering several most prominent tags, such as M7 "Climate & Environment," M9 "Nature," and M12 "Arts" (See the first column in Table 7 for the full list of the classes). The topics that include the highest percentage of tags are M2 "Science" (56%), followed by M5 "Business and Economy" (19.9%), M3 "Technology" (17.41%), and M1 "Health" (10.95%) (See Fig. 3 for the network visualization).

3.2 LDA Topic Modeling

The ten most probable terms in each topic were used to help us interpret its meaning. In addition, the top five talks most associated with a topic were also used for disambiguation. For example, Table 4 shows the five typical talks, their related tags, and the ten high probability terms in T0 in the transcript corpus, which was thus labeled "genetics."

Fig. 3. Visualization of the modularity classes and their associated tags

Table 4. Information about the top five typical talks and the most probable terms in T0

talk id	Talk name	Related tags	Ten high-probability terms within the topic
39689	Can we cure genetic diseases by rewriting DNA?	['biology', 'CRISPR', 'genetics', 'science', 'disease', 'technology', 'future', 'molecular biology', 'innovation', 'biotech', 'health', 'virus']	gene, dna, genetic, evolution, sequence, biology, genome, protein, DNA, code
331	DNA folding, in detail	['DNA', 'MacArthur grant', 'creativity', 'nanoscale', 'physics', 'science', 'technology']	
1213	DNA clues to our inner Neanderthal	['DNA', 'biology', 'evolution', 'science']	

(*continued*)

Table 4. (*continued*)

talk id	Talk name	Related tags	Ten high-probability terms within the topic
23058	The radical possibilities of man-made DNA	['genetics', 'science', 'invention', 'technology', 'biology', 'DNA', 'future', 'synthetic biology', 'medicine', 'medical research', 'bioethics', 'biotech', 'molecular biology', 'nature']	
863	Watch me unveil "synthetic life"	['DNA', 'biology', 'business', 'genetics', 'life', 'science', 'technology']	

Following the same heuristics, we could interpret and label the topics present in the description (See Table 5) and transcript corpora (Table 6).

Table 5. Naming of the topics modeled on talk description

Topic	Most probable terms
D0 Space exploration	Earth, mission, space, planet, map, universe, NASA, fly, Australia, Robert
D1 Human right	woman, activist, justice, movement, Women, violence, law, campaign, gender, race
D2 Green architecture	city, building, urban, architecture, architect, small, side, local, large, Green
D3[a] TV & Films	film, series, David, documentary, produce, tv, BBC, John, television, production
D4 Global Economy	member, Global, World, president, President, list, International, board, chair, Economic
D5 Education	teach, school, student, education, program, teacher, kid, college, class, passionate
D6 Letters	Times, language, publish, journalist, writing, writer, Journal, Street, magazine, editor
D7[b] Scientific research	National, receive, Center, Institute, Science, Research, Fellow, School, Harvard, American
D8 Psychology	question, Anderson, Addison, society, answer, psychology, Studio, choice, shape, psychologist

(*continued*)

Table 5. (*continued*)

Topic	Most probable terms
D9 Arts	art, artist, Museum, Art, visual, document, collection, photo, culture, Arts
D10 Data Sciences	datum, number, fact, present, set, step, animation, data, point, trend
D11 Sustainability	climate, energy, water, problem, sustainable, solution, solve, low, cost, plan
D12 Management	business, innovation, entrepreneur, strategy, management, Business, industry, ceo, firm, China
D13 Engineering	MIT, computer, designer, material, engineering, device, Lab, kind, engineer, form
D14 Family	child, family, love, speak, age, mother, bear, struggle, parent, father
D15 uninterpretable_1*	Chris, Paul, success, Bill, individual, foundation, lay, early, organize, radical
D16 uninterpretable_2*	day, death, funny, read, single, hard, history, person, rule, dream
D17 Refugee	food, open, war, attention, deliver, end, Scott, eye, pay, refugee
D18 Gaming animation	digital, video, game, internet, online, reality, platform, web, medium, user
D19 Health	health, disease, care, Health, medical, cancer, medicine, patient, drug, doctor
D20 Music	play, performance, sound, perform, song, tour, stage, audience, musician, album
D21 Neurosciences	brain, scientific, behavior, understand, lab, physics, memory, experiment, scientist, biology
D22 AI	robot, machine, car, intelligence, thing, knowledge, AI, robotic, analysis, artificial
D23 Poverty	economic, Africa, government, policy, financial, poverty, security, conflict, poor, East
D24 Preservation	animal, ocean, protect, specie, natural, plant, de, ancient, land, nature

[a] The typical talks in D3 are about documentaries of nature preservation
[b] The typical talks in D7 are about the scientific research on mental health and sleep

Table 7 juxtaposes the resultant topics produced by the three procedures. Topic modeling tends to extract topics of higher granularity, consistent with previous findings using subject headings and free-text side-by-side for the purpose of mapping sciences [11]. In most cases, a tree-like hierarchy can be drawn from the broader category generated by co-word analysis to the narrower, more specific subjects extracted by topic modeling. For example, M2 "Science" can be further split into D7 "Scientific research," D0/T19" Space exploration," T0 "Genetics," T13 "Biochemistry" and T21 "Meteorology," Similarly, topics modeled on the transcripts also tended to more precise than those modeled on the description. For example, the topic most associated with the talk entitled "A new way to remove CO_2 from the atmosphere" in the description model was

Table 6. Naming of the topics modeled on transcript

Topic	
T0 Genetics	gene, dna, genetic, evolution, sequence, biology, genome, protein, DNA, code
T1 Development	China, India, chinese, West, Asia, East, western, indian, North, Brazil
T2 Math & puzzle	puzzle, complexity, boundary, metaphor, flow, communicate, interaction, sum, dynamic, graph
T3 Emotion/Personality	emotion, happiness, desire, emotional, personality, joy, meaning, regret, negative, cheat
T4 Cancer cure	cancer, tissue, medicine, organ, surgery, tumor, stem, skin, breast, cure
T5 Family	mom, sister, dad, wife, cry, brother, husband, hate, anger, angry
T6[a] Mythology	ancient, king, hero, stone, gold, myth, god, Rome, greek, narrative
T7 Conflicts	military, weapon, conflict, soldier, War, security, gun, bomb, Afghanistan, Iraq
T8 Mental health	stress, pain, mental, depression, disorder, anxiety, illness, loss, medication, symptom
T9 Diet	coffee, drink, taste, cook, milk, meat, restaurant, meal, diet, cup
T10 Plants/animals	bird, bee, insect, ant, elephant, flower, monkey, cat, dinosaur, wild
T11 Management	meeting, failure, manager, confidence, leadership, lesson, mistake, executive, advice, employee
T12 Traffic safety	roll, ride, seat, finger, plane, flight, ball, airplane, shoe, crash
T13 Biochemistry	chemical, bacteria, molecule, oxygen, microbe, compound, gut, spider, chemistry, enzyme
T14 Infectious diseases	virus, vaccine, HIV, infection, infect, malaria, epidemic, pandemic, antibiotic, flu
T15 uninterpretable_1*	tv, John, player, sport, television, show, Paul, David, laugh, athlete
T16 Ocean	ocean, fish, sea, boat, ship, coral, shark, swim, whale, reef
T17 Arts	film, artist, paint, photograph, painting, beauty, museum, theater, photo, scene
T18 Human rights	violence, gender, gay, identity, immigrant, racism, racial, racist, slavery, abuse
T19 Space exploration	universe, star, particle, galaxy, hole, physics, quantum, telescope, sky, atom
T20 Internet	Google, web, Facebook, content, online, website, digital, app, user, Twitter

(continued)

Table 6. (continued)

Topic	
T21 Meteorology	ice, Mars, mountain, satellite, cloud, temperature, cold, cave, weather, heat
T22 Letters	English, letter, card, speaker, text, page, sentence, author, cartoon, translate
T23 Renewable energy	CA, nuclear, electricity, solar, vehicle, battery, electric, engine, Chris, wind
T24 Crimes	police, prison, crime, legal, court, criminal, justice, jail, officer, arrest
T25 Reasoning	argument, moral, bias, argue, participant, assumption, opinion, debate, FALSE, objective
T26 Learning	learning, classroom, grade, university, professor, graduate, University, School, training, peer
T27 Carbon emission	carbon, oil, gas, plastic, emission, environmental, waste, CO2, forest, coal
T28 Democracy	democracy, vote, election, citizen, politic, president, politician, campaign, flag, democratic
T29 Economy	financial, income, bank, invest, fund, wealth, profit, investment, capital, economist
T30 Engineering	designer, print, fold, electronic, 3d, glass, chip, camera, prototype, laser
T31 Performance	♫, song, dance, noise, poem, sing, instrument, musician, poetry, ear
T32 Robots	robot, leg, robotic, muscle, limb, cord, wheelchair, drone, soft, joint
T33 Agriculture	innovation, farmer, farm, worker, factory, production, brand, patent, consumer, supply
T34 Poverty	poverty, refugee, african, village, aid, continent, Kenya, Nigeria, Africans, corruption
T35 Sex/Reproduction	sex, male, female, %, birth, smell, sexual, hair, mosquito, pregnant
T36 Neuroscience	signal, neuron, dog, consciousness, electrical, perception, cortex, visual, neural, activate
T37 AI	artificial, intelligence, AI, algorithm, mathematic, predict, math, mathematical, square, match
T38 Religions	religion, compassion, religious, faith, honor, grandmother, tribe, tradition, soul, muslim
T39 Architecture	architecture, park, urban, neighborhood, City, San, architect, bus, Francisco, housing

[a] The talks on the topic of mythology mostly belong to TED-ED series

D11"sustainability." IIn contrast, in the transcript model, it was T27 "Carbon emission," which is more specific to the subject of the talk. There were also cases where a talk was assigned different topics when different corpus was used. For example, for the talk "The best stats you've ever seen," the transcript model identified T2 "Development" as the most important topic. In the description model, it was more correctly assigned the topic of D10 "Data Sciences," as the speaker Hans Rosling was using demographic data of countries at different development stages to demonstrate the power of data visualization.

Cosine similarities between topics uncovered by the three procedures were calculated and visualized in Fig. 4, 5, and 6.

Table 7. The topics generated by three topic extraction procedures

Co-word network	Topic modeling (description)	Topic modeling (transcript)
M0 Design & architecture	D0 Space exploration	T0 Genetics
M1 Health	**D1 Human rights**	**T1 Development**
M2 Science	D2 Green architecture	T2 Math and puzzles
M3 Technology	D3 TV and films	T3 Emotions & Personality
M4 TED-ED animation	D4 Global Economy	T4 Cancer cure
M5 Business & Economy	D5 Education	**T5 Family**
M6 Neurology	D6 Letters	T6 Mythology
M7 Climate & Environment	D7 Scientific research	**T7 Conflicts**
M8 Music	D8 Psychology	T8 Mental health
M9 Nature	D9 Arts	T9 Diet
M10 Oceans	D10 Data sciences	T10 Plants & Animals
M11 Education	D11 Sustainability	T11 Management
M12 Arts	D12 Management	T12 Traffic safety
	D13 Engineering	T13 Biochemistry
	D14 Family	T14 Infectious diseases
	D15 *uninterpretable_1*	T15 *uninterpretable*
	D16 *uninterpretable_2*	T16 Ocean
	D17 Refugee	T17 Arts
	D18 Gaming and animation	**T18 Human rights**
	D19 Health	T19 Space exploration
	D20 Music	T20 Internet
	D21 Neurosciences	T21 Meteorology
	D22 Artificial intelligence	T22 Letters
	D23 Poverty	T23 Renewable energy
	D24 Preservation	**T24 Crimes**

(*continued*)

Table 7. (*continued*)

Co-word network	Topic modeling (description)	Topic modeling (transcript)
		T25 Reasoning
		T26 Learning
		T27 Carbon emission
		T28 Democracy
		T29 Economy
		T30 Engineering
		T31 Performance
		T32 Robots
		T33 Agriculture
		T34 Poverty
		T35 Sex/Reproduction
		T36 Neurosciences
		T37 Artificial intelligence
		T38 Religions
		T39 Architecture

Fig. 4. The topic similarity between co-word analysis and topic modeling on description

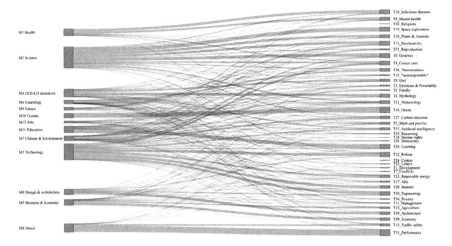

Fig. 5. The topic similarity between co-word analysis and topic modeling on transcripts

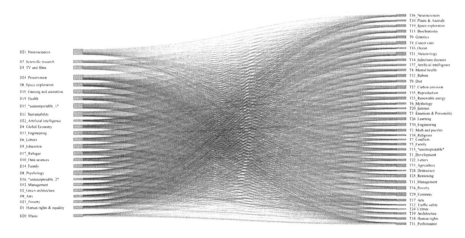

Fig. 6. The topic similarity between talk description and transcript models

However, some topics extracted by topic modeling cannot be readily subsumed under the broader categories. Topics uncovered uniquely by topic modeling were highlighted in Table 7, which shows that, firstly, topic modeling on transcript produced the highest term specificity. Secondly, these uniquely identified terms are mostly about social issues and humanities, for example, D1/T18 "Human rights," D14/T5″ Family," "D23 Poverty", D17 "Refugee," T7 "Conflicts," T28 "Democracy," and T38 "Religions."

4 Discussion and Conclusion

In this study, we experimented with different automatic topic discovery methods on the TED Talks. The presence of subject terms and free-text contents of the TED Talks affords us a unique opportunity to explore different topic-extracting techniques on different elements of its metadata. Specifically, co-word network analysis was applied to the human-assigned tags. This is done by, firstly, constructing a co-word network based on the cooccurrence patterns of the tags. Modularity analysis, an algorithm that detects groups of densely connected items, was then performed on the resulting network to identify the main themes. Tags with high centrality within each modularity class were used for interpreting the meaning of the class. For the free-text, LDA topic modeling was applied separately to the description and transcript to reveal topics in the contents. Besides highly probable terms in the topic, the talks that included the topic as its major topic were also used to help interpret its meaning. Both co-word network analysis and LDA topic modeling proved to be effective in identifying major themes in the Ted Talks. While both methods require human interpretation and naming of the topics produced, topic modeling was more efficient and identified topics with greater specificity. In addition, topic modeling was able to identify topics beyond the broad categories generated by the co-word analysis. This is even more apparent in the transcript model, which uncovered the most unique topics. Interestingly, the topics uniquely identified by topic modeling were mostly related to social issues and humanities. This might be due to TED's expanding its scope beyond its earlier emphasis on technology, entertainment, and design to social, political, and humanitarian issues. One of the possible reasons is that terms contained in the free-texts were more current and specific than the human-assigned tags. It is not to say that co-word analysis does not have its merits. Though less precise, it gives a bird-eye view of the domain, which affords fast comprehension of its major topics and their relationships. Another interesting aspect of our finding is that although models built on the description and the transcript generated comparable results, there were also discrepancies between the two. In scientometrics, the difference between the description and the transcript of a talk is akin to that between the abstract and the full-text of a journal article. Future studies can be done to examine how comparable the topics extracted from the abstract and the full-text are.

References

1. Callon, M., Courtial, J.P., Laville, F.: Co-word analysis as a tool for describing the network of interactions between basic and technological research: the case of polymer chemistry. Scientometrics **22**(1), 155–205 (1991)
2. Rip, A., Courtial, J.P.: Co-word maps of biotechnology: an example of cognitive scientometrics. Scientometrics **6**(6), 381–400 (1984)
3. Leydesdorff, L., Nerghes, A.: Co-word maps and topic modeling: a comparison using small and medium-sized corpora (N < 1,000). J. Am. Soc. Inf. Sci. **68**(4), 1024–1035 (2017)
4. Ding, Y., Chowdhury, G.G., Foo, S.: Bibliometric cartography of information retrieval research by using co-word analysis. Inform. Process. Manag. **37**(6), 817–842 (2001)
5. Hu, J., Zhang, Y.: Research patterns and trends of recommendation system in China using co-word analysis. Inform. Process. Manag. **51**(4), 329–339 (2015)

6. Tang, M.C., Cheng, Y.J., Chen, K.H.: A longitudinal study of intellectual cohesion in digital humanities using bibliometric analyses. Scientometrics **113**(2), 985–1008 (2017)

7. Tang, M.C., Teng, W., Lin, M.: Determining the critical thresholds for co-word network based on the theory of percolation transition: a case study in Buddhist studies. J. Documentation **76**(2), 462–483 (2019)

8. Blei, D.M., Ng, A.Y., Jordan, M.I.: Latent dirichlet allocation. J. Mach. Learn. Res. **3**(Jan), 993–1022 (2003)

9. Blei, D.M., Lafferty, J.D.: Dynamic topic models. In: Proceedings of the 23rd international conference on Machine learning, pp. 113–120, Association for Computing Machinery, New York (2006)

10. Fan, W.M., Jeng, W., Tang, M.C.: Using data citation to define a knowledge domain: a case study of the Add-Health dataset. J. Assoc. Inform. Sci. Technol. **74**(1), 81–98 (2023)

11. Suominen, A., Toivanen, H.: Map of science with topic modeling: comparison of unsupervised learning and human-assigned subject classification. J. Am. Soc. Inf. Sci. **67**(10), 2464–2476 (2016)

12. Callon, M., Law, J., Rip, A.: Qualitative scientometrics. In: Callon, M., Law, J., Rip, A. (eds.) Mapping the Dynamics of Science and Technology, pp. 103–123. Palgrave Macmillan, London (1986)

13. Blondel, V.D., Guillaume, J.L., Lambiotte, R., Lefebvre, E.: Fast unfolding of communities in large networks. J. Stat. Mech.: Theory Exp. **2008**(10), 10008 (2008)

Emotional Analysis through EEG on In-Store Journey

Pilot Methodology for Evocation of Emotions Through Video Stimuli to Measure Performance Metrics Using EEG Emotiv EPOC+ on In-Store Experiences

Fernando U. Osornio García⬭, Gilberto A. Fragoso González⬭,
Mayté V. Martínez Pérez⬭, Fernando Báez Martínez⬭,
Mario H. Salas Barraza(✉) ⬭, and Víctor M. González⬭

Sperientia:[Studio+Lab]®, San Andrés Cholula, 72810 Puebla, Mexico
{fernando.osornio,gilberto.fragoso,mayte.martinez,fernando.baez,
mario.salas}@sperientia.com

Abstract. This work presents the results of a pilot study for emotion evocation to explore the customer journey in a departmental store to detect possible friction points using biosignals and a retrospective verbalization. Our proposed methodology relies on the assumption that states how episodic memory elicits recollection of past events, thus, the previous emotional state of that specific event, providing a more efficient and vivid approach to measure their experience in a controlled environment. The present methodology is based on video-recording the participant's purchase journeys at the store, subsequently using the mentioned recording as visual stimuli to measure and report the emotional journey (as-is) based on quantitative data acquired by the use of an electroencephalograph (EEG). This data was complemented by retrospective interviews in order to build a solid comprehension of the user's journeys and their interactions with the services. The results obtained through these tests (n = 3) offer measurable genuine reactions towards their personal journey through the store, areas of friction, and key interactions during critical stages of their journey such as the payment system failures, crucial interactions, and the presence of the store employees, along with the imperative store's infrastructure to offer a homogeneous in-store experience. We present these results despite the limitations of the study (sample size) due to the potential of the methodology and expect that more researchers can expand our findings.

Keywords: UX Research · EEG · Emotional Journey · Customer Journey · Episodic Memory · Human computer interaction (HCI) · Bioinformatics · Laboratory experiments

1 Introduction

One of the main objectives of business organizations is that customers can access their value proposition. However, they can only access them through the interaction of the different spheres that conform to the offered service [21]. These spheres may include

© Springer Nature Switzerland AG 2023
F. Fui-Hoon Nah and K. Siau (Eds.): HCII 2023, LNCS 14039, pp. 155–169, 2023.
https://doi.org/10.1007/978-3-031-36049-7_12

interactions with technological interfaces such as digital applications or web pages, and human-human interactions, with systems and operational processes to complete a certain task. The integration of all interactions in these spheres and their consequences for the individuals result in the user experience [15, 17]. Therefore, it is necessary to ensure that all of them are working properly. To achieve this, it is indispensable to know which are the key points of friction or rupture through all interactions employing research processes.

The user experience can be categorized into different levels of an individual's relationship with a company; the single interaction level reflects the experience of an individual who has used a single device to perform a specific task. The journey level captures the complete experience of a user to achieve a goal and can be made up of multiple interactions between channels or devices. The main phases of a typical journey are; the need for recognition, information search, evaluation of alternatives, purchase decision, and post-purchase behavior [21]. The last level, the relationship level, refers to the interactions between a person and the company throughout the life of the customer-company relationship [19]. To create memorable experiences and ensure that they not only meet but exceed user expectations, at any level of relationship with companies, they must be known and understood in their entirety. Understanding the experience in this way will help us to identify the needs of stakeholders and not just customers [10].

Among the methods used for this purpose, it can be found effective tools such as the so-called journey maps. This type of human-centered tool (HCT) helps to understand the interactions that users have with other stakeholders by visualizing the interrelationships of people over time between the service channels and the users. A journey map not only includes the steps where individuals interact with the services but also reveals the key steps of an experience. Journey maps help to find gaps in user experience and explore potential solutions. This tool can be used to visualize existing processes (as-is) like a sequence of events, moments, experiences, interactions, or activities. Journey Maps make intangible experiences visible and facilitate a common understanding point and offer possible solutions [21]. Nevertheless, data collection methods for these types of tools usually include interview methods and retrospective verbalization protocols, so that the information collected may be biased by the user. That is, for researchers to fully understand users' interactions with the services they offer, they rely on the user to consciously recall and acknowledge the processes they underwent. This can be difficult when talking about a journey in a physical store, where there is not a completely controlled environment for observation.

This pilot study aims to offer an alternative to end-to-end knowledge through the analysis of psychophysiological data acquired through biosensors. The advantage of these is that the psychophysiological reactions elicited by the interaction of the users in the journey offer quantitative information that does not rely on the consciousness of the actions of each individual. The methods presented are based, like the retrospective verbalization protocols, on the evocation of memories and therefore emotions for the tracking of the journey through biosensors.

Thanks to the evocation of emotions and interactions it could be possible not only to measure in a more reliable way the experience in a physical place, but on digital screens by placing users in a specific context. Therefore, now we could evaluate more complex interactions and elicited emotions by the users due to cultural, social or individual contexts with the EEG. These evoked emotions could be reported robustly without the traditional worries that come with EEG signals like mobility of the equipment, travel costs, BCI Illiteracy [3] and those derived from the active interaction: user's movement, noise in the signals and artifacts, which need to be removed by the researchers; this is our main contribution to the field of HCI.

1.1 A Glimpse into Human Memory

Human memory can be divided into two main categories; short-term memory, which is defined as that memory that we can access only when we are aware of specific actions or events that have occurred within a short period, and declarative memory or long-term memory, which is that which we can access even when we are not aware of the actions performed and the period of occurrence of the events may dwell in a place far away from the present [24].

Declarative memory is categorized into three areas; semantic, which refers to the association of events with meanings, this kind of long-term memory can be abstract and are not context-dependent [4]; procedural, which has to do with that type of memory that we acquire through the constant practice of some activity [12] and episodic memory. In this case, the study was oriented toward the third area of long-term memory, episodic memory, which is defined as that type of memory that consciously elicits past events conceptualized in the first person (events personally experienced by the individual) in a specific Spatio-temporal context [18], that is, the evocation of past events conscious defined by a specific cognitive state; by temporal circumstances, by spatial circumstances, and by the affective state. The importance of the emotional state in which individuals are at the moment of memory fixation is evident when it is noted that people do not remember just any events, but those that carry an intrinsic emotion [2, 7, 24, 25].

1.2 Episodic Memory and Emotional State

Some authors show evidence of the close relationship between memory and cognition with the emotional state of individuals [2, 8, 18]. In this sense, the emotional state can be understood as the sum of complex psychophysiological patterns in the brain and body derived from the response to external events or objects perceived through sensory activity, whether this response occurs in real-time or is retrieved from memory [5, 13]. The existence of the emotional state and its change endows individuals with the ability to assign values, either positive or negative, concerning a psychosocial belief system to the events they experience, and this in turn mediates the way they interact with their environment [7].

Thus, cognitive processes, and memory, rarely occur completely independently of emotional processes. Memory and emotions are affected in both directions, in other words, cognitive processes can mediate and evoke emotions, just as an emotional response can evoke memories. In the case of episodic memory, this type of memory mediates complex emotions. Recalling past events in a specific context should evoke the emotional experiences associated with this event, which in addition would generate the emotional state that the subject experienced from the event in question. This is because the limbic system, the one that processes emotions, is related to memory processes through two types of somatic markers [2].

Therefore, two types of somatic markers contribute to the experience of emotions, the first are those markers that are learned or are innate, the second, and the most relevant for this pilot study, are those markers that cause emotional states thanks to the retrieval of information stored as memories [5].

1.3 Evoking Emotions Through Episodic Memory

According to Ellard [8] it is not enough that an emotional state is recalled, but the environment and the context by which it is recalled influence the quality of the experience. In this sense, the evoked episodes are more vivid when the same contextual conditions are available to the individual during the memory creation process [2]. There is evidence that recall of personal experiences through stimuli that are recognized as personal evoke more complex emotions [8].

Likewise, the methods for evoking episodic memories and emotional states directly affect the quality of the information collected for study. Hu [13] divides the two types of stimuli used to evoke these, internal stimuli where there are no external stimuli to induce a memory, for example, guided recall practice, and, on the other hand, external or media-based stimulation. The second category refers to those stimuli that come from a source outside the individual, such as photographs, audio, or video [18]. When talking about the study of emotions at the laboratory level, media-based studies are usually chosen because these types of stimuli evoke responses automatically and the bias of the participants tends to be lower than in guided recall activities [8].

Therefore, if it is desired to elicit memories and emotions as closely as possible to the moment in which the event was experienced, it is important to achieve an ecological environment that emulates as closely as possible the environment experienced by the individuals. Several authors have suggested that the best source to recall emotions is video, since this creates environments of high intensity and emotional complexity, offering greater stimulation in a more efficient manner and with a much more vivid approach in a laboratory environment; where participants have a greater sense of substitution of the environment and a faster response to the materials presented [8, 13, 18]. In addition, the video material can be complemented with meaningful audio, congruent with the recording, to increase the effectiveness of the evocation, creating a conducive, ecological and familiar environment for the participant to recall these events with the emotional states involved.

1.4 Methods for Emotion Quantification

The method by which emotional information is obtained also plays an important role in ensuring the quality of the data. Usually, the analysis of the emotional state of individuals is based on self-reported information that gives researchers an idea of what is happening as long as the user is aware of this.

In addition to the self-reported emotional state, there is another way to obtain information about the emotional state of individuals, which refers to the measurement and analysis of physiological indices through biosensors. In general, biosensors are used to measure spontaneous physiological changes due to the psychophysiological change of individuals in reaction to external stimuli, and these measurements have the advantage of greater consistency between individuals [13].

Finally, some of the most commonly used physiological parameters for the quantification of psychophysiological and emotional states refer to the change in cardiac variability (HRV), the change in respiratory rate (RFV), or the change in electrodermal skin response (EDA). Spontaneous physiological changes in these indices are often related to cognitive and emotional responses [1, 6, 16]. However, from a neurophysiological perspective, electroencephalography (EEG), a method mediated by a non-invasive device, can provide information about electrical changes in the brain more accurately and in real-time, without being mediated by some other type of limbic system mechanism by measuring the electrical activity of the cerebral cortex under different emotional states [16]. The information retrieved from the EEG can provide information about the mental and emotional processes that lead to certain behaviors.

1.5 Similar Studies Measuring Emotions

An emerging need for product and service evaluations has been measuring emotions in real time, with short times being the most accurate because of the difficulty involved in measuring them beyond a few hours [20]. The Emotiv EPOC+, biosensor used in this pilot, has been used in different studies: such as the one by Zabcikova in 2018 [28] to measure the performance metrics, using the software company, in relation to visual and auditory stimuli; in addition to the use in measuring stress in architectural virtual environments [9].

Other EEGs have also been used to localize frontal brain regions during the memory development process, as well as to explore other important factors related to long-term memory and its remembering [11]. Also using these devices, it has been found that there is a greater effect on user experience when combining sounds and video, compared to using only one type of stimulus [26].

2 Materials and Methodology

2.1 Participants

The sampling method used for this pilot study was non-probabilistic, by convenience. This study was open to any individual who was a customer of the evaluated department store, in this case being residents of the city of Puebla, Mexico, in an age range between 18 to 60 years old; it was also required that they had the availability to attend the different stages of the study in person. The exclusion criteria for the study were that participants must not have any self-reported neurological condition or be under any medical treatment that would interfere with the acquisition of electroencephalographic signals and that they should not present BCI Illiteracy [3]. From this sampling method and due to the pilot characteristics of the study the sample size consisted of two females and one male (n = 3, avg 31.4 years old) in which an in-store test and a biometric test were applied.

2.2 Procedure

Recording of Stimuli. Prior to the in-store trial, a technical test was carried out. It was verified that each one of the participants did not present BCI Illiteracy [3], meaning they were not part of the 20% of the population whose neurological signals could not be recorded by an EEG. This was performed in order to avoid complications of readings in the second part of the trial and ensure reliable signals during the interviews. In case of having valid readings, the test was continued, in case of not having readings, the participants could not be considered within the sample, since there would be no EEG readings.

Once the technical test with EEG was performed and the signals obtained were validated, their experience in the store was recorded using a GoPro HERO9 Black with a harness on the chest, as evidence of their journey through the store, which would be used the next day to evaluate their journey biometrically. The participants' journey consisted of typical journey that many clients do on a regular basis, which consisted of the following activities:

- Need recognition: The participants looked around the store for products they wanted without exceeding 500 Mexican pesos.
- Information search: During their search, they were allowed to interact with every information point, including the store employees.
- Evaluation alternatives and Purchase decision (user's choice): The participants indicated when they were ready to pay for the items selected through the product search.
- Product Payment: The participants pay their items by interacting with the store employees and ended when the articles were given to them.
- Post-purchase behavior: the participants interacted with the systems and employees after the purchase.

The participant's journey could be summarized in three broad moments which include the before-paying phase (Need recognition, Information search, Evaluating alternatives, and Purchase decision), the payment phase (Product payment), and the after-payment phase (post-purchase behaviors).

After the observation test in the department store, the videos of the three participants were transferred to a computer, to be used in the next phase of the study. The second phase of the study was conducted at the offices of Sperientia: [Studio + Lab]® a day after the in-store trial, where the participants were fitted with the EEG headset model EPOC+ of the EMOTIV brand, where they watched the aforementioned videos, to measure performance metrics obtained through the EEG. In addition, retrospective interviews were also conducted to complement the information retrieved from the readings obtained by the biometric device in order to understand the complete customer journey of the participant.

Retrospective Interview with EEG about the Store Experience. Interviews were conducted along with the observation test with EEG model EPOC+ by Emotiv. The observation test used as stimuli the video recordings of each participant along their purchase journey. The EEG was used to observe the emotional reactions of their journey.

The data obtained was stored at an internal server from Emotiv, for future processing, lecture, and interpretation.

To ensure the reliability of the signals, participants were asked to go to the test without hair products, not wearing any metal accessories on their ears or neck; keep the cell phone and any electronic device in airplane mode, or turned off, as far away as possible from the EEG. Participants were also asked to limit their head movements to reduce and/or eliminate noise in the signals caused by electronic devices or movement of the participant, respectively.

A baseline was used in the study, as a reference, to measure the relative changes in emotional states concerning the participant's baseline. During the baseline, participants were asked to try to relax as much as possible for 60 s with their eyes open and to repeat this process with their eyes closed for the same period.

The main analysis tool was the performance metrics graphs associated with emotional states, especially those related to task stress. These are provided by EmotivPRO software; the emotional spectrum obtained is explained below:

- Stress (ST) is a metric of comfort with the current task. High stress can result in the inability to complete a difficult task, feeling overwhelmed, and fearing the negative consequences of failing to meet task requirements.
- Engagement (ENG) is perceived as awareness and conscious attention to task-related stimuli. It measures the level of immersion at the moment and is a mixture of attention, and concentration that contrasts with boredom.
- Interest (VAL) is the degree of attraction or aversion to the current stimulus, environment, or activity and is commonly referred to as valence.
- Excitement (EXC) is a feeling with a positive value. It is characterized by activation of the sympathetic nervous system which results in a range of physiological responses including pupil dilation, eye-opening, sweat gland stimulation, heart rate, increased muscle tension, blood redirection, and digestive inhibition.
- Concentration (FOC) is a measure of attention to a specific task. It measures the level of depth of attention, as well as the frequencies at which attention shifts between tasks.
- Relaxation (MED) also known as meditation, is a measure of the ability to shift concentration and recover from intense concentration.

3   EEG Data Interpretation

The data collected by the EMOTIV PRO software is represented in a simplified scale with a range of 0 to 100 points. There is also the baseline, which refers to the lack of stimulation or conscious rest of an individual, to be compared with situations by measuring the change between the signal at the desired time for the initial normal values marked by the baseline for each human being.

Each individual presents different performance metrics, either by their mental state on that day, or session, up to the moment of obtaining data, so the baseline allows an understanding of the current resting state of each participant and a comparison. Therefore, due to the different metrics in each individual, it is not possible to have a standardization of performance metrics. However, it is possible to measure the change in these metrics over time when different stimuli are presented, then a comparison is made between the baseline and some other relevant event to observe the change in the metrics.

4 Results

Applying the emotion evocation pilot methodology, the customer experience at the store could be measured by the aggregation of Emotiv's performance metrics through the shopping journey. Since these results only use information from 3 participants, it allowed the researchers to identify problems on the customer journey experience on the retail store on one of its locations. The results are depicted like radar charts, which are formed by 6 axes. Each axis represents a performance metric variable, and each set of data is represented by a color. These figures are used for a better lecture.

4.1 Payment System Importance for the Customer Experience

It was observed that while the participants recalled their payment journey, the three observed their problems when they tried to pay by credit card. The performance metrics showed a stress increase after the store staff stated the issue with their payment system (yellow color and squares in Figs. 1 and 2).

Additionally, relaxation, which represents the ability to change focus and switch between tasks; all the participants upon completion of the payment journey (green color and circles) had a low level of this metric compared to the point of the statement of the crash pay system. This could indicate that the experience of this failure generates stress that remains after the payment journey is completed (Figs. 1 and 2). Without this methodology, it would have been difficult to find this reaction to the error or crash in the payment system.

On the other hand, the three participants showed an increase in their interest levels while the payment task in the journey was recalled. This metric can note aversion or attraction to a stimulus. In this case, it could be taken on aversion, as indicated by the aforementioned metrics.

As shown in Figs. 1 and 2, it is possible to observe an increase in stress along all three (3) participants. This increase was observed during the payment procedure, where in all three participants, the system failed and didn't allow them to pay with anything except cash (yellow color and squares).

Furthermore, the performance metrics observed in Fig. 3, excitement, interest, and relaxation (green color and circles) increased compared with the start levels before the payment of the items (blue color and triangles). This could mean that the participant had a conscious positive emotion at the time of the purchase, even though they had to pay for the items in different store departments. This positive emotion can be related to a decrease in the excitement level at the moment before and over the purchase. It's possible meaning, regardless of the outcome, the excitement still increased, even with the payment failure.

Participant 2 during crash in payment system

Fig. 1. Performance metrics along the payment journey, with system failures on participant 2. Showing an increase in stress and interest; with stress levels from 38 to 39 points and interest levels from 47 to 48 points.

Participant 3 during crash in payment system

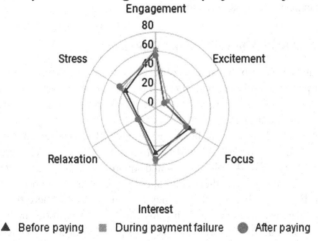

Fig. 2. Performance metrics along the payment journey, with system failures on participant 3. Showing an increase in stress and interest among the participant 3, showing stress levels from 37 to 45 points and interest levels from 47 to 56 points.

Therefore, by using the emotion evocation methodology it was possible to observe the psychological effects along the journey in the product payment. For this instance, the negative effects of a payment system failure lead to an increase in stress. This allows

Fig. 3. Performance metrics along the payment journey, with system failures on participant 1. Showing an increase in stress and interest among the participant 3, showing stress levels from 37 to 45 points and interest levels from 47 to 56 points.

us to find a possible problem that the chain store can fix and prevent further instances of bad payment experience for their customers.

4.2 Shortage of Staff Affects Customer Experience

On top of that, it was also possible to observe how the performance metrics change with the staff presence and their help through the shopping journey. This change in the performance metrics was seen when the staff attendant brought assistance to a participant (Fig. 4, blue color and triangles).

Also, it was noted in the data how the presence of staff affected the participants at an emotional level; a decrease could be observed when activities were carried out without collaborators (Fig. 4, color yellow and square), and an increase when there was the opportunity to interact with them (Fig. 5, color yellow and square).

Firstly, in Fig. 4, we can see the journey of participant 1, where the performance metrics of her sensory reactions to being lost in the store without having a specific goal (blue color and triangle) were observed. It is possible to see how the lack of store personnel affects the values of the performance metrics by not observing a collaborator and reacting to her journey in the store.

The increase in the levels of engagement can be attributed to the concentration on remembering the task of his trip the day before as the participant observes on the screen, making him aware of his behavior in the recorded video. Likewise, should also be emphasized the metrics of their concentration and interest when browsing the store; they have values that correspond to the interaction with a collaborator (blue color and triangle), the search for items without a collaborator (yellow color), and the continuation of their search without a collaborator (green color and circle), respectively. There it's

Participant 1 Searching with Employee and after

▲ With interaction   ■ Searching for a product   ● Continuing searching

Fig. 4. Performance metrics along the searching journey, with contributor assistance and after the assistance on participant 1. It is possible to observe an increase in engagement by 13 points compared to their initial level during their search in the store; a decrease in excitement from 96 to 37 points and in concentration from 82 to 53 points.

Participant 2 Finding an employee

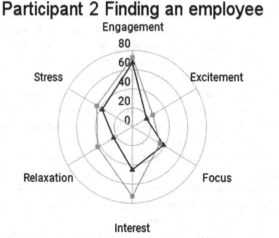

▲ Searching for a product   ■ Finding an employee/staff member

Fig. 5. Performance metrics along the searching journey, with contributor assistance and after the assistance on participant 2. Showing an increase in excitement, from 18 to 25 points; interest from 45 to 73 points; relaxation from 23 to 43 points.

noted how excitement and focus decrease, having a peak value when the participant interacts with a staff member and decreasing for the rest of the journey.

The changes in excitement and concentration allow us to infer how the lack of personnel caused a loss of excitement and concentration in their search for products, which could be the cause of a potential purchase at risk of not being completed. The participants (P1 and P2) commented on the possibility of having help but not receiving it due either to the lack of presence of collaborators or available personnel. On the other hand, excitement measures the levels of emotion that physiologically show high attention to a task, generalizing along with the exponential loss of concentration, where we can observe that P1 had a decrease in attention in the search for products when he started a part of his journey without attention to the customer or showing interest in his present needs.

In addition, another important sensory metric is the increase in relaxation of more than half, increasing compared to the baseline when initiating a search for your products. If possible, the user witnessed an event where their cognitive load was low by not having something that caught their attention. This could be processed as a positive metric, but the emphasis should also be given to the timely cognitive load for decision-making explained by Victor Yocco [27] in his book Design for the Mind: Seven Psychological Principles of Persuasive Design; in this case, the selection of items with control reasoning and normative beliefs could lead them to make a decision which is supported by the context of being assisted by a collaborator previously knowledgeable in the area of children.

Secondly, Fig. 5 shows how participant 2 was performing tasks 1 and 2, needing recognition and information search, for about ~40 min, during which no staff interacted with him. As the participant was passing through several departmental areas and observed several employees, who were seen to be available but not interacting with him. These conditions give rise to the episodic memory reactions when showing the video to participant 2, where the values of when he goes through the shoe department (blue color and triangle) and when he observes a collaborator (yellow color and square).

These increased values of excitement, interest, and relaxation could show that the cause of these increments is due to the possibility of help being present in the area; causing the participant to feel more excited or interested in interacting with a staff member. This may be due to a person's tendency to have an idea of what they are looking for, allowing the employee to inquire and present options or recommendations as they see fit. During this event described and seen in Fig. 5, participant 2 appears to have an increase in their interest, meaning attention focused on the employee, accompanied by a relaxation regarding possible confidence in having a person available for his or her search. The potential of offering options from an experienced or knowledgeable store employee could allow the customer a short and concise trip.

To conclude, the described results have the potential found with the emotional evocation methodology, offer new ways to explore the customer experience, to find significant insights into critical points of the journey, and help to show some emotions difficult to express to the participants in the cultural or educational context. In the pilot study, it was possible to observe the emotional reaction to the crash of the payment system during task 4, product payment, and on tasks 1 and 2, need for recognition and information search, how the employees play an important part in the experience with their presence and aid to the customers.

5 Discussion

First, it was observed that the participants had common friction points throughout the shopping journey. The inability to pay by credit card, possibly derivatives in increases of stress and aversion to the experience. This could mean a possible negative emotion, that would result in an unpleasant experience and a potential dropout of the purchase or loss of loyalty from regular customers.

Another observation was the possible effect of the presence and companionship of the store staff throughout the shopping journey. The absence and lack of aid from them, generate a general decrease in the performance metrics, while the opposite stimuli increase them. Highlighting the importance of the store staff for the customer journey is vital for a better shopping experience.

The implementation of performance metrics with EEG (Emotiv EPOC+) had a positive impact on the analysis, as it was complemented with more usual methods to measure customer experience. This is a major element of change because it doesn't rely only on the self-report of experience of the participants [13]. The information collected by the proposed research method is relevant to building a journey map as described by Stickdorn, Hormess, Lawrence, and Schneider [21], specifically to know precisely all the stages and steps of the journey as well as the construction of an emotional journey and a quantitative dramatic arc and to observe and identify the channels and stakeholders involved in order to achieve a thoroughly understanding of the user experience. This information allowed us to conclude how the in-person shopping experience is still critical for department store customers.

Finally, it is important to note that although the results obtained are not generalizable, due to the sample size and the contextual specificity of the present pilot, they are relevant because the proposed methodology allows the detection of possible friction areas to explore different scenarios, thanks to the combination of different research methods; usual verbalization, video analysis and analysis of biosignals. Through which issues could be identified, without the presence of EEG would have been hard or even impossible to find, since the information obtained couldn't be expressed through conventional methods like interviews or surveys. This kind of approach could help the difficult population externalize their thoughts and feelings with service and products; being an inversion to customer experience research, making room for the improvement of the store experience. Furthermore, the methods used to identify the population with BCI Illiteracy [3] before starting a participants' journeys benefiting from the Emotiv EPOC+'s portability, could save time and efforts on research expenses.

We assume that the combination of the aforementioned methods and the implementation of biosignal analysis provide the study with a greater capacity for discoverability of critical points on the journey, making this method a viable and useful alternative for the task of user experience research.

References

1. Agrafioti, F., Hatzinakos, D., Anderson, A.K.: ECG pattern analysis for emotion detection. IEEE Trans. Affect. Comput. **3**, 102–115 (2012)

2. Allen, P.A., Kaut, K.P., Lord, R.R.: Emotion and episodic memory. Handb. Behav. Neurosci. **18**, 115–132 (2008)
3. Allison, B.Z., Neuper, C.: Could Anyone Use a BCI? In Brain-Computer Interfaces, pp. 35–54. Springer, London (2010)
4. Antonucci, S., Reilly, J.: Semantic memory and language processing: a primer. Semin. Speech Lang. **29**(1), 005–017 (2008)
5. Bechara, A., Damasio, H., Tranel, D., Damasio, A.: The Iowa gambling task and the somatic marker hypothesis: some questions and answers. Trends Cogn. Sci. **9**, 159–162 (2005)
6. Das, P., Khasnobish, A., Tibarewala, D.N.: Emotion recognition employing ECG and GSR signals as markers of ANS. In: 2016 Conference on Advances in Signal Processing (CASP), pp. 37–42 (2016). https://doi.org/10.1109/CASP.2016.7746134
7. Dolan, R.J.: Emotion, cognition, and behavior. Science **298**(5596), 1191–1194 (2002)
8. Ellard, K.K., Farchione, T.J., Barlow, D.H.: Relative effectiveness of emotion induction procedures and the role of personal relevance in a clinical sample: a comparison of film, images, and music. J. Psychopathol. Behav. Assess. **34**(2), 232–243 (2012)
9. Ergan, S., Radwan, A., Zou, Z., Tseng, H.A., Han, X.: Quantifying human experience in architectural spaces with integrated virtual reality and body sensor networks. J. Comput. Civ. Eng. **33**(2), 04018062 (2019)
10. González, V.M.: Ensayos de Sperientia. Sperientia Studio + Lab (2018)
11. Hanouneh, S., Amin, H.U., Saad, N.M., Malik, A.S.: EEG power and functional connectivity correlates with semantic long-term memory retrieval. IEEE Access **6**, 8695–8703 (2018)
12. Horikawa, E., et al.: "[Procedural memory]" Nihon rinsho. Jpn. J. Clin. Med. **69**(Suppl 8), 331–336 (2011)
13. Hu, W., Huang, G., Li, L., Zhang, L., Zhang, Z., Liang, Z.: Video-triggered EEG-emotion public databases and current methods: a survey. Brain Sci. Adv. **6**(3), 255–287 (2020)
14. Kellog, R.T.: Fundamentals of Cognitive Psychology. SAGE Publications, Inc. (2012)
15. Law, L.-C., Roto, V., Hassenzahl, M., Vermeeren, A., Kort, J.: Understanding, scoping and defining user experience: a survey approach. In: Proceedings of the CHI'09, pp. 719–728 (2009). https://doi.org/10.1145/1518701.1518813
16. Mauss, I.B., Robinson, M.D.: Measures of emotion: a review. Cogn. Emot. **23**(2), 209–237 (2009)
17. Nielsen, J.: Enhancing the explanatory power of usability heuristics. In: Proceedings of the SIGCHI conference on Human Factors in Computing Systems, pp. 152–158 (1994)
18. Repetto, C., et al.: Immersive episodic memory assessment with 360° videos: the protocol and a case study. In: Cipresso, P., Serino, S., Villani, D. (eds.) Pervasive Computing Paradigms for Mental Health: 9th International Conference, MindCare 2019, Buenos Aires, Argentina, April 23–24, 2019, Proceedings, pp. 117–128. Springer International Publishing, Cham (2019). https://doi.org/10.1007/978-3-030-25872-6_9
19. Salazar, K.: User Experience vs. Customer Experience: What's The Difference? Nielsen Norman Group (2019). https://www.nngroup.com/articles/ux-vs-cx/
20. Shu, L., et al.: A review of emotion recognition using physiological signals. Sensors **18**(7), 2074 (2018). https://doi.org/10.3390/s18072074
21. Stickdorn, M., Hormess, M., Lawrence, A., Schneider, J.: This Is Service Design Doing: Applying Service Design Thinking in the Real World. O'Reilly Media (2021)
22. Thaler, H.R., Sustein, C.R.: Un pequeño empujón, 2da edición. Penguin Random House Grupo Editorial (2017)
23. The Definition of User Experience (UX): Nielsen Norman Group. https://www.nngroup.com/articles/definition-user-experience/ (2013)
24. Tulving, E.: Episodic and semantic memory. In: Tulving, E., Donaldson, W. (eds.) Organization and Memory, pp. 381–403. Academic Press, New York, NY (1972)

25. Tulving, E.: Elements of Episodic Memory. Clarendon, Oxford (1983)
26. Van Camp, M., De Boeck, M., Verwulgen, S., De Bruyne, G.: EEG technology for UX evaluation: a multisensory perspective. In: Advances in Neuroergonomics and Cognitive Engineering: Proceedings of the AHFE 2018 International Conference on Neuroergonomics and Cognitive Engineering, July 21–25, 2018, Loews Sapphire Falls Resort at Universal Studios, Orlando, Florida USA, vol. 9, pp. 337–343. Springer International Publishing (2019)
27. Yocco, V.: Design for the mind: seven psychological principles of persuasive design. Simon and Schuster (2016)
28. Zabcikova, M.: Measurement of visual and auditory stimuli using EEG headset Emotiv Epoc+. MATEC Web of Conferences **292**, 01023 (2019)

A Machine Learning Model for Predicting a Movie Sequel's Revenue Based on the Sentiment Analysis of Consumers' Reviews

Suyanee Polsri[1], Ya-Wen Chang Chien[2], and Li-Chen Cheng[1](✉)

[1] National Taipei University of Technology, Taipei 106, Taiwan
jessicacheng@mail.ntut.edu.tw
[2] Huafan University, New Taipei 223, Taiwan

Abstract. The relationship between the performance of movie sequels, the performance of the corresponding original movies and the users' review sentiments is actively studied in the scientific community. However, the precise constitution of this relationship remains unclear due to the complex multidimensional nature of the problem. In particular, the precise correspondence between the users' review sentiments and the topic structure of the reviews (that represents the aspects of the movie that impacted the sentiment the most) is yet to be fully understood. In this study, a machine learning topic modeling algorithm (Latent Dirichlet Analysis, LDA) is performed on the three movies from the Jurassic World franchise. The analysis is performed on a dataset of reviews gathered from the IMDB website. The reviews are separated into six datasets – a positive and a negative subset for each of the three movies. The outputs of the topic modeling are represented as word clouds of the most salient terms. The subsequent analysis of the word clouds demonstrates the heterogeneity of the topics within reviews and the nature of the ambiguity that often complicates the vocabulary-based sentiment analysis. Based on the results of the topic modeling, using comparative methods we determine the possible reasons behind the significant decline of the box office performance for "Jurassic World: Dominion" and the franchise in general.

Our result illustrated that successful sequel would have to be consistent with other movies of the franchise and to have enough originality at the same time to receive positive feedback. Future works includes developing an approach that can leverage the heterogeneity of the LDA-produced topic representations, applying roBERTa model to handle sentimental analysis, and predicting movie sequel's revenue based on machine learning models.

Keywords: Data mining · Topic modeling · Movie review · Machine learning

1 Introduction

Better performance of movie sequels compared to the original movies is a well-established fact that drives the production of franchises and whole cinematic universes (i.e., Marvel, DC etc.). However, the number of studies on sequel movie revenue prediction is still limited. Additionally, factors such as the changing preferences of the

© Springer Nature Switzerland AG 2023
F. Fui-Hoon Nah and K. Siau (Eds.): HCII 2023, LNCS 14039, pp. 170–180, 2023.
https://doi.org/10.1007/978-3-031-36049-7_13

audience, the level of competition in the film market, and the different approaches taken by filmmakers and studios to create sequels may also impact the popularity and financial success of movie sequels. The reasons for the success of movie sequels are complex and multifaceted, and more research is needed to gain a deeper understanding of these factors.

It has been shown in research that such variables as the release date, runtime, number of stars of the original movie on the review websites such as IMDB are relevant in terms of sequel movies' revenue prediction. For example, the study [1] demonstrated the impact of the users' reviews on the sequel movie by applying the accessibility-diagnosis framework combined with log-log modeling. Along with those and similar features, that can be denoted as "definitive" for shortness, textual features have also been shown to positively affect the revenue prediction accuracy. Namely, sentiment analysis has been proven to be an effective tool for this and similar tasks. For instance, the authors of the paper [2] collected movie-related information from IMDb using web crawlers. Based on applying the sentiment analysis tools to the gathered data, the movie reviews were quantified to forecast the box office success of a film by combining the quantified data with the fundamental film information and ambient variables. The authors of the study [3] also utilized a combination of sentiment analysis with machine learning to forecast the movies' box office performance.

Nevertheless, to that end there is still a demand for better variable sets, as well as predictive methods, that would demonstrate even higher prediction capability. The machine learning algorithms typically used for the sequel performance prediction are also limited in literature: the authors mostly consider linear models, support vector machines or simple fully connected neural networks, without applying more complex and potentially more performant techniques (recurrent neural networks, ensemble methods etc.).

We propose the following complete processing and exploration pipeline:

1. Gather the reviews from IMDB using the provided API.
2. Separate the positive and negative reviews by the movie's rating.
3. Preprocess the reviews by removing punctuation and lemmatizing the words.
4. Create a dictionary of stop words by excluding "non-meaningful" word classes (i.e. prepositions) and several specific words that introduce noise into the topic allocation model.
5. Using the created dictionary, train the sentence encoder and the downstream unsupervised machine learning model – LDA.
6. Explore the topic composition of the reviews and combine them into meaningful aspects.
7. Using the Robustly Optimized BERT Pretraining Approach (roBERTa) sentiment prediction model, obtain the review sentiments.
8. Apply some machine learning models to identify the causal configurations of the sequel movies' performance.

In this study, an experiment is conducted following processing steps from 1 to 6. Our aim is to explore the potential of the LDA algorithm in the context of movie sequel performance prediction and to shed some light on the factors that contribute to the popularity and financial success of movie sequels. Further research will continue to be conducted to complete steps 7 and 8 to gain a deeper and more comprehensive understanding of the correlations between movie success and specific factors, and to further predict the box office revenue of movie sequels.

The rest of the paper is organized as follows: Sect. 2 provides a review of related work; Sect. 3 provides an introduction of the proposed method; Sect. 4 presents the experimental results; and Sect. 5 summarizes the major contributions and future research directions.

2 Related Work

Sentiment analysis of the Internet users' reviews is a major part of many box office prediction frameworks represented in the literature. For example, The study [4] proposes a movie revenue prediction model based on sentiment analysis combined with machine learning. Three main sources were used by the authors to gather data: IMDB, Box Office Mojo, and YouTube. First, excerpts from the reviews of the first five films in each of five different movie franchises—a total of 25 separate films—were taken. Second, the reviews were preprocessed by removing the stop words, spelling correction, and de-noising. The final step was to extract the movies' budgets, box office receipts, and release dates from IMDB and Box Office Mojo, as well as to perform sentiment analysis on the preprocessed reviews. Multiple experiments were further performed on the resulting dataset, including missing value imputation and revenue forecasting. Four different imputation techniques were considered: hot-Desk, kNN, IRMI, and the novel method based on inter-sample differences. The latter was shown to have superior performance compared to the former three methods. The algorithms used in the revenue prediction included a multilayer perceptron neural network (MLP), support vector machine (SVM) and multiple linear regression (MLR). It was demonstrated that MLP does not perform similarly well due to the small size of the dataset compared to SVM and MLR.

The study by [1] is aimed at determining the causes of sequels' success, focusing on the importance of reviews of the first movie. The model of sequel box office performance proposed there is based on the accessibility-diagnosis framework and the log-log modeling method to evaluate the original movie's impact on the sequel. The authors demonstrated that the performance of the sequel is directly influenced by the critical reception of the original movie.

The effects of online reviews of concurrently released competing movies on the box office and attendance of a focal film were also examined by [5] using field data. The empirical findings imply that the review rating of a competitor, which is featured on the same ticketing page as a focal movie, has a negative impact on the box office success of the focal movie. This work further investigated the heterogeneous effect in the target movies (sequels vs. non-sequels) and discovered that non-sequels are more strongly impacted by competitor reviews.

The authors of the study [6] examined the relationships between movie names and the success of the associated movies using text mining and exploratory factor analysis (EFA). Two movie markets were considered, namely the USA and China. The results showed that popular movies with remarkable total lifetime grosses had names with contrasting and comparable patterns of the most commonly used terms, providing important insights into the preferences of audiences in these nations.

The paper [7] proposes a component focusing multi-head co-attention network model for determining the aspects' sentiments. The model consists of three modules: (1) extended context, (2) component focusing, and (3) multi-headed co-attention. The extended context module improves the ability of the transformer encoder to handle aspect-based sentiment analysis tasks, and the component focusing module improves the weighting of adjectives and adverbs. This module helps to address the problem of average pooling, which treats every word as an equally important term. The multi-head co-attention network combines the multi-word target with the context and builds a strong correspondence between them. The performance of the proposed model is evaluated in extensive experiments on publicly available datasets, including the data on restaurant and laptop reviews from SemEval 2014 and the ACL 14 Twitter dataset. The authors show that the performance of the proposed model is on par or better than that of the recent state-of-the-art models.

The authors of the paper [8] investigate customer satisfaction through aspect-level sentiment analysis and visual analytics using the impact of COVID-19 on the airline industry as a case study. In that regard the deep learning based sentiment analysis techniques are of special interest. The authors utilize the BERT transformer neural network in conjunction with discriminative linguistic features. The latter refers to a widely used set of features used in sentiment analysis, while the former represents a state-of-the-art deep learning model for many natural language processing (NLP) tasks. The proposed method was shown to outperform the baseline models (ABSA) and therefore proved prospective in the review analysis tasks.

The paper [9] proposes a framework for studying the effect of online reviews on movie box office sales based on aspect-based sentiment analysis and economic modeling. The authors gathered reviews and box office statistics from the Box Office Mojo and IMDb websites. After that, the aspect-opinion pairs were extracted from the reviews, where aspect is a word or a word combination carrying a meaningful latent feature, i.e., "director (PDRP)", "special effects (SEP)" etc. The extracted aspects were quantified based on frequency of occurrence and processed with a linear regression model along with numerical rating features. It was shown that additional textual reviews, both positive and negative, positively impact the movie revenues, even when the ratings do not have a strong effect on the box office performance.

In current literature, the use of machine learning algorithms for predicting the performance of movie sequels is largely limited to a few basic techniques, such as linear models, support vector machines, and simple fully connected neural networks. The authors rarely explore more advanced and potentially more accurate techniques, such as recurrent neural networks and ensemble methods.

3   Data and Methodology

In this study, three sequel movies from the Jurassic World franchise were chosen: Jurassic World, Jurassic World: Fallen Kingdom and Jurassic World: Dominion. This choice was motivated by the overall popularity of the franchise as well as by the mixed reception of the movies. The latter facilitates splitting the reviews into positive and negative subsets of equal volume. This is important for the LDA convergence since this algorithm is unsupervised and therefore sensitive to data imbalance.

The movies were retrieved from IMDB using the built-in API provided by IMDB itself (see Fig. 1.) and numerous reviews will be crawled. Only reviews written in English will be considered. The dataset was separated by review's rating. Positive reviews are defined as those with at least 5 stars, and negative reviews are those with less than 5. So far, steps 1 and 2 have been completed.

3.1 Preprocessing

Details of the implementation of steps 3 and 4 are described below. Python NLTK (Natural Language Toolkit) package was used to preprocess the review data. First, each review was tokenized. After that, special characters were removed, including punctuation. Additionally, all uppercase characters were converted to lowercase. Next, the "stop words" that were gathered from all reviews and unnecessary words (i.e., conjugations, prepositions, pronouns etc.) were removed to improve signal-to-noise ratio in the dataset. Finally, the words were transformed into their dictionary forms via lemmatization.

Fig. 1. An example of the review from IMDB

3.2 LDA Topic Modeling

In step 5, LDA topic modeling was performed separately on the six corpora (positive and negative reviews for each of the three movies) after tokenization and lemmatization. We have set the number of topics to four and trained the LDA model using the TF-IDF representation of the movie reviews. Both the LDA and TF-IDF models were taken from the scikit-learn library [10]. In step 6, the outcomes were visualized using the pyLDAvis library and word cloud tools.

Tables 1 and 2 demonstrate the statistics of the collected datasets. From Jurassic World to Jurassic World: Dominion, although the cost of investment showed an upward trend, the box office revenue showed a decreasing trend. It can be concluded from Table 1 that a higher budget does not necessarily lead to a higher box office revenue.

Table 1. General information on the movies used in the study.

	Year	Director	Box-office revenue, $	Budget, $ mil
Jurassic World	2015	Colin Trevorrow	208,806,270	150
Jurassic World: Fallen Kingdom	2018	J.A. Bayona	148,024,610	170–187
Jurassic World: Dominion	2022	Colin Trevorrow	145,075,625	165–185

Table 2. The quantitative statistics of the gathered datasets.

	Number of reviews	Positive	Negative	Rating
Jurassic World	1,559	840	719	6.9
Jurassic World: Fallen Kingdom	1,900	642	1,258	6.1
Jurassic World: Dominion	2,502	750	1,752	5.6

From the data in Table 2, it was observed that Jurassic World: Dominion had a higher number of negative reviews and a lower consumer rating. It can also be seen from Table 2 that the resulting positive and negative corpora are roughly equal in volume. However, for Jurassic World: Dominion the ratio between positive and negative reviews is equal to 0.4, which means that the negative LDA model will have a significantly larger set of tokens to train on. Arguably, for the current study this issue is not critical since the LDA models are trained separately for each subset, i.e. the imbalance will not affect the results as long as the number of reviews for each model is large enough.

4 Results and Discussion

From the results of the three sequel movie topic modeling experiments, it was determined that the LDA algorithm can successfully identify the key topics that consumers value in the data set, regardless of whether the reviews are positive or negative. To clearly showcase the results of the LDA algorithm for topic modeling, the top 20 salient terms for each topic in each of the review corpora were presented in a word cloud format. The word clouds are shown in Tables 3 and 4. Several conclusions can be drawn from the results.

Firstly, unlike handcrafted aspects frequently used in semantic analysis of the reviews (i.e. Story, Visuals etc.), the extracted topics are much more heterogeneous. For instance, the positive topic 1 for the Jurassic World movie mostly contains terms related to dinosaurs ("indominus", "animal", "t-rex" etc.), but it also contains more general terms, e.g. "original", "major" and "action". The same observation is true for all topics. This is caused by the unsupervised nature of the LDA algorithm which does not make any model assumptions regarding the topic composition and semantic structure of the data. We argue that this is not a drawback of the method, however, since it captures the real structure of the latent topic space without imposing any artificial constraints, therefore making the topics more representative in practice.

Table 3. Topics extracted from the positive subsets of the review dataset.

For the same reason positive and negative topics are intertwined in terms of vocabulary: for example, topic 2 for the Jurassic World: Fallen Kingdom movie contains such terms as "dull" and "unnecessary", while topic 1 for Jurassic World: Dominion contains the word "interesting".

Secondly, it is evident that overall the extracted topics can be unambiguously differentiated based on the majority of the related terms. Positive topic 1 for Jurassic World mostly covers dinosaurs, topic 2 – visual effects and action, topic 3 – actors and CGI, topic 4 – overall originality and technical excellence of the movie. Based on this, the following set of vocabulary-based aspects can be derived from the obtained word clouds for each movie:

1. Story (including overall originality and novelty along with the writing quality)
2. Action
3. Actors and characters
4. Technical realization

Table 4. Topics extracted from the negative subsets of the review dataset.

	Jurassic World	Jurassic World: Fallen Kingdom	Jurassic World: Dominion
Topic 1			
Topic 2			
Topic 3			
Topic 4			

5. Production (i.e. how the users characterize the effort conducted by producers, is the movie worth the ticket price etc.).

Each extracted topic is mostly focused on one of these aspects, while containing elements of others. For example, such words as "story", "original" and "stupid" can be attributed to the aspect 1; "action", "scary" and "bland" – to the aspect 2; "actor", "animal", "chris" – to the aspect 3; "cgi", "visuals" and "special" – to the aspect 4; "franchise", "money" and "studio" – to the aspect 5. In other words, the extracted topics represent a fuzzy composition of the aspects with different weights assigned to them.

Thirdly, to investigate the connection between the movies' box office performance and the discovered topics several sentences containing the most salient terms were extracted from the reviews and gathered in Table 5.

The first movie is criticized mainly for the special effects (including dinosaurs specifically) and the story (both in terms of plot and originality). The second movie is additionally criticized for poor acting. Finally, the third movie receives critique across the board: for the story, the writing (including the inconsistency relative to the previous movies), the actors and the CGI; however, the former was positively evaluated by some users, resulting in an ambivalent reception.

Starting from the second movie, the term "money" plays an important role in the topic construction. However, its use is not identical for the second and the third movies. In reviews on Fallen Kingdom, the users mainly criticize the producers for not delivering a movie with a level of quality adequate to its budget; in reviews on Dominion, some users explicitly refer to watching the movie as a waste of money.

Table 5. Selected sentences containing the most salient terms from the negative reviews.

Jurassic World	
Animal	Real <u>animals</u> don't move every single joint with every step, like robots
Original	Because this movie is an <u>original</u> follow up to the <u>original</u> much more thought provoking movie, and it should at least try and live up to its intellectual standard
Writer	I hope the <u>writer</u> and director never make another film again
Jurassic World: Fallen Kingdom	
Money	They must have run out <u>money</u> to pay for good writing
Story	No <u>story</u>, no innovation after the last episodes
Actor	No chemistry between the <u>actors</u>
Terrible	Such a <u>terrible</u>, depressing awfully acted film
Jurassic World: Dominion	
Actor	I feel bad for the <u>actors</u>, their potential is being squandered on a waste of CGI!
Franchise	Ever since the original Jurassic Park this <u>franchise</u> has gone downhill
	This is the worst movie in the <u>franchise</u>
Story	The film begins by leaving out its own <u>story</u> plots from the last two films
	The dialogue, <u>story</u>, everything was just an abomination and disrespectful to the original Jurassic Park
Money	It's a waste of your <u>money</u>
	So much money spend on so much nonsense
CGI	It has bad acting, bad story, bad continuity, bad characters and even bad <u>cgi</u>/models
	Some <u>cgi</u> is bad

Therefore, if for a given franchise an overall increase in different aspects featured in the negative reviews is observed, the box office is likely to decrease. Additionally, reviewers deterring readers from watching the movie for financial reasons is the major indicator of poor box office performance, as observed for the Jurassic World: Dominion movie. In our future work, this comparative analysis will be replaced by more advanced machine learning techniques.

Finally, the following specific conclusions can be drawn for the movies under consideration:

1. For the Jurassic World movie, the users mostly positively characterized originality, visuals and other special effects, while negatively referring to the story in general and certain specific actors or characters.
2. For the Jurassic World: Fallen Kingdom movie, the users positively rate the story, action and CGI, while criticizing the story, the actors and the production. This suggests that different users can have polar opinions on certain aspects of the movie, which should be taken into consideration when building sentiment analysis-based models.

3. For Jurassic World: Dominion the users enjoy actors and action. The most criticized aspects include the story, production and originality.

Overall, it can be concluded that the viewers judge sequel movies in the context of their predecessors and pay attention not only to visuals, acting and story, but also to the originality and production aspects. Therefore, a successful sequel in that scenario would have to be consistent with other movies of the franchise and to have enough originality at the same time to receive positive feedback.

5 Conclusion

In this study, the reviews on three movies from the Jurassic World franchise were studied via the LDA topic modeling. Based on the word clouds build from the topics extracted from the positive and negative reviews, the following conclusions can be made:

1. The topic composition of the reviews is heterogeneous and can be characterized as a fuzzy mixture of vocabulary-based aspects with different prevalence of each aspect for each topic. Consequently, this study shows that vocabulary-based sentiment analysis techniques do not fully capture the relations between aspects and sentiments, and a more complex approach is required to effectively leverage topics as features for the downstream analysis.
2. Overall, five distinct aspects can be extracted from the topics, namely "story", "action", "actors and characters", "technical realization" and "production".
3. The ambivalence of users' opinions on such aspects as story and technical realization must be taken into account when performing the sentiment analysis to adequately describe the feature space.
4. In terms of sequel performance, the originality and consistency with the previous movies of the franchise seems to play an important role in the reception of the movies. Users do not judge a sequel solely on its own merits and evaluate its quality in the context of the previous movies.
5. In particular, the movies under consideration are most often criticized for the quality of the story, originality and production, while being positively evaluated in terms of technical realization, action and actors' performance.

In summary, the main result of the study so far is the demonstration of the complex nature of the relationship between the review sentiments and the topics covered in them. This relationship is likely to be effectively covered by fuzzy and/or trainable models as opposed to vocabulary-based approaches. Therefore, in our future work we aim to achieve three major tasks. Firstly, we will develop an approach that can leverage the heterogeneity of the LDA-produced topic representations to better model the correspondence between the users' review content and the overall review sentiments and movie ratings. Secondly, we will perform sentimental analysis with the roBERTa model which will be trained on a labeled dataset of positive and negative movie reviews. Finally, we will apply some models to identify the causal configurations of the sequel movies' performance.

The outcome of future work will be used to build up a deeper understanding of the correlations between review sentiments, specific aspects of the movies, and the movies' success rate. This will provide valuable insights into the sequel movie revenue prediction, and help stakeholders in the film industry make better informed decisions. The results of this study can also be used as a foundation for future research in the field of movie sequel performance prediction.

Acknowledgements. This study was supported in part by the Ministry of Science and Technology of Taiwan under grants MOST 109-2410-H-027-009-MY2 and MOST 111-2410-H-027-011-MY3.

References

1. Belvaux, B., Mencarelli, R.: Prevision model and empirical test of box office results for sequels. J. Bus. Res. **130**, 38–48 (2021)
2. Hu, Y.-H., et al.: Considering online consumer reviews to predict movie box-office performance between the years 2009 and 2014 in the US. Electron. Libr. **36**(6), 1010–1026 (2018)
3. Hur, M., Kang, P., Cho, S.: Box-office forecasting based on sentiments of movie reviews and Independent subspace method. Inf. Sci. **372**, 608–624 (2016)
4. Ahmad, I.S., et al.: Sequel movie revenue prediction model based on sentiment analysis. Data Technol. Appl. **54**(5), 665–683 (2020)
5. Xu, Y., Wu, H., Li, R.: How the competitors' online reviews affect product performance in the dynamic competitive market: an empirical analysis of the movie industry. Appl. Econ. Lett. 1–6 (2022)
6. Xiao, X., Cheng, Y., Kim, J.-M.: Movie title keywords: a text mining and exploratory factor analysis of popular movies in the United States and China. J. Risk Financ. Manag. **14**(2), 68 (2021)
7. Cheng, L.-C., Chen, Y.-L., Liao, Y.-Y.: Aspect-based sentiment analysis with component focusing multi-head co-attention networks. Neurocomputing **489**, 9–17 (2022)
8. Chang, Y.-C., Ku, C.-H., Nguyen, D.-D.L.: Predicting aspect-based sentiment using deep learning and information visualization: the impact of COVID-19 on the airline industry. Inform. Manag. **59**(2), 103587 (2022)
9. Cheng, L.-C., Yang, Y.: The effect of online reviews on movie box office sales. J. Glob. Inf. Manag. **30**(1), 1–16 (2022)
10. Pedregosa, F.: Scikit-learn: machine learning in Python. J. Mach. Learn. Res. **12**, 2825–2830 (2011)

Participative Process Model for the Introduction of AI-Based Knowledge Management in Production

Tobias Rusch[✉], Nicole Ottersböck, and Sascha Stowasser

Institute of Applied Industrial Engineering and Ergonomics, Uerdinger Str. 56, 40474 Dusseldorf, Germany
t.rusch@ifaa-mail.de

Abstract. The growing shortage of skilled workers and the large number of employees from the 1960s so-called baby boom generation who will be retiring in the next few years are current and growing major challenges for companies in Germany. When people retire, companies can lose a lot of valuable know-how and experience. Experienced professionals enjoy a special status in companies because of their extensive and in-depth knowledge of work processes. Salaries and positions in companies are often linked to knowledge. The fear of losing status or even financial loss by sharing knowledge can be a reason for employees to keep their knowledge to themselves. The use of new technologies such as artificial intelligence in companies can also trigger several fears among employees like fears regarding job loss, new skill requirements or surveillance. An innovative work culture, where knowledge sharing is actively encouraged, required and rewarded, is not successfully implemented everywhere. A particular challenge in the KI_eeper project is to convince experienced employees in the participating organizations of the importance of sharing and preserving knowledge to ensure the company's long-term competitiveness. The aim is to encourage them to provide their knowledge to other employees with the help of an AI assistance system. In addition to the development of an assistance system, an important task of the project is the development of methods promoting knowledge transfer and contributing to the acceptance of the use of the new technology. Information and awareness-raising workshops with managers and employees have been designed and tested in the companies. Employees are involved in the development of the system in a participatory approach. This ensures that the system can be developed in a user- and needs-oriented manner and thus the acceptance of the system's use in future everyday work can be promoted, which is crucial for long-term successful application.

Keywords: AI · knowledge transfer · skills shortage

© Springer Nature Switzerland AG 2023
F. Fui-Hoon Nah and K. Siau (Eds.): HCII 2023, LNCS 14039, pp. 181–191, 2023.
https://doi.org/10.1007/978-3-031-36049-7_14

1 Motivation for Knowledge Transfer

1.1 Challenges in Knowledge Transfer

In the coming years, an increasing number of employees born in the 1960s will retire. According to the German Federal Statistical Office, by 2036, 12.9 million workers, or about 30% of the workforce in 2021, will have been reaching retirement age. [1] Subsequent generations with lower birth rates will not be able to meet the resulting demands for skilled workers. The experience of long-serving employees, built up over decades, will be lost to companies unless action is taken to preserve it. In particular, the loss of the tacit knowledge that employees have built up through experience over the course of their working lives is a major challenge for companies. This knowledge often exists unconsciously in the minds of employees and is difficult to capture. This experience is particularly crucial when mistakes, unforeseen events or changes occur during the work processes. Problems that arise can usually be solved by long-serving employees based on their previous experience with similar cases. Organizations now face the challenge of capturing this knowledge and preserving it for future generations. For experienced employees preferred method of capturing and transferring knowledge is to pass on their experience regarding the work process to new, inexperienced employees. Due to increasing labor shortages, vacancies cannot be filled immediately and often not with equally qualified employees. This may result in major problems regarding knowledge transfer processes such as mentoring. This is increasingly motivating companies to recruit workers from abroad. Language barriers and different education standards complicate the integration of foreign workers into companies [2]. Another growing challenge for companies is the increasing individualization of products and services which leads to an ever-increasing variety of products and a growing product and service portfolio. The variety of variants leads to an increased complexity which can lead to errors in the production processes. Therefore, these challenges can often result in increased physical and mental strain for employees [3].

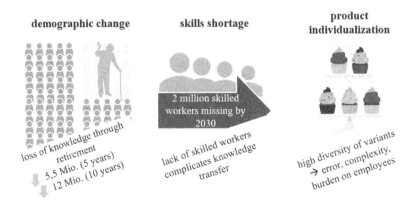

Fig. 1. Challenges in knowledge transfer for German companies [6]

New technologies are now being used to find innovative solutions to the challenges of knowledge transfer and knowledge retention (see Fig. 1) to support companies. One solution could be an AI-based assistance system that automatically captures and stores both empirical and tacit knowledge in the ongoing work process. The aim is to make the captured knowledge available to less experienced employees during the planning process and respectively during the work process via an information-based assistance system.

1.2 Project KI_eeper

The research project "KI_eeper – Know how to keep" addresses the growing need of companies to efficiently keep the experiential knowledge of experts in the company. Funded by the German Federal Ministry of Education and Research (BMBF), the project aims to develop a technical solution to the challenges of knowledge transfer and retention. This includes the design and implementation of a pre-competitive demonstrator for the acquisition, processing, storage, and application of knowledge at production-related workstations. In cooperation with several system developers, an AI-based assistance system will be developed that automatically captures and stores knowledge during the ongoing work process. The consortium includes two companies from the metal and electrical industries, in which the demonstrator is to be developed, tested, and evaluated. The aim is to make the captured knowledge available to less experienced employees during the induction phase as well as during the work process via an information-based assistance system. By this, new employees could be qualified by using the stored knowledge of previous employees. This requires participation and intense cooperation with employees. Promoting and ensuring this is also a central goal of the project. From the consortium's point of view, this includes considering the requirements of the respective workplace as well as the employees' point of view, see also Fig. 2, and the acceptance and willingness of the employees to cooperate with the project staff and later with the technical system. Various further stakeholders and decision-makers, such as the works council, are involved in the system development as well. The development process of the demonstrator is designed for practical use cases at knowledge-intensive workplaces in manufacturing and assembly like metal processing in the case of the participating companies.

In addition to an AI-based assistance system, a set of socio-technical methods is being developed that will enable a broader range of companies to successfully design the implementation of this system in their organization while ensuring the highest possible level of acceptance among employees. This set of methods is based on socio-technical system design and concretizes the involvement of employees in a participatory introduction process of an AI-based assistance system at the workplace [8, 9]. The process model focuses on a universal approach with the aim of successfully and economically deploying the demonstrator beyond the use cases. To promote a broad range of applicability a consortium of so-called transfer partners is interested in participating in the project. For this purpose, an accompanying network of companies will be established within the project. The companies in the network will be able to evaluate the project results for their own production processes. Suitable other use cases can also be found

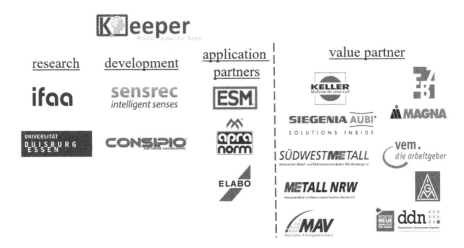

Fig. 2. Project consortium KI_eeper [6]

among this network and used for further development of the technology. The participating associations and organizations also intend to make the knowledge available to their member companies. The validation of technical subsystems and the prototype demonstrator at these corporate partners is part of this project. In addition, the consortium is accompanied by several social partners who, beyond the generality of the project, are monitoring compliance with relevant legislation, such as the GDPR, and contributing their experience in change management.

1.3 Use Case: Surface Technology

The company apra-norm Elektromechanik GmbH, located in the rural Vulkaneifel region, is a medium-sized company that manufactures a wide variety of enclosures and enclosure parts. About 200 employees produce a portfolio of almost 3000 different products and customized product solutions, see Fig. 3.

Apra-norm maps the entire manufacturing process from the processing of sheet metal plates to the assembly of individual components. An important part is the department of surface technology (OFT). In this workplace area, the manufactured products are cleaned, painted, and subjected to initial quality control. The multi-layered processes take place at several, predominantly manual workstations, which demand a high level of process understanding and experience knowledge from the employees, see Fig. 4.

The complexity of the work process and the need for a high level of unrecorded empirical knowledge were the key criteria for the selection of a specific use case for the project which was evaluated in a workshop with the company's management. Apra-norms choice fell on surface technology because it is the company's most cost-intensive and error-prone production area. Each of the products in the company's portfolio must pass through this division before assembly and delivery [5]. The entire process has three work areas. In the receiving area, all products have to be hung up on a conveyor system via traverses. In a subsequent step, the components are cleaned. Depending on the

Fig. 3. Company headquarters apra-norm, manufacturer of enclosures and switch boxes [4]

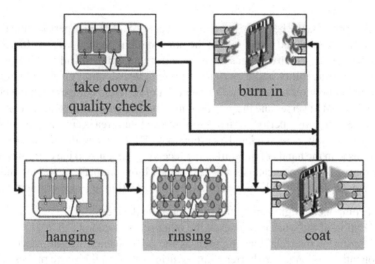

Fig. 4. Stations in the OFT process at the company apra-norm [5]

requirements, water is used for rinsing only, or all possible residues on the components are completely removed by chemical treatment in various baths. After drying and possible reworking of the components, the next process step takes place. The conveyor system then moves the components through the various workstations in the paint stop which include several painting stations and a large gas oven in which the applied coating is burned. This oven has a certain capacity, that should be fully exploited, given the high costs involved. In the application area, the finished, coated products are checked for quality, removed from the conveyor system and packaged for further processing or shipment. Products with quality defects remain on the conveyor and are returned to the receiving station for finishing. Various products hanging from a cross beam are depicted in Fig. 5.

This simple appearing process is very complex and knowledge-intensive. For this reason, a focus is on the receiving station as an example. The work area of the surface

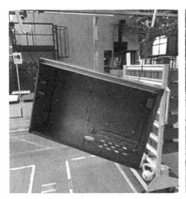

Fig. 5. Products hanging on a traverse [4]

technology represents the beginning of the whole process. Workers' performance at this station has an impact on the workflows of the other stations and the overall efficiency of the whole process especially the gas oven. To ensure an efficient and reliable process, a high degree of flexibility and experience is required. Attention must be paid to the correct alignment of components, the combination of different components, the correct sequence of loaded traverses and the path planning for the conveyor system. According to management, the training period for new employees in this area is currently more than a year, even for employees with related qualifications. The responsibility for the timing of all work areas is in the hands of the staff at the receiving station, and the current required turnaround time for a crosshead is set at 2:35 min [5]. In summary, the requirements in this work area include the following aspects correct selection of transportation devices:

- correct component alignment on the transportation devices
- consideration of the work processes at the other process stations
- consideration of interrelationships such as the consequences of the processing of components with longer or shorter painting/firing times on the entire process flow

Due to the high age structure in the company, apra-norm faces the challenge that many experienced employees throughout the organization will retire in the next few years. The severe shortage of skilled workers makes it difficult for the company to train new employees at an early stage. The increasing integration of employees with immigrant backgrounds poses new challenges, such as language barriers and a country-specific, very different educational background of the employees. At present, it is not possible to assess which types of experience particularly influence the efficiency of the entire surface treatment process. Based on multimodal data collection and evaluation by an AI algorithm to be developed, conclusions will be drawn about the knowledge that is critical for success. On the one hand, these challenges which are faced by the employees slow down the learning process. On the other hand, these challenges make it difficult to optimize the entire processes of the plant on the technical side. To be able to better deal with these challenges, apra-norm is participating as an application partner in the KI_eeper project. [6].

2 Process Model for the Participatory Involvement of Employees

The first work package of the research project aims at a detailed analysis of processes and requirements and a resulting socio-technical specification. This specification will provide the technical partners with all the necessary information about the work process. Based on this information, the development process will begin. An important focus of the process analysis was also the involvement of the employees at the receive station in the process intake. For this reason, the requirements were grouped into functional, technical and qualitative dimensions, as well as organizational, ethical and legal implications (ELSI). In this way, the methods to be used and involved internal stakeholders can be assigned to specific clusters. Figure 6 shows a structured scheme of these groups for determining the respective requirements. The selection of the methods and the corresponding allocation of internal stakeholders must be considered individually for each company. The procedure can be individually adapted and designed depending on the chosen application.

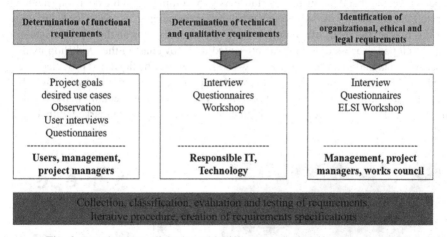

Fig. 6. Requirements elicitation with different methods and target groups [7]

The recording of the process as a basis for the requirements elicitation for the AI system was established within the combination of different methods of field research in the social sciences. The implementation of the individual methods, an interview, a participatory observation, and a silent observation, took place successively with each (individual) employee. The amount of information gained about the work process can thus be increased by each method applied. In the form of a control loop, this procedure is applied in the companies with the corresponding employees and thus the process is iteratively refined from both the explicit and implicit sides, see Fig. 5 (Fig. 7).

At the start of the requirements analysis, a workshop was conducted with managers and leading surface technology specialists to develop a rough process flow in the form of a flowchart. In addition to the initial process analysis, the aim was also to enable the company management to select and define concrete project goals for the company. In groups corresponding to all three areas of the OFT process, the workflows were examined

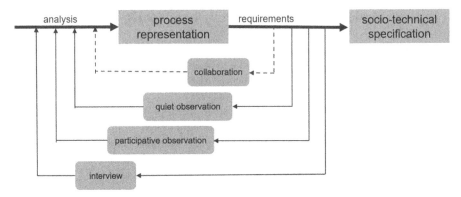

Fig. 7. Control loop for a participative process model by combining different methods [7]

in detail and the interdependencies between them were highlighted. Discussions with the experts revealed that the registration station is very complex, and the decisions made by the staff there influence the subsequent processes in the other areas of surface technology. For this reason, the focus of the initial requirements analysis is on the receiving station. The developed process diagram was presented as a flowchart in the workshop using a flipchart and later digitized. Figure 8 shows both versions. The flowchart provided a very good overview of the interlinking of the work steps. Once the individual work steps had been structured, those with a high susceptibility to error were specially marked with a 'lightning bolt'. All steps requiring a high level of experience were marked with a 'light bulb' [7].

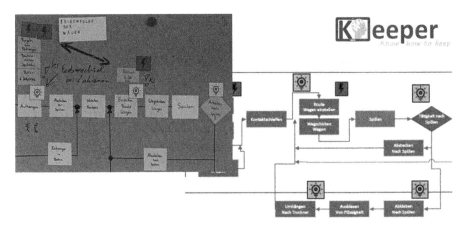

Fig. 8. Flow chart as a result of the workshop [7]

The definitions of the project goals as well as the process diagram as the result of this workshop have been the starting point for the following steps. This information base was successively further expanded and supplemented in collaboration with the admissions department staff. Other departments were also consulted as sources of information. For example, the internal IT systems and databases were examined for available data and relevant information that could be used for the development of the application. All measures implemented are constantly coordinated with the Works Council to ensure maximum transparency and acceptance among the employees. The most important stakeholders regarding the project process are the employees which are directly involved in the processes related use case. Their knowledge and experience must be captured as the project progresses. In a kick-off meeting with all employees, the project and company goals were presented. The purpose of this workshop was to inform about the project plan and to provide employees with the opportunity to address their needs and fears about the project. The workshop's main objective was to ensure the participation of the employees and to increase the acceptance of the project goals. In several steps, a personal interview was conducted with each employee to determine their view of the process, their individual approach and their personal circumstances. By conducting a guideline-based interview, socio-demographic data, affinity for technology, possible migration background/language barriers and the employees' ideas and perceptions of the work process were recorded. The results of the analysis of the interviews were combined with the results of silent and participant observation of each employee during the work at the station. The data and information that were gathered by these methods were used to complete the process map. The map was then discussed and verified in a consolidation workshop with all employees involved in the entire OFT process. This also provided additional process information. The resulting model of the entire process is the basis for the specifications, which can be used by the development partners to design an AI-based assistance system tailored to the selected use case [8]. The necessary process information, which is currently recorded by the project staff, should then be automatically recorded and processed by the developed technical system. In addition to the purely technical process flow, the system image will also contain the recorded work steps that result from the employees' experiential knowledge. This can be supplemented by further data from the general conditions of the OFT system, e. g. temperature etc.

3 Summary and Outlook

The participatory process model enables a comprehensive analysis of the selected work processes by involving all stakeholders from the management level to the employees at the workplace. In addition to technical data, it also includes socio-technical aspects. During the development process, even the process experts were surprised about how complex the work processes in the receive department are. Before it had previously been considered simple. It was not expected that the staff would need such a high level of experience. An important criterion for detailed analysis was the cooperation of the various process experts with external facilitators who had no insight into the work process. The external moderators' questions about the process made the experts aware of the diversity and complexity of the process. This approach helped to create a detailed picture of the process. Through the resulting need to explain the process, the process participants became aware of many aspects of their intuitive work and their existing implicit experiential knowledge. It is planned to allow scientific staff to participate to gain an even deeper insight into the processes. This could lead to even deeper insights into the working process as a method. The socio-technical specifications resulting from the analysis will be further elaborated by the technical developers and form the basis for the design of the AI-based assistance system. The networking of the two systems for capturing and displaying knowledge can also create the possibility of not only displaying existing experiential knowledge but also comparing different experiential knowledge with each other. This could then be used for process optimization in companies. The collaboration and joint development of process knowledge also led to an increased appreciation of surface engineering work on all sides. In the further course of the project, the gained acceptance, and cooperation of the employees with the project team will be further promoted. An ELSI workshop will be developed to provide employees with a deeper insight into the technology and the planned data collection. This workshop will be an opportunity for the employees to provide direct feedback on the planned technology and increase acceptance of the use of the system. A corporate network will be established through the participating value creation partners to validate and evaluate the developed process model and the technical implementation in other companies to provide a broad base of other organizations with the opportunity to benefit from the methods developed within the research project KI_eeper.

Acknowledgements. This research and development project is funded by the German Federal Ministry of Education and Research (BMBF) within the framework of the program "Zukunft der Arbeit" (funding codes 02L20C500- 02L20C505) and is supervised by the Project Management Agency Karlsruhe (PTKA). The responsibility for the content of this publication lies with the authors.

References

1. German Federal Statistical Office: 12,9 Millionen Erwerbspersonen erreichen in den nächsten 15 Jahren das gesetzliche Rentenalter. Pressemitteilung Nr. 330 vom 4. August 2022. https://www.destatis.de/DE/Presse/Pressemitteilungen/2022/08/PD22_330_13.html (2022). Accessed 4 Jan 2023
2. Schultz, S.: Fachkräftemonitor 2022. Fachkräfteengpässe von Unternehmen in Deutschland, Trends und Potenziale zum Zuzug ausländischer Fachkräfte. Bertelsmann Stiftung (Hrsg) (2022). https://www.bertelsmann-stiftung.de/de/publikationen/publikation/did/fachkraefte-migrationsmonitor-2022. Accessed 4 Jan 2023
3. Bornewasser, M., Bläsing, D.: Humanorientierte Beurteilungskriterien zur Implementation informatorischer Assistenzsysteme in der manuellen Montage. In: Arbeit interdisziplinär analysieren – bewerten – gestalten. Gesellschaft für Arbeitswissenschaft (Hrsg): Frühjahrskongress 2019, Dresden (2019)
4. apra-norm Elektromechanik GmbH. b. d. untersten Mühle 5, 54552 Mehren (2023). https://leanbase.de/awards/die-apra-gruppe
5. Ottersböck, N., Prange, C., Dander, H., Ochterbeck, J., Peters, S.:Flexibler Arbeitskräfteeinsatz durch KI-basierten Wissenstransfer. Den Herausforderungen der demographischen Entwicklung und des Fachkräftemangels mit innovativen Technologien begegnen. In: ifaa – Institut für angewandte Arbeitswissenschaft (Hrsg) Werkwandel – Zeitschrift für angewandte Arbeitswissenschaft 1/23 (in print) (2023)

6. Ottersböck, N., Rusch, T.: Babyboomer weg – Wissen weg. In: ifaa – Institut für angewandte Arbeitswissenschaft (Hrsg) Werkwandel – Zeitschrift für angewandte Arbeitswissenschaft 3/22:28–32 (2022)

7. Rusch, T., Ottersböck, N., Ternes, J.: Partizipative Prozessaufnahme als Grundlage eines KI-basierten Assistenzsystems für den Wissenstransfer im Produktionsbetrieb. In: GfA–Frühjahrskongress 2023, Beitrag C.6.19 (in print) (2023)

8. Stowasser, S., et al. (Hrsg): Einführung von KI-Systemen in Unternehmen. Gestaltungsansätze für das Change-Management. Whitepaper aus der Plattform Lernende Systeme, München https://www.plattform-lernendesyteme.de/files/Downloads/Publikationen/AG2_Whitepaper_Change_Management.pdf (2020). Accessed 3 Jan 2023

9. Pokorni, B., Braun, M., Knecht, C.: Menschenzentrierte KI-Anwendungen in der Produktion. In: Bauer, W., Riedel, O., Renner, T., Peissner, M. (Hrsg.) Praxiserfahrungen und Leitfaden zu betrieblichen Einführungsstrategien. https://www.ki-fortschrittszentrum.de/content/dam/iao/ki-fortschrittszentrum/documents/studien/Menschzentrierte-KI-Anwendungen-in-der-Produktion.pdf (2021). Accessed 3 Jan 2023

Automating Data Personas for Designing Health Interventions

Gaayathri Sankar[1], Soussan Djamasbi[1(✉)], Daniel J. Amante[2(✉)],
Adarsha S. Bajracharya[2(✉)], Qiming Shi[2(✉)], Yunus Dogan Telliel[1(✉)],
and Torumoy Ghoshal[1(✉)]

[1] User Experience and Decision Making (UXDM) Laboratory, Worcester Polytechnic Institute, Worcester, MA 01609, USA
{gsankar,djamasbi,ydtelliel,tghoshal}@wpi.edu
[2] University of Massachusetts Chan Medical School, 55 Lake Avenue North, Worcester, MA 01655, USA
{daniel.amante,qiming.shi}@umassmed.edu,
adarsha.bajracharya@umassmemorial.org

Abstract. In user experience (UX) research, persona development can be particularly beneficial in designing digital health interventions. From our previous work, we identified challenges that diabetic patients face in adhering to guideline-recommended care through persona development. We developed data personas from Electronic Health Records (EHR) using two-step clustering and combined them with proto personas that were generated by a group of medical experts for the same patient population. Not only did our results prove beneficial for intervention design, but also highlighted fairness issues that may result from the underrepresentation of certain populations in EHR datasets. In our current paper, we validate the results of our prior work and we extend our previous work on persona development to build a model that can automatically associate a patient record to its representative data persona. The model was built using the results of the two-step clustering as the inputs. We believe that building such a model will help clinicians quickly trace new patient records to already developed personas, thereby assisting in immediate health interventions based on the health risk characteristics of the patient.

Keywords: Persona development · Predictive modeling · Intervention design · Electronic Health Records (EHR) · Type 2 diabetes · Cluster analysis · Proto personas · Data personas

1 Introduction

Today, the availability of an enormous amount of data has allowed for automation in various industries and in different capacities. The integration of automation into the workplace has been revolutionary. Automation processes have the potential to transform the way we work, saving us time, money, and effort, thus helping to improve efficiency and productivity. Automation can be used to avoid mundane and repetitive tasks, thus

© Springer Nature Switzerland AG 2023
F. Fui-Hoon Nah and K. Siau (Eds.): HCII 2023, LNCS 14039, pp. 192–201, 2023.
https://doi.org/10.1007/978-3-031-36049-7_15

freeing up employees' time and energy for more creative and innovative tasks, such as strategic decision-making. It can also be used to improve accuracy, as well as to improve customer service. In the current digital era, good customer service relies on truly understanding the needs, goals, and challenges of the customer. One method to achieving this is through persona development.

Persona development is an important tool in the creation and execution of successful information systems. It involves creating detailed representations of target customers, also known as personas, to better understand the needs and wants of the individuals that make up the target users. Personas can be created in various ways. They can be created from existing user information such as data from EHR (data personas), by collecting and synthesizing information from actual and/or potential users (user personas), or by gathering information from a group of experts who have knowledge about the users (proto personas). Combining more than one type of persona development method is known to enhance knowledge about user needs [1, 2].

The objective of this paper is to extend our previous work on persona development. While our previous work focused on the creation of proto-data personas to understand the needs of Type 2 diabetes patients receiving treatment from the UMass Memorial Accountable Care Organizations (ACOs), our current work focuses on validating and automating the generated data personas, thus enabling ACOs to easily associate the data from a diabetic patient's EHR in their network to its associated persona cluster for designing self-care intervention decisions.

2 Background

In UX research, personas provide designers with a framework to guide their design policies and better connect them with their target users. By understanding their personas, innovators are able to create more meaningful products/services that are more likely to resonate with their target audience. Personas can also help inform business decisions and can be used by decision-makers to create value with customized experiences meeting customers' needs. Developing personas is an essential part of UX-driven innovation today that should not be overlooked [3].

As automation is set to become a staple in the workplace in the near future, the automation of persona development is revolutionizing the way businesses interact with their customers. By using artificial intelligence and machine learning techniques, companies are increasingly able to create accurate data personas of their target audiences. These personas can then be used to guide their design decisions, improve user experiences, and personalize product offerings. Automation of persona development is also enabling companies to become more data-driven and make informed decisions about their customers, thus creating a more efficient and effective user experience [4].

Good user experiences should account for the inclusion of all users. Automation is rapidly becoming an integral part of our lives, and it is important that fairness is taken into consideration when creating these systems. Responsible automation can be used to reduce bias and create a level playing field for all. Fairness and transparency should be considered when designing and implementing automation systems. This can help to ensure that everyone, regardless of their background or identity, is treated equally and given the same opportunities. Automation should be used to provide equitable and accessible services for all, so that everyone is able to benefit from technological advances. By taking into account fairness and transparency, automation can become a source of social good for everyone. This is particularly important in the field of medicine and healthcare since reducing inequalities, and good health and well-being of all are important Sustainable Development Goals (SDGs) laid out by the United Nations.

As stated earlier, in our previous work, the results from combining proto-data personas not only proved beneficial for intervention design but also highlighted fairness issues that may have resulted from the underrepresentation of certain populations in EHR datasets. Thus, using the clustering technique to create the data personas and then combining them with the proto personas not only helped us understand these fairness issues but also helped avoid a blind automation of the created clusters/data personas.

3 Methodology

As mentioned previously, our prior work involved investigating UMass ACOs' type 2 diabetes patients' needs by combining data and proto personas. To achieve this goal, we first developed a set of data personas from electronic health records (EHR data) for this population. Next, we mapped the generated data personas to proto personas that were developed in a prior study for the same patient population [5]. In this section, we explain how we validated the data personas generated in our previous work. Then we explain the process we used for automating them. As explained earlier, by automation, we mean enabling ACOs to easily associate the data from a diabetic patient's EHR in their network to its associated persona cluster for designing self-care intervention decisions.

In our previous work, the data personas from the EHR data were created using two-step hierarchical cluster analysis in IBM SPSS Statistics v26. The original dataset contained both continuous and categorical variables. NULL values (or missing data) were embedded in all of the variables in the dataset. Because we wanted to preserve the presence of missing information in the continuous variables, we converted the continuous variables into categorical variables as outlined in [1]. This in turn also guided our choice of algorithm.

While the above-mentioned analysis proved to be a satisfactory methodology for analyzing the EHR data, we proceeded with validating our results with other clustering methods before moving on to the automation stage. The reason behind validation through other clustering algorithms was to examine whether these algorithms could possibly have better results for our dataset than the two-step hierarchical clustering. To address this question, we used k-means (a centroid-based clustering algorithm) as well as Gaussian Mixture Modeling (GMM) (a density-based clustering algorithm). Both the algorithms were run using sci-kit learn machine learning library on Python.

Since our focus was on validation, we used the same three input variables as we used in the prior two-step hierarchical clustering, i.e., the 3 major health risk variables: HbA1C, LACE+, and BMI. Hemoglobin A1C (HbA1C) levels refer to blood glucose levels. HbA1C is considered a direct measure of risk for developing type 2 diabetes. LACE+, the second health risk variable we used in our study, is a predictor of mortality rates. LACE+ scores are calculated by considering factors such as age, sex, comorbidities, whether the patient was admitted through the emergency department (ED), the number of ED visits in the six months prior to admission, and the number of days the patient was in an alternative level of care during admission [6]. Obesity, the third health risk variable in our study, is considered a global pandemic. Studies show that a majority of people with type 2 diabetes are obese [7]. Obesity is measured by a score representing a person's body mass index (BMI) [8].

After the validation process using k-means and GMM, the results of which are discussed in Sect. 4 below, we went forward with automation by building a predictive model using the Classification and Regression Trees (CART) algorithm in IBM SPSS Modeler v.18.2.1. CART is a powerful machine learning algorithm used for both classification and regression tasks. CART is a decision tree-based algorithm that uses a divide-and-conquer approach to identify the optimal split points for the independent variables. It is a supervised learning method that builds a decision tree model by identifying the best split points in the data. The tree is then used to make predictions about the dependent variable, (in this case, our data personas). CART can handle both numerical and categorical data, making it a powerful approach for a variety of predictive tasks. Thus, this algorithm was chosen based on the same principle chosen for our two-step clustering, i.e., we wanted to preserve the presence of missing information in the continuous variables and therefore converted the continuous variables into categorical variables. The clusters generated in our prior work, i.e., the data personas, were used as inputs in our model. The dataset was partitioned into training and test sets in the 80:20 ratio. The CART method is known for its high variance [9], and therefore, to avoid overfitting, we pruned the trees as well as applied bagging (bootstrap aggregation) to enhance model stability.

The results from the validation process and the automation are outlined in the next section.

4   Results

In the following section, we briefly describe the results from the data preparation and then explain the results from the validation process and subsequent automation (building the predictive model).

4.1   Data Preparation

As the above-mentioned clustering algorithms (k-means and GMM) consider only continuous variables, we imputed the NULL values using the MICE imputer, commonly used for EHR datasets [10]. Since each of the three variables explained previously (HbA1C, LACE+, and BMI) have varying units, we also standardized the data. We then created boxplots to visualize outliers (see Fig. 1). Although the box plots show outliers for BMI and HbA1C, none of these data points were beyond the acceptable range for these variables. So, we did not delete any of these outliers and kept all of the data points. Then we drew up a correlation matrix and pair plots since these graphs examine possible correlations between these variables (see Figs. 2 and 3). This is important since if the variables correlated, it would mean that they would probably cluster together too. From Figs. 2 and 3, we see that none of the variables were correlated with the other.

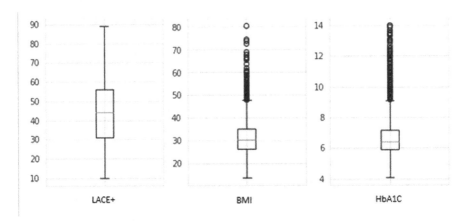

Fig. 1. Box plots of the input variables.

4.2 Validation Process

From expert opinion, we had developed 8 proto personas. Our prior work using two-step clustering for the EHR data resulted in 9 data personas. Therefore, keeping this as the maximum limit, for k-means, we ran the algorithm for cluster sizes ranging from 2 to 9. Among these trials, the best silhouette score was 0.3 at a cluster size of n = 4 (Fig. 4). However, using the two-step clustering approach we were able to produce a higher (i.e., better) silhouette score (0.5) for the same data for a cluster size of n = 9 [1]. We also compared the average sum of squared errors (SSE) between the results of k-means and two-step clustering. At the same n = 9, for two-step clustering, we found a lower SSE (0.06) compared to SSE for k-means (0.83). Based on these findings (higher silhouette score and lower SSE), we chose the two-step clustering in favor of k-means.

GMM requires a normal distribution in order to produce the most accurate results. For the GMM, we first drew up histograms for each of the variables. The histograms are a determination of the distribution of the variable based on the Kolmogorov-Smirnov test (KS-test). This test is used to assess the goodness of fit of any given set of data to a certain theoretical distribution [11]. Conducting the test was important for the GMM analysis since variables that do not exhibit a Gaussian (normal) distribution may not produce optimum results. As shown in Fig. 5, the results of KS-test showed that the data for the variables used in the analysis was not normally distributed. Nevertheless, for exploratory purposes, we continued with the GMM cluster analysis to explore possible results using this algorithm.

As per the sci-kit learn library, for GMM, we used the Bayes Information Criterion (BIC) for model selection (see Fig. 6). BIC is more useful in selecting a correct model while the Akaike Information Criterion (AIC) is more appropriate in finding the best model for predicting future observations [12]. Since our focus is on the former, we used BIC to help identify the true model.

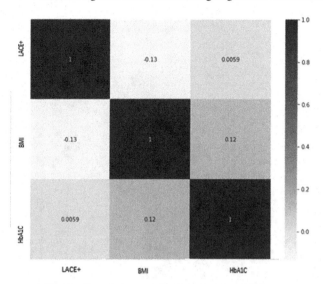

Fig. 2. Correlation matrix of the input variables.

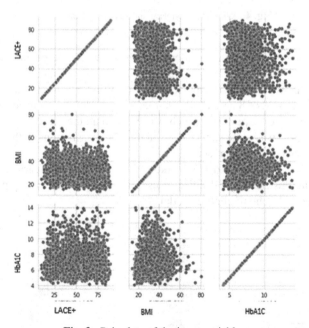

Fig. 3. Pair plots of the input variables.

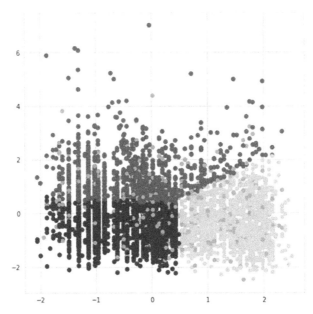

Fig. 4. k-means clustering at cluster size or n=4

In this case, model selection considers both the covariance type and the number of components used in the model. We generated the BIC score per model to select the best cluster number as well as the covariance type. From Fig. 6, we see that the algorithm suggested a cluster size of 8 and a full covariance type. However, clustering using n = 8 returned a silhouette score of 0.1 which was worse than the score from k-means (see Fig. 7). Besides, this was quite expected given that the distributions of the variables did not follow a normal distribution (see Fig. 5).

Since both the k-means and GMM performed badly in comparison to the two-step hierarchical clustering, we decided to use the results of the two-step hierarchical clustering for building the predictive model.

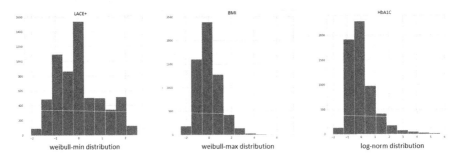

Fig. 5. Distribution of the three input variables for GMM based on Kolmogorov-Smirnoff test.

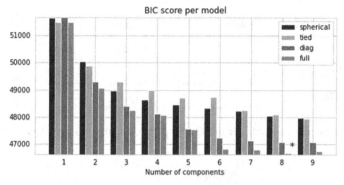

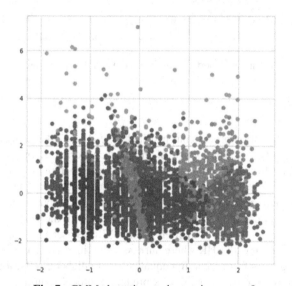

Fig. 6. The asterisk shows the model selection using the BIC score, in this case, cluster size or n = 8 and covariance type as full.

Fig. 7. GMM clustering at cluster size or n = 8

4.3 Automation: Building the Predictive Model

Based on the results of the validation process discussed in the previous section, we used the results obtained from the two-step hierarchical clustering process [1] to automate the identification of a persona cluster for an unseen EHR. The variables used as inputs included the 3 health risk variables mentioned earlier – HbA1C, LACE+, and BMI. The 9 data personas named alphabetically from A to I from our prior study were used as the target variable [1]. The predictive model generated by the CART algorithm resulted in a training accuracy of 93.18% and a test accuracy of 92.44% (see Fig. 8). All the three predictors were assigned an equal importance of 0.33 by the algorithm signaling the relative importance of each predictor in estimating the model (see Fig. 9).

'Partition'	Training		Testing	
Correct	4, 534	93.18%	1, 125	92.44%
Wrong	332	6.82%	92	7.56%
Total	4,866		1, 217	

Fig. 8. Results from our predictive model after partitioning the dataset in the ratio of 80-20.

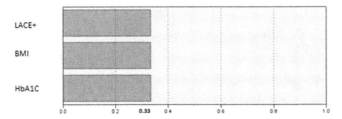

Fig. 9. Predictor Importance of the input variables during prediction.

5 Discussion

In this study, our validation process has justified the choice of clustering algorithm used in our previous study [1]. The choice of the best clustering algorithm for the given EHR dataset came from hierarchical-type clustering rather than from density-based or distribution-based clustering. The two-step hierarchical clustering not only helped identify patient groups based on their health risk characteristics but combining the results of the two-step hierarchical clustering (data personas) with the proto personas helped to identify problems of fairness inherent in the EHR dataset [1]. This is particularly beneficial when dealing with EHR where fairness issues inherent in the data may prevent minority patient groups from receiving equal access to diabetic self-care interventions. Once we are aware of this, by building a predictive model (automation) with a high accuracy, we can facilitate ACOs to easily associate the new data from a diabetic patient's EHR in their network to its associated persona cluster for designing self-care intervention decisions. Thus, if the new data happens to belong to Spanish-speaking females (young or old) [1], then the medical expert would know that this associated cluster given by the predictive model, however accurate, reflects the majority, and therefore, customization of health interventions can be done for these minority patient groups as well. Additionally, from building this predictive model, we observed that conducting the cluster analysis prior to prediction tended to yield a more accurate model [13, 14].

Thus, through our validation process and automation, we emphasize the need to include proto persona development before clustering, classification, and prediction.

6 Limitations and Future Research

This study was an attempt at addressing one of the future research suggestions identified in our prior study. All of the other limitations and future research possibilities outlined in the prior study apply to our current paper too. These include the development of user personas, using geospatial data along with median household income to understand the socio-economic status of patients thereby assisting health officials and the local government in improving access to care, and using data science methods such as text mining to further understand patients' emotional needs and challenges by analyzing their comments and feedback not necessarily included in the EHR datasets.

References

1. Sankar, G., Djamasbi, S., Telliel, Y.D., Bajracharya, A.S., Amante, D.J., Shi, Q.: Developing personas for designing health interventions. In: Nah, F.F.-H., Siau, K. (eds.) HCI in Business, Government and Organizations: 9th International Conference, HCIBGO 2022, Held as Part of the 24th HCI International Conference, HCII 2022, Virtual Event, June 26 – July 1, 2022, Proceedings, pp. 325–336. Springer International Publishing, Cham (2022). https://doi.org/10.1007/978-3-031-05544-7_25
2. Jain, P., Djamasbi, S., Wyatt, J.: Creating value with proto-research persona development. In: Nah, F.-H., Siau, K. (eds.) HCII 2019. LNCS, vol. 11589, pp. 72–82. Springer, Cham (2019). https://doi.org/10.1007/978-3-030-22338-0_6
3. Djamasbi, S., Strong, D.: User experience-driven innovation in smart and connected worlds. AIS Trans. Hum.-Comput. Interact. 11(4), 215–231 (2019). https://doi.org/10.17705/1thci.00121
4. Djamasbi, S.: Presentation at UX Symposium. Worcester Polytechnic Institute, Worcester, MA, USA (2022)
5. Bajracharya, A., McDonald, A., Girardi, C., Ahumada-Zonin, G., Djamasbi, S., Amante, D.: Proto-research persona development: a user experience design approach. In: Third Annual Diabetes Center of Excellence, UMass Medical School, Diabetes Day (2019)
6. Weiss, A.P.: Morning CMO Report. Upstate University Hospital (2017)
7. Eckel, R.H., et al.: Obesity and type 2 diabetes: what can be unified and what needs to be individualized? Diab. Care 34(6), 1424–1430 (2011). https://doi.org/10.2337/dc11-0447
8. Body Mass Index (BMI): Centers for Disease Control and Prevention. https://www.cdc.gov/healthyweight/assessing/bmi/index.html. 7 June 2021
9. Hayes, T., Usami, S., Jacobucci, R., McArdle, J.J.: Using Classification and Regression Trees (CART) and random forests to analyze attrition: results from two simulations. Psychol. Aging 30(4), 911–929 (2015)
10. Wells, B.J., Chagin, K.M., Nowacki, A.S., Kattan, M.W.: Strategies for handling missing data in electronic health record derived data. Egems 1(3), 7 (2013)
11. Berger, V.W., Zhou, Y.: Kolmogorov–smirnov test: Overview. Wiley statsref: Statistics reference online (2014)
12. Chakrabarti, A., Ghosh, J.K.: AIC, BIC and recent advances in model selection. In: Philosophy of Statistics, vol. 7, 583–605 (2011)
13. Karpati, T., Leventer-Roberts, M., Feldman, B., Cohen-Stavi, C., Raz, I., Balicer, R.: Patient clusters based on HbA1c trajectories: a step toward individualized medicine in type 2 diabetes. PLoS ONE 13(11), e0207096 (2018). https://doi.org/10.1371/journal.pone.0207096
14. Trivedi, S., Pardos, Z.A., Heffernan, N.T.: The utility of clustering in prediction tasks. arXiv preprint (2015). arXiv:1509.06163

Change Management Process and People's Involvement when Introducing AI Systems in Companies

Sascha Stowasser[1]([⊠]), Oliver Suchy[2], Sebastian Terstegen[1], and Alexander Mihatsch[3]

[1] ifaa – Institut für angewandte Arbeitswissenschaft e. V., Uerdinger Str. 56, 40474 Duesseldorf, Germany
s.stowasser@ifaa-mail.de
[2] Deutscher Gewerkschaftsbund, Henriette-Herz-Platz 2, 10178 Berlin, Germany
[3] Plattform Lernende Systeme, c/o acatech – Deutsche Akademie der Technikwissenschaften, Karolinenplatz 4, 80333 Munich, Germany

Abstract. Artificial intelligence (AI) offers great potential for companies and their employees – whether through improved work processes or digital business model innovations. At the same time, the change in companies must – and can – be shaped together and the challenges in the use of AI systems must be solved. This is the only way to overcome challenges and negative side effects in the use of AI systems. The overall aim is to create a new relationship between humans and machine, in which people and AI systems work together productively and the respective strengths are emphasized.

The paper shows the possibilities of how AI systems can be introduced successfully and in the interests of the employees in the company. The challenges and design options for companies are based on the phases of the change process: starting from the goal setting and impact assessment, planning and design, preparation, and implementation, right through to evaluation and continuous adaptation. The process was derived by the working group Future of Work and Human-Machine-Interaction of the Plattform Lernende Systeme – Germany's Platform for Artificial Intelligence and want to sensitize for the requirements of the change management in AI [1].

Keywords: Artificial Intelligence · Change Management · Introduction Process

1 Artificial Intelligence Changes the Work

AI not only offers all kinds of opportunities for innovative business models for companies and institutions, the working world in companies is also undergoing revolutionary changes. AI instruments or learning (work) systems are developing the world of work 4.0 into the world of work 5.0. While the world of work 4.0 has so far been the focus of networked digitization and the flexibilization of work location, time, organization and freedom of action, the world of work 5.0 will be enriched with intelligent assistance, learning robots and user-optimized information provision. For employees, the use of AI means even more flexibility, carrying out more demanding activities, individually adapted information and facilitation of monotonous routine mental activities [2].

© Springer Nature Switzerland AG 2023
F. Fui-Hoon Nah and K. Siau (Eds.): HCII 2023, LNCS 14039, pp. 202–213, 2023.
https://doi.org/10.1007/978-3-031-36049-7_16

Since AI technology and learning systems have the potential to reduce or channel the flood of work-relevant information in knowledge work and learning robot systems or AI-based automation solutions in production work can take over physically demanding parts of the activity the prospect that employees in a working world shaped by AI will tend to experience reduced stress. In the best-case scenario, this has a positive effect on the risk analyzes for the effects of stress on employees.

The use and implementation of new technologies are a well-known factor in companies and in the work environment, which is based on a familiar set of change instruments and the legal rules - for example on co-determination and data protection. Nevertheless, new challenges for change processes arise from the specifics of AI, such as the learning aspect of machines, robots and software systems, the use of large amounts of data as a basis for learning or predictive analytics using AI systems. In addition, questions of discrimination through data and algorithms, personal rights or the relationship between man and machine - including the scope for action and the attribution of responsibility - are increasingly coming into focus [3].

2   Employees and AI Implementation

2.1   Fears and Necessities in the Introduction of AI

The world of work is changing. Our own project analyzes show that around 75% of the work systems will change. Incidentally, this applies to all types of work systems, i.e., both production and knowledge work.

A survey within the project "Digital Mentor - Model and testing of a preventive AI helper (en[AI]ble for short)" (September 2020 – September 2023, the project is funded by the Federal Ministry of Labor and Social Affairs (BMAS) as part of the New Quality of Work (INQA) initiative and is professionally supported by the Federal Institute for Occupational Safety and Health (BAuA). The project management organisation is Gesellschaft für soziale Unternehmensberatung mbH (gsub). Funding number: EXP.01.00008.20.) reflects the assessment of AI technologies from various business stakeholders [4]. While people are willing to use AI in the private sphere without any major reservations, for example as a navigation aid or when selecting music with their smartphone, many at work see things very differently. They fear that their personal data will be misused and are afraid of being controlled by the new technologies. In the project and in operational implementation projects, we register three basic fears that are addressed again and again:

- What will happen to my job?
- What happens to my personal data?
- Can I keep up with digitization? Am I competent enough to work with an AI?

It is essential to take these fears seriously, which is why clarity about the use of AI must be created. The actionist use of AI technologies, unstructured implementation processes of new instruments and the lack of consideration for the people affected are of little help here. Rather, an employee-oriented, participatory introductory process emerges as a success factor. An intensive change participation of the employees and the works council includes.

- Demonstrating the advantages and benefits of AI,
- the determination of the need for qualification and its implementation,
- helping to shape the new work systems.

Finally, the design of a relationship between humans and technology that is changing due to AI is in the foreground. It is important to emphasize the respective strengths of man and machine to enable productive interaction and to support people in their work.

2.2   Development of a Change Management Process in the Plattform Lernende Systeme – Working Group "Future of Work and Human-Machine Interaction"

Designing AI and self-learning systems for the benefit of society is the goal pursued by the Plattform Lernende Systeme which was launched by the Federal Ministry of Education and Research (BMBF) in 2017 at the suggestion of acatech - National Academy of Science and Engineering. Learning systems "made in Germany" should improve people's quality of life, strengthen good working practices, and ensure sustainable growth and prosperity. This involves translating the know-how available in Germany into marketable applications and enabling individuals to take a considered, autonomous approach to handling learning systems in their everyday lives and in the workplace.

Plattform Lernende Systeme brings together leading experts in self-learning systems and AI from science, industry, politics, and civic organizations. In specialized working groups, they discuss the opportunities, challenges, and parameters for developing self-learning systems and using them responsibly. They derive scenarios, recommendations, design options and road maps from the results (for more information see [5]).

Working group (WG 2) - Future of Work and Human-Machine Interaction - of the Plattform Lernende Systeme focuses on human-centered design of the future working world and on human-machine interaction issues (HMI). At the same time WG 2 serves as the interface between HMI and the area of Manufacturing and Industry 4.0. WG 2 identify action areas and develop joint positions, concrete application scenarios, and practical recommendations [6].

A major focus of the WG 2 is the consideration of the socio-technical work system. Appropriate and employee-oriented change management in the company is decisive for the success of technical AI systems. The experts of the working group developed a practical approach and process steps of change management for the introduction of AI systems. The members of the WG 2 want to sensitize for the requirements of the change management in Artificial Intelligence. Therefore, the experts have developed a practice-oriented catalogue of requirements that is intended to provide orientation for the practical implementation of the introduction of AI systems in the various phases of the change process.

This publication is a summary of the white paper "Introduction of AI systems in companies – design approaches for change management" of the WG 2 [1].

3 Framework of Change Management for AI Implementation

In the following, the framework conditions for change in AI and the challenges it faces will be outlined and possible solutions and best practice examples will be presented. The aim is to provide orientation for a practical implementation and design of the introduction of Artificial Intelligence. Different phases of change processes are taken as a basis and the requirements including the practice-oriented solution approaches are explained (see Fig. 1). The challenges and design options for companies are based on the phases of the change process: starting from the *objective and impact assessment, planning and design, preparation and implementation*, right through to *evaluation and adaptation*.

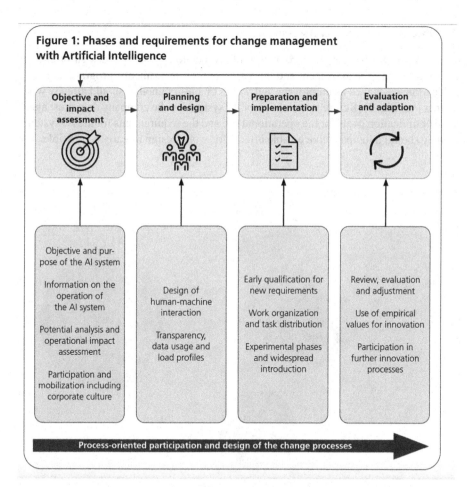

Fig. 1. Phases and requirements for change management with AI [1].

An important success factor for the productive use of AI in companies is the acceptance of AI systems by employees. This is even more important because the introduction of AI is often accompanied by concerns regarding the handling of personal data and surveillance, the impact on employment, the design of humane working conditions and the training and further education for AI technologies. A transparent and comprehensible design of AI deployment and a process-oriented participation of employees and company representatives in the change process are a central element in this respect – to be able to address reservations at an early stage or to find and negotiate constructive solutions for conflicting objectives – for example regarding data use. Ethical guidelines of the companies within the framework of the existing recommendations – such as the EU High-Level Expert Group on Artificial Intelligence (2018) – and along the existing (legal) framework for the use, introduction and handling of AI systems can also be a suitable instrument [7].

3.1 Objective and Impact Assessment

All actors responsible for the introduction of AI technology and the design of change processes - such as management and human resources departments, programmers, and IT departments as well as employees and works or staff councils - should already work together (see Fig. 2). Before introduction of AI systems in the company they must agree on the optimization goals for the operational use and the requirements for the AI systems and they should anticipate conceivable effects for work design as early as possible.

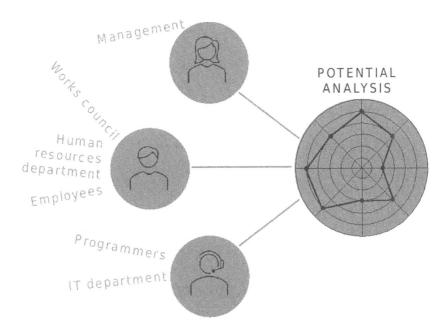

Fig. 2. Relevant stakeholder in the change management process.

When introducing AI systems, a careful analysis of potential and an operational impact assessment are necessary to optimally utilize potential, anticipate risks, develop design solutions, and gain the acceptance of employees. Nevertheless, even when AI systems are introduced, a careful analysis of potential and an operational impact assessment are necessary to make optimum use of potential, anticipate risks, develop design solutions, and gain acceptance among employees. When designing AI-based work systems, several aspects such as (health) impact assessment, technical and social impact assessment, workplace perspectives and, finally, the employees' scope for action must be considered.

The impact assessment in connection with the use of AI-based applications should consider the usual criteria of ergonomics – from feasibility to work-related psychological aspects. This assessment must also be made regarding long-term effects. In the technical and social impact assessment, one criterion should be that the use of AI systems leads to an improved work-life balance rather than to the removal of boundaries from work by exploiting the potential for flexibility.

In general, the use of AI systems should make work much easier for all employees by transferring routine activities and by increasing flexibility in terms of place, time, and work content. Attention should be paid to a healthy level of challenging and routine activities for individuals and the preservation of their sovereignty: On the one hand, flexibility should not exceed a healthy level, and, on the other hand, it should also be possible for employees to determine their own working hours.

It is important to have sufficient transparency about the impact of the AI application before it is introduced. Since there is no such thing as "the" standard AI, it makes sense to take an application-oriented approach to be able to assess sensitive aspects (criticalities) and develop specific measures for implementation. When designing AI-based work systems, several aspects, such as the (health) impact assessment, the technical and social impact assessment, the workplace perspectives and, finally, the employees' scope for action, must be considered.

Workforces must not only be "taken along" with change processes in AI but must also be actively involved in the change process – because: The agility and innovative ability of a company is largely determined by its employees. In addition, important impulses for the innovative ability of companies can also emanate from interest groups. The success of many change processes depends not least on a change of culture in which the employees concerned are involved. The introduction of AI technologies alone does not create added value: it is therefore important to keep in mind the questions of how processes and organizational structures in the company, but also the skills and competencies of the employees, should be designed in connection with the introduction of AI systems. This applies not least to the central questions of leadership and cultural change in companies.

3.2 Planning and Design of the AI-System

In a second step, the focus is on the design of the AI systems themselves. This is primarily about the design of the interface between humans and AI systems along criteria for the humane and productive implementation of human-machine interaction in the work environment. A balanced relationship between the requirements for good and conducive working conditions on the one hand and the technological and economic potentials of AI on the other increases the chances for the acceptance of AI systems in change processes.

[8] provide more detailed information on the design of human-machine interaction in AI systems (see Fig. 3).

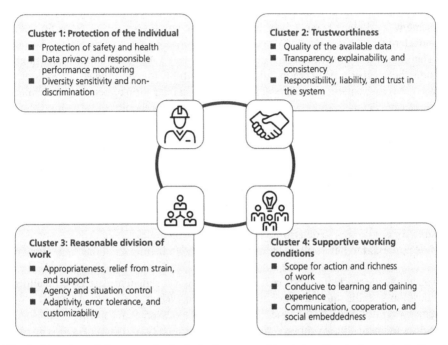

Cluster 1: Protection of the individual
- Protection of safety and health
- Data privacy and responsible performance monitoring
- Diversity sensitivity and non-discrimination

Cluster 2: Trustworthiness
- Quality of the available data
- Transparency, explainability, and consistency
- Responsibility, liability, and trust in the system

Cluster 3: Reasonable division of work
- Appropriateness, relief from strain, and support
- Agency and situation control
- Adaptivity, error tolerance, and customizability

Cluster 4: Supportive working conditions
- Scope for action and richness of work
- Conducive to learning and gaining experience
- Communication, cooperation, and social embeddedness

Fig. 3. Overview of clusters and criteria for human-machine interaction in the context of work with AI [8].

A first cluster of criteria deals with questions of individual protection. This concerns fundamental questions of attack and operational security, robustness, and health protection. At the same time, questions of data protection, responsible performance recording and non-discrimination of AI systems are also relevant in the development. A second cluster of criteria deals with the issue of trustworthiness. The focus here is on data quality, the issues of transparency, explainability and consistency, as well as questions of responsibility, liability, and system trust. The third cluster of criteria is devoted to the relationship between human and machine and the question of a sensible division of work. This involves issues of relief and support by AI systems, situation control and the adaptivity and fault tolerance of AI systems. The last cluster of criteria in turn focuses on working conditions: Important here are, for example, the scope for action for employees, the learning and experience-promoting nature of the work or the appropriate integration of AI-based systems in an existing social and operational context.

In change processes in AI, the questions of transparency and explainability of the systems, the processing of the necessary data – especially personal data – and the determination of load profiles appear to be an essential key factor – for the functionality, reliability, and performance of AI systems on the one hand, and for successful implementation in the companies and gaining acceptance among users on the other. For this reason, these issues should be considered with special care in the change process and in development and planning:

- Strengthen explainability, graded transparency and traceability of AI systems
- Implement privacy-by-design and anticipate and resolve data protection issues
- Create load profiles and design human-machine interaction without contradictions

3.3 Preparation and Implementation

The implementation of AI in companies requires not only a careful and well-thought-out planning and development of the AI system, but especially a suitable integration into existing or new work processes and possibly changed organizational structures. For this purpose, it is necessary, on the one hand, to prepare employees and their representatives for their new tasks at an early stage and to raise the full potential of AI technologies and strengthen their acceptance with suitable qualification measures. Closely linked to this are the changes in the task and activity profiles of employees, which are often associated with the introduction of new technologies. This question is particularly relevant for AI, as the relationship between human and machine can also change in the long term. For many companies, the switch to self-learning systems will not happen overnight. Especially in view of the major changes that self-learning systems can mean for many processes, it is advisable to test the new possibilities in pilot projects, involve employees and gather experience before the systems are introduced across the board.

Companies should create a framework and concrete training, or further education offers for their employees in dealing with AI systems, which the employees can actively and responsibly use. The acceptance of AI technology and the acceptance of the training or further education offers can be further increased by motivating employees to contribute their experience knowledge to the design processes. In addition to technical topics, interdisciplinary skills such as analytical and critical thinking, the ability to judge, creativity, complex problem solving, project and customer management will become increasingly important in the future. Against the background of partly new forms of cooperation in teams and networks, the ability to collaborate, communication skills and conflict management will become even more important in the future. In addition, self-management and self-directed learning, willingness to change and independent thinking and decision-making will gain in importance [9, 10].

AI can – depending on the system design and the area of application – require changes in the organization of work and in the distribution of tasks within the company, but also in the relationship between human and machine. For this reason, these issues should be considered at an early stage and those affected, who usually have valuable expert and process knowledge, should be involved. In this way, frictional losses during the introduction of AI systems can be reduced, possible conflicts and reservations of employees can be better anticipated and moderated and the change process in AI can be made more sustainable. An early risk assessment can also help to correct possible undesirable developments in the early stages. All in all, the aim is not to set up hurdles for AI systems, but to describe sensible instruments for productive and safe use and to define the scope of action for employees.

A good practice to be recommended for conducting a risk assessment also for AI-based work processes is to first.

1. identify affected work areas and activities,
2. identify risks and pressures caused by the AI system in these areas and activities, and identify changed communication, organization, and cooperation processes, including necessary aspects of data security and data protection,
3. to demand transparent, comprehensible information from the manufacturer or provider of the AI system about which data is collected and processed, where the data is stored and who has access to the data,
4. assess the identified pressures to determine whether they cause hazards,
5. define necessary protective measures,
6. take the defined measures into account when procuring or programming the AI system,
7. to investigate the possibilities of effectiveness control by the AI system itself and by responsible persons and to control the effectiveness of the measures accordingly, and finally
8. to document the processes of risk assessment (possibly by the AI system itself) – such as access regulations and times of data access, responsible decisions of the AI system and of persons in the process etc.

The change processes in AI should include pilot and experimental phases in which experience and best practice examples can be gathered. These phases can make it possible to check the effects and interfaces of the AI systems regarding the objectives and with a view to humane work design, as well as to exclude undesirable effects as far as possible and to gain positive experience with the systems in the working environment – for example about the benefits. Within the framework of these pilot and experimental phases, employers and employees can find a suitable balance to reconcile the protection needs of the employees with new flexibility requirements of the company in transformation.

Pilot and experimental phases are of great importance due to the high dynamics of technological development in the field of Artificial Intelligence: trying, experimenting, evaluating, accepting – but also rejecting and learning as early as possible – should also be part of the change process in Artificial Intelligence. These phases of practical testing open – also in the sense of a process-oriented participation of employees and their representatives – the possibility to make necessary changes before a comprehensive introduction and to integrate the experience and process knowledge of the employees even better – for example in the planning of work processes, activity profiles or organizational structures.

Trial phases also allow for a broader information base on how an AI system works and a more thorough assessment of its potential and impact. In addition, it will also make it possible to better assess competence requirements and to implement qualification measures in a more targeted manner. Pilot projects and applied research in real operational environmental can thus help to minimize frictional losses during the introduction in the company, offer an important basis for the change process and process-oriented participation and at the same time contribute to the acceptance of AI systems based on practical experience among employees.

3.4   Evaluation and Adaption

Once self-learning systems have been introduced, a company faces a phase that is almost as important for the success of the transformation process as the planning and introduction (see Fig. 4). Plans that have been made, criteria that have been defined and tasks that have been assigned must now prove themselves in practical everyday life in the most diverse company processes. The use of AI systems and especially the defined criteria and standards for human-machine interaction geared to good work must be evaluated with the systematic involvement of the employees and adapted if necessary – initial conclusions have already been drawn in the experimental phases and have been incorporated into the design of the nationwide introduction.

Evaluation and Adaption	
Review, evaluation and adjustment	• Use pilot phases for evaluation and make necessary adjustments • Consider further development of self-learning systems and accompany it with examinations • Create continuous evaluation processes and establish a feedback culture
Participation in further innovation processes	• Using the experience and ideas of employees and interest groups as a motor for further innovations • Establish an open corporate culture and establish incubation tools • Promote interdisciplinarity and strengthen exchange between company divisions

Fig. 4. Overview of tasks in phase "Evaluation and Adaption"

Over time, companies are gaining more and more experience with the use of AI systems in their products and internal processes. Making the best possible use of this experience means incorporating it into future innovation processes using suitable methods. Here too, the company can and should pay attention to the rich experience of those who are constantly working with the learning systems: The participation of the workforce in further innovation processes thus rounds off the cycle of the transformation process. At the same time, the participation of employees in further innovation processes offers the added value that they can participate in the design of new systems right from the start. If the ideas for an AI system, which is to optimize work processes used within the company, are based on the working reality of the employees, the developed system can ultimately be better integrated into this working reality. In this way, it is possible to avoid the need for costly adjustments.

4   Summary and Practical Recommendations for Action

How the forms of employment and work activities will develop in detail in the working world of the future cannot yet be clearly determined. Without a doubt, the successful implementation of new working environments requires the appropriate corporate culture for a positive use of AI technologies.

AI introduction means: An entrepreneur should always ask himself why an AI process should be used in his company. So, what benefits does the company expect from using an AI process or what added value does the use of one create for customers and employees? Once this question has been answered clearly, it is a matter of introducing the procedures in a structured manner. As with every technological innovation, questions about the corporate strategy, planning of the processes, handling of data and the procurement of the technologies play a role.

The following are proving to be promising elements in change management:

- Experimental rooms and pilot projects, which then radiate as beacons, turn out to be useful to a) test and evaluate new technologies in the field and b) achieve comprehensive willingness to use and accept AI solutions.
- Transparent and legally secure company agreements that include the data protection officer make sense, which clarify the use of the AI and, above all, the handling of employee data in a binding manner.
- A sensitization offensive for an active data protection culture with the employees or the works council and the data protection officer prove to be positive.
- Early qualification offensives to impart the necessary skills to employees and managers are to be carried out.

When introducing the AI process into the company, the following applies, among other things: also, to keep an eye on the topic of corporate culture and leadership. The four phases outlined here are action-bearing and should be gone through extensively by the company.

Acknowledgement. This publication is reusing text portions (and figures) that are written for the (German) white paper "Introduction of AI systems in companies – design approaches for change management", Munich, 2020 [1]. The original version of this publication is available at: https://www.plattform-lernende-systeme.de/publikationen.html. The authors are members of the WG 2 of Plattform Lernende Systeme [6].

References

1. Stowasser, S., Suchy, O., et al.: Introduction of AI systems in companies, Working Group Future of Work and Human-Machine-Interaction, Plattform Lernende Systeme, München (2020) https://www.plattform-lernende-systeme.de/files/Downloads/Publikationen_EN/AG2_Whitepaper_Change_Management_EN.pdf, last accessed 19 January 2023
2. Stowasser, S.: Erfolgreiche Einführung von KI im Unternehmen. In: Knappertsbusch, I., Gondlach, K. (eds.) Arbeitswelt und KI 2030, pp. 145–153. Springer Gabler, Wiesbaden (2021)
3. Terstegen, S., Suchy, O., Stowasser, S., Heindl, A.: In: GfA (ed.) Arbeit HumAIne Gestalten, Beitrag B.9.4. GfA-Press, Dortmund (2021)
4. ifaa, Stowasser, S. (eds.): KI-Zusatzqualifizierung – Für eine produktive und menschengerechte Arbeitsgestaltung. ifaa – Institut für angewandte Arbeitswissenschaft e. V., Duesseldorf (2021). https://www.arbeitswissenschaft.net/fileadmin/Downloads/Angebote_und_Produkte/Broschueren/enAIble_Broschuere.pdf, last accessed 19 January 2023

5. Plattform Lernende Systeme: Homepage, https://www.plattform-lernende-systeme.de/about-the-platform.html, last accessed 19 January 2023
6. Plattform Lernende Systeme: WG 2 Future of Work and Human-Comupter Interaction, Homepage, https://www.plattform-lernende-systeme.de/wg-2.html, last accessed 06 February 2023
7. EU High-Level Expert Group on Artificial Intelligence (ed.): Ethical guidelines for a trustworthy AI. European Commission, Brussels (2018). https://ec.europa.eu/newsroom/dae/document.cfm?doc_id=60419, last accessed 19 January 2023
8. Huchler, N., et al.: Criteria for Human- Machine Interaction When Using AI – Approaches to its humane design in the realm of work. Working Group Future of Work and Human-Machine-Interaction, Plattform Lernende Systeme, München (2020). https://www.plattform-lernende-systeme.de/files/Downloads/Publikationen_EN/AG2%20_WP_MMI_Englisch.pdf, last accessed 19 January 2023
9. Jacobs, J.C., Kagermann, H., Spath, D. (eds.): Promoting Lifelong Learning – Good Practical Examples. A good practice report by the Human Resources Circle of acatech. Lessons Learned, scientific analyses and options for action. acatech, München (2020). https://www.acatech.de/wp-content/uploads/2020/02/acatech-DISKUSSION-Lebenslanges-Lernen-fördern.pdf, last accessed 19 January 2023
10. André, E., Bauer, W., et al. (eds.): Competence development for AI – Changes, needs and options for action. White paper from Plattform Lernende Systeme, München (2021). https://doi.org/10.48669/pls_2021-2, last accessed 06 February 2023

Option Pricing Using Machine Learning with Intraday Data of TAIEX Option

Chou-Wen Wang[1], Chin-Wen Wu[2]([✉]), and Po-Lin Chen[1]

[1] National Sun Yat-Sen University, Kaohsiung, Taiwan
[2] Nanhua University, Chiayi County, Taiwan
cwwu@nhu.edu.tw

Abstract. The use of artificial intelligence (AI) in the financial sector has become increasingly popular in recent years. This study focuses on the application of machine learning (ML) for option pricing using intraday data on Taiwan's Capitalization Weighted Stock Index (TAIEX). This study compares this method with the traditional Black-Scholes option pricing model to determine if the results of ML are more accurate in predicting market prices. The empirical results show that ML can provide more accurate option pricing than the BS model, particularly when training the model with option volatility. Moreover, the pricing ability of the model is positively correlated with the frequency of data used in this study. However, when predicting prices for the next six months, machine learning does not outperform a BS model using lagged prices.

Keywords: Option pricing · Implied volatility · Machine Learning · Intraday data · XGBoost · CatBoost

1 Introduction

As technology advances, artificial intelligence (AI) is becoming more and more popular in our lives. In recent years, there has been a growing trend of applying AI in finance, where it is used for a wide range of purposes, including studying consumer behavior and financial products. AI can help digital finance departments to identify consumer habits or attributes based on their information, and this can also help them achieve more accurate marketing, save costs, and improve the product being purchased. In terms of financial products, AI is mainly used for predicting prices, returns, volatility, or fluctuations, as well as for asset allocation and investment portfolio weighting to maximize returns or Sharpe Ratio.

Option pricing has been a popular topic since the Black-Scholes (BS) model was developed in 1973. Since then, scholars have been refining the model by breaking some of its assumptions. For example, in 1988, Dumas et al. developed the ad hoc BS model, which improved the accuracy of the model by changing the volatility function from a constant function of historical volatility to a more complex function involving the squared terms of maturity and strike price and their intersection. In 1993, Heston introduced the

© Springer Nature Switzerland AG 2023
F. Fui-Hoon Nah and K. Siau (Eds.): HCII 2023, LNCS 14039, pp. 214–224, 2023.
https://doi.org/10.1007/978-3-031-36049-7_17

concept of stochastic volatility (SV) to option volatility forecasting, which led to the development of the famous Heston Model (SV model).

With recent advances in technology and hardware, scholars are now able to use machine learning to price options using large amounts of data to validate their models. For example, Liang et al. (2009). Used Support Vector Regression (SVR) and Multilayer Perception (MLP) to price options, and more recently, the Gradient Boosting method has emerged as a popular method. In 2020, Ivaşcu used Gradient Boosting algorithms such as XGBoost and LightGBM to price options, and achieved better results than the traditional BS model.

This study applies machine learning methods, such as XGBoost and the latest algorithm CatBoost, to option pricing. The goal is to use machine learning to find the relationship between option prices and related variables from a large amount of historical intra-day data, and then compare the results with the traditional BS model to see if those models are more accurate or not.

2 Literature Review

2.1 Option Pricing Model

Hull and White (1987) used BS model determining call option price with stochastic volatility which is uncorrelated with the price of securities. The result shows that the price of BS model overvalues At-The-Money (ATM) options and undervalues deep In-The-Money (ITM) and Out-of-The-Money (OTM) options. Dumas et al. (1998) continued the previous method, that is, the volatility of asset return is a deterministic function of assets and prices and the time. They further developed the Deterministic Volatility Function (DVF) as option pricing model so as to find out the historical volatility in the BS model, which is called ad hoc BS model.

Yung et al. (2003) used GARCH model to compare with the ad hoc BS model. The option pricing model they used is based on the exponential GARCH (E-GARCH) model proposed by Nelson. The result shows E-GARCH model outperforms the ad hoc BS model both in-sample and out-of-sample. Berkowitz (2009a) mentions that previous empirical results indicating ad hoc BS model performs better than BS model. Wang et al. (2012) used ARMA combining with BS model and ad hoc BS model for the prediction of option volatility and option pricing. The empirical result shows that the AR effect is more significant than the MA effect, and the best model is the ad hoc ARMA(1,1), which outperforms the BS model and the ad hoc BS model.

Kim (2021) compares option pricing models from both pricing and hedging perspectives. In this study, the SVJ model performs the best in after-one-day pricing, and the ad hoc BS performs the best when the options priced one week later. From hedging perspective, the ad hoc BS model outperforms BS model, CS (Corrado and Su) model, SV model and SVJ model.

2.2 XGBoost and CatBoost for Financial Applications

Al Daoud (2019) used three gradient boosting algorithms, XGBoost, LightGBM (Light Gradient Boosting Machine) and CatBoost to predict housing credit. Basak et al. (2019) used random forest (RF) and XGBoost algorithms to predict the direction of stocks.

Jha et al. (2020) used Linear Regression, Support Vector Regression (SVR), Decision Tree, Random Forest (RF), XGBoost, CatBoost, Lasso, Voting Regressor to predict housing price. Among them, XGBoost performs the best. Chowdhury et al. (2020) used Call and Put price to predict the stock price (closing price). This study uses Decision Tree, Neural Network, and Ensemble method to predict stock prices and the empirical results showing that all three machine learning methods outperform the BS model, with the Ensemble method performing best.

Jabeur (2021) compared the accuracy of predicting corporate bankruptcy using Cat-Boost with other older machine learning algorithms including discriminant analysis (DA), logistic regression (LR), SVM, NN, RF, GBM, DNN, XGBoost, and Catboost. The empirical results show that CatBoost outperforms all previous models, when applying to classification model.

2.3 Option Pricing Using Machine Learning

Andreou et al. (2009) used å-insensitive SVR and Least Square SVR for option pricing. The result shows that two methods outperform BS model. Abdelmalek et al. (2009) use genetic programming (GP) to predict implied volatility of S&P500 index options and compare between time series samples and moneyness time to maturity classes. This study suggests that the genetic programming is good at solving financial problems.

Liang et al. (2009) first used three conventional parametric methods, the binomial tree method, the finite difference method and the Monte Carlo method for option pricing. They used LNN, MLP and SVR model to raise the accuracy of these results. The results show that the MLP approach is slightly better than the LNN approach, and the SVRs outperforms the MLPs. Culkin and Das (2017) use fully-connected feed-forward deep learning neural network (Deep FNN) to make a better option pricing model than BS model. The results show that Deep Learning can be used to learn option pricing models from the market. Han et al. (2019) used machine learning models such as K-Nearest Neighbor (KNN), SVM, RF, Gradient Boosting (GB), extreme random tree (ET), and deep learning with selective learning. The results show that Gradient Boosting is the best model and performs better than ATM both in Out-of-The-Money (OTM) and In-The-Money (ITM). Liu et al. (2019) used Artificial Neural Network (ANN) to determine options and calculating Implied Volatility. The results show that the ANN solver can accelerate the calculation of numerical method significantly. Ivaşcu (2021) use SVR, RF, XGBoost, LightGBM, NN, Genetic algorithms (GA) machine learning algorithms to determine the price of an option and compared these models to traditional BS models. The best performance among all algorithms are XGBoost and LightGBM.

3 Methodology

3.1 XGBoost

Chen and Guestrin propose the eXtreme Gradient Boosting (XGBoost) algorithm in 2016. XGBoost is a supervised machine learning model that can be used for classification or regression prediction. This method is based on Gradient Boosting Decision Tree

(GBDT) algorithm of Friedman (2001). The main idea of XGBoost is ensemble learning, which combines many previous models to form a better model, so that the model is more stable. Compared with GBDT, XGBoost expands a drift function through a Taylor second-order expansion makes it outperform than most of other machine learning algorithms in many predicts.

3.2 CatBoost

CatBoost (Categorical boosting), proposed by Prokhorenkova et al. (2018) and Dorogush et al. (2018), is an algorithm based on the oblivious trees in GBDT, which is good at solving categorical features. It also solves the problems of Gradient Bias and Prediction shift, and reducing the overfitting problem to improve the accuracy of the algorithm. This makes CatBoost performing well in recent years.

In gradient boosting algorithm, one commonly used method is to group according to Target Statistics (TS). TS is used to calculate the expected value of the target feature for each category, and some people even directly use TS as a new numerical feature to replace the original category feature. Threshold value is important for TS value based on the Gini Coefficient or Mean Square Error(MSE). The purpose is to obtain a number which is the best among all possible divisions of the class feature into two slices for the training data.

Any combination of several different categorical features can be considered as a new feature. Therefore, if there is a data set has multiple class features, then the total number of these features will grow exponentially. Hence, it isn't possible to get all combinations in the algorithm. When constructing new split points for the current tree, CatBoost uses a greedy strategy to consider combinations. For the first partition of the tree, no combination is considered. For the next split, CatBoost combines all combinations of the current tree, categorical features and all categorical features in the dataset. The newly obtained categorical features dynamically convert to be numerical features.

3.3 Model Comparison

In this study, the Root Mean Square Error (RMSE) is used as the basis for model comparison, which is the same as Ivaşcu (2021). The smaller the RMSE of the test data, the more accurate the model can be predicted. The RMSE of the BS model is 1.3598, The three stages of the model comparison are shown below.

Different labels (y), features (x), algorithms, data splitting methods, and data frequency are presented in three stages for model comparisons. The first stage (Fig. 1) is grouped into predicting option prices directly or predicting option volatility. Then, the four sets of features in Table 1 are used to predict the above two sets of labels. Then, the four sets of features in Table 1 are used to predict the above two sets of forecasting targets. The 1st set is traditional BS variables. The 2nd set is 1st set adding the lagged option price. The 3rd set is the 1st set without considering Implied Volatility, which is also used in Ivaşcu (2021). The fourth set is the variables of ad-hoc BS model. The algorithm uses the XGBoost and CatBoost methods to compare with the traditional BS model. The data splitting method is random cutting, and the data frequency is one second. In the second stage, (Fig. 2), 10 s, 20 s, 30 s, and 60 s are used as data frequency to compare with one second.

4 Empirical Results

4.1 Data

This study uses the futures contracts (TX) and option contracts (TXO) of the Taiwan Stock Exchange Capitalization Weighted Stock Index (TAIEX) as research data. TXO have the following characteristics: (1) TXO are European options; (2) TX and TXO are both cash settled; (3) TX and TXO have the same expiration date, which is the third Wednesday of each month.

The data for this study was collected from July 2017 to December 2021 from the Taiwan Futures Exchange (TAIFEX). The data includes price information for each transaction with a minimum time unit of seconds. To align the futures and options data, only the last transaction data per second is used in this study.

The raw data include monthly and weekly options which are new commodities that only traded at every Wednesday except the third week of the month and expire on the following Wednesday with a one-week trading period. Weekly options have greater trading volumes and liquidity than monthly options which have longer expiration date. However, the volume of weekly futures is quite low due to lack of futures data satisfying the no-arbitrage boundary condition. Therefore, we cannot use the weekly option data but can only use the monthly option data as a sample for this study.

After cleaning and filtering the data, only 758,138 items remain. Table 2 shows the descriptive statistics about important variables in this study. The option price (Opt) is between 1 and 2100, mostly within 100, with a median of 59 and a mean of 73.22, which means that there are many extreme values that make the mean higher. The futures price (Fut) is between 8345 and 18267, and the strike price (K) is between 8700 and 18300. The value of futures are divided by performance price (Fut/K) is between 0.8407 and 1.221, which represents the moneyness of options. The first quartile and third quartile are 0.9966 and 1.0028 respectively, which means that most of the data are around 1. This means that most of the data used in this study are At-The-Money (ATM) options data. The Implied Volatility (IV) is between 0.0123 and 0.8144. From the third quartile of 0.1555, most of the values are less than 0.2. The number of time to maturity has been removed from the data within 1 day during data cleaning.

The remaining data ranges from 1 to 160.104 days, most of which are within one week. There are two main reasons for this. One is that most of the volume over one week will be in the weekly options, and the second is that the longer the time to expiry, the more likely the data will have price bias causing it to fail to satisfy the no-arbitrage boundary condition of the put-call parity. Finally, this study uses the Bank of Taiwan one-year time deposit rate as the risk-free rate, which has a value between 0.0076 and 0.0104 with no significant difference during the study period.

Table 1. Features Sets Table

	Features(x)		Features(x)
The first set	Futures Price(F),	The second set	Futures Price(F),
	Strike Price(K),		Strike Price(K),
	Time to Maturity(T),		Time to Maturity(T),
	Risk-free Interest Rate(rf),		Risk-free Interest Rate(rf),
	Lag Implied Volatility(lag_iv)		Lag Implied Volatility(lag_iv),
			Lag Options price(lag_C)
The third set	Futures Price(F),	The fourth set	Futures Price(F),
	Strike Price(K),		Strike Price(K),
	Time to Maturity(T),		Squared Strike Price,
	Risk-free Interest Rate(rf)		Time to Maturity(T),
			Squared Time to Maturity,
			Interaction of Strike Price and Time to Maturity(K*T),
			Risk-free Interest Rate(rf),
			Lag Implied Volatility(lag_iv)

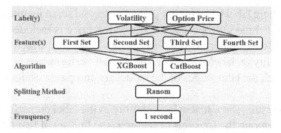

Fig. 1. Model Comparison with the Four Sets of Features

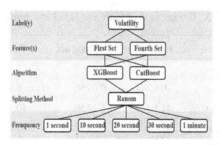

Fig. 2. Model Comparison with Different Frequencies

Table 2. Descriptive Statistics of the Important Variables

Stat	Opt	Fut	K	F/K	IV	Days	r
Min.	1	8345	8700	0.8407	0.0123	1	0.0076
1st Qu.	36.5	10512	10500	0.9966	0.0954	1.14	0.0076
Median	59	10867	10900	0.9996	0.1223	2.134	0.0104
Mean	73.22	11938	11943	0.9996	0.1329	3.947	0.0095
3rd Qu.	93	12177	12200	1.0028	0.1555	5.15	0.0104
Max.	2100	18267	18300	1.221	0.8144	160.104	0.0104

4.2 Comparison of Different Label and Features

First, as shown in Table 3, the performance of the label is volatility, regardless of which set of features is much better than directly using the option price as the forecast target. In the model with label being the volatility, the fourth group x performs the best. In other words, the performance of the model with ad hoc BS variables will be better than that of the BS model. The inclusion of the lag option prices in the second set slightly reduces the accuracy of the model. The third set has substantially lower, predictive accuracy due to the absence of Implied Volatility as an important variable of the option, same as the results of Ivaşcu (2021). In the algorithm, XGBoost is better than CatBoost in all four sets.

In the model where label is the option price, the second set of features performs much better than the other three sets because the lag option price is included as a variable. In the algorithm, CatBoost outperforms XGBoost in all four sets, which is the opposite of the result for volatility as label. However, the overall performance is much worse than the volatility as label, the later stages of the model are compared using only the volatility as label.

The empirical results show that using volatility as the label outperforms using option price as the label. Among the four sets of features, the first set of features and the fourth set of features performed better, with the fourth set of features performing the best.

Table 3. Comparison of Different Label and Features

label	Algorithm	First set	Second set	Third set	Forth set
Volatility	XGBoost	0.9034	0.9067	7.5329	0.8793
	CatBoost	0.9241	0.9250	8.8944	0.9026
Option price	XGBoost	41.6727	3.3450	40.8657	41.6776
	CatBoost	37.0605	2.4282	39.6685	37.0251

Table 4 shows that the results of the first set of features cut by rolling scheme are much worse than the results of the previous section randomly cut by 8:2. In Model 2, the testing data is the first half of 2020. Because of the plunge of Taiwan stocks in the Pandemic, the option prices changed dramatically and it wasn't possible to predict the prices accurately. In Model 4, as the Pandemic entered Alert Level 3 in Taiwan, the TAIEX was highly fluctuating and dropping 8% to 9% on one day, so was the forecast.

Table 4. First Set of Features Cut by Rolling Scheme

Model	Training/Testing Data	Start Data	End Date	BS	XGBoost	CatBoost
1	Training Data	20170630	20190630	0.9871	0.6116	0.6889
	Testing Data	20190701	20191231	0.8808	3.2518	2.3893
2	Training Data	20180101	20191231	1.0161	0.6406	0.7201
	Testing Data	20200101	20200630	1.7264	27.0307	26.5047
3	Training Data	20180701	20200630	1.2360	0.7322	0.8003
	Testing Data	20200701	20201231	1.4193	7.1065	6.5739
4	Training Data	20190101	20201231	1.3139	0.7984	0.8572
	Testing Data	20210101	20210630	2.6582	11.7189	18.4433
5	Training Data	20190701	20210630	1.6761	0.8629	0.9418
	Testing Data	20210701	20211231	1.8628	6.2971	3.2969

In Table 5, the fourth set of features is compared with the first set of features by using the rolling scheme. For model 1, the fourth set of features outperforms the first set of features in both XGBoost and CatBoost, while for models 2 to 5, the first set of features outperforms the fourth set of features. Comparing the two algorithms, CatBoost outperforms XGBoost in most cases.

Table 5. Fourth Set of Features Cut by Rolling Scheme

Model	Training/Testing Data	Start Data	End Date	BS	XGBoost	CatBoost
1	Training Data	20170630	20190630	0.8269	0.6219	0.7387
	Testing Data	20190701	20191231	0.7108	2.5511	1.8214
2	Training Data	20180101	20191231	0.8522	0.6614	0.7843
	Testing Data	20200101	20200630	1.3504	33.1735	30.7476
3	Training Data	20180701	20200630	0.9758	0.7865	0.9186
	Testing Data	20200701	20201231	1.1257	13.5309	9.5118
4	Training Data	20190101	20201231	1.0398	0.8164	0.9324
	Testing Data	20210101	20210630	1.9452	23.3439	19.5805
5	Training Data	20190701	20210630	1.3080	0.9158	1.0498
	Testing Data	20210701	20211231	1.4269	23.6758	6.3183

In this study, the data frequency of all the previous models are 1 s. In the following, we use volatility as label and the first set and the fourth set of features for prediction with frequencies of 10 s, 20 s, 30 s and 60 s, respectively. As shown in Table 6, the lower the frequency, the less data can be used. The data for 1 s, 10 s, 20 s, 30 s, and 60 s are 758,138, 287,420, 252,318, 233,947, and 200,934 respectively. Looking at different time frequency, the lower the frequency, the longer it takes to predict the option price, the larger the error will be. Comparing two different sets of features, the fourth set of features predicted better than the first set, except for XGBoost with a frequency of 10 s and CatBoost with a frequency of 60 s. Simply by comparing the two algorithms, XGBoost is slightly better than CatBoost. Probably because the features in this study are all numerical variables rather than categorical variables, the results of CatBoost are not so outstanding. The results obtained using both algorithms are better than the BS model.

Table 6. Comparison of different data frequencies

Frequency No. of data		1 s 758,138	10 s 287,420	20 s 252,318	30 s 233,947	1 min 200,934
First Set	XGBoost	0.9034	0.9778	1.0403	1.1021	1.2345
	CatBoost	0.9241	0.9619	1.0561	1.1136	1.2464
	BS	1.3598	1.9577	2.6224	3.2729	4.4662
Fourth Set	XGBoost	0.8793	0.9762	1.0262	1.0939	1.2268
	CatBoost	0.9026	0.9660	1.0509	1.1081	1.2491
	BS	1.3598	1.9577	2.6224	3.2729	4.4662

Based on the empirical results, the random split method has better model performance than the rolling scheme. Additionally, higher data frequency leads to more accurate predictions. Comparing the models, the best performance is achieved when using volatility as the label, the fourth set of features as the predictor, XGBoost as the algorithm, random split as the data splitting method, and using the highest data frequency. The obtained RMSE of 0.8793 is significantly lower than the BS model's RMSE of 1.3598.

5 Discussion and Conclusion

This study used TAIEX underlying futures and options data from July 2017 to December 2021 for morning and evening trading as samples. Machine learning algorithms were used to predict option prices, and the models were compared based on their labels, features, algorithms, data splitting methods, and data frequencies.

The result indicates that volatility as a label performed better than option price. The first set of features (features of BS model) and the fourth set (features of ad hoc BS model) show better performance among the four sets of features. Data splitting through random split provided the best performance in comparison to the rolling scheme. XGBoost and CatBoost outperform the traditional BS model, with XGBoost performing slightly better than CatBoost. Training is conducted at different data frequencies (1 s, 10 s, 20 s, 30 s, and 60 s), and the results revealed that higher data frequency lead to more accurate predictions.

Future research will contain other variables that influence option prices, such as technical and macroeconomic indicators, and explore alternative algorithms like Long Short Term Memory (LSTM) and so on.

References

Abdelmalek, W., Hamida, S.B., Abid, F.: Selecting the best forecasting-implied volatility model using genetic programming. Advances in Decision Sciences (2009)

Al Daoud, E.: Comparison between XGBoost, LightGBM and CatBoost using a home credit dataset. Int. J. Comp. Info. Eng. **13**(1), 6–10 (2019)

Andreou, P.C., Charalambous, C., Martzoukos, S.H.: European option pricing by using the support vector regression approach. In: International Conference on Artificial Neural Networks, pp. 874–883. Springer, Berlin, Heidelberg (2009, September)

Basak, S., Kar, S., Saha, S., Khaidem, L., Dey, S.R.: Predicting the direction of stock market prices using tree-based classifiers. The North American Journal of Economics and Finance **47**, 552–567 (2019)

Berkowitz, J.: On justifications for the ad hoc Black-Scholes method of option pricing. Studies in Nonlinear Dynamics & Econometrics **14**(1) (2009a)

Black, F., Scholes, M.: The pricing of options and corporate liabilities. J. Polit. Econ. **81**(3), 637–654 (1973)

Chen, T., Guestrin, C.: XGBoost: a scalable tree boosting system. Paper presented at the Proceedings of the 22nd acm sigkdd international conference on knowledge discovery and data mining (2016)

Chowdhury, R., Mahdy, M.R.C., Alam, T.N., Al Quaderi, G.D., Rahman, M.A.: Predicting the stock price of frontier markets using machine learning and modified black-scholes option pricing model. Physica A **555**, 124444 (2020)

Culkin, R., Das, S.R.: Machine learning in finance: the case of deep learning for option pricing. Journal of Investment Management **15**(4), 92–100 (2017)

Dumas, B.J., Fleming, J., Whaley, R.E.: Implied volitility functions: empirical tests. Journal of finance **53**, 2059–2106 (1998)

Han, H., Huang, H., Hu, J., Kuang, F.: Implied volatility pricing with selective learning. In: International conference on Data Science, Medicine and Bioinformatics, pp. 18–34. Springer, Singapore (2019, June)

Heston, S.L.: A closed-form solution for options with stochastic volatility with applications to bond and currency options. The review of financial studies **6**(2), 327–343 (1993)

Hull, J., White, A.: The pricing of options on assets with stochastic volatilities. J. Financ. **42**(2), 281–300 (1987)

Ivaşcu, C.F.: Option pricing using machine learning. Expert Syst. Appl. **163**, 113799 (2021)

Jabeur, S.B., Gharib, C., Mefteh-Wali, S., Arfi, W.B.: CatBoost model and artificial intelligence techniques for corporate failure prediction. Technol. Forecast. Soc. Chang. **166**, 120658 (2021)

Berkowitz, J.: On Justifications for the ad hoc Black-Scholes Method of Option Pricing. Department of Finance University of Houston (2009b)

Jha, S.B., Babiceanu, R.F., Pandey, V., Jha, R.K.: Housing Market Prediction Problem using Different Machine Learning Algorithms: A Case Study. arXiv preprint arXiv:2006.10092 (2020)

Kim, S.: Portfolio of volatility smiles versus volatility surface: implications for pricing and hedging options. J. Futur. Mark. **41**(7), 1154–1176 (2021)

Liang, X., Zhang, H., Xiao, J., Chen, Y.: Improving option price forecasts with neural networks and support vector regressions. Neurocomputing **72**(13–15), 3055–3065 (2009)

Prokhorenkova, L., Gusev, G., Vorobev, A., Dorogush, A.V., Gulin, A.: CatBoost: unbiased boosting with categorical features. Advances in Neural Information Processing Systems 31 (NeurIPS 2018) (2018)

Liu, S., Oosterlee, C.W., Bohte, S.M.: Pricing options and computing implied volatilities using neural networks. Risks **7**(1), 16 (2019)

Merton, R.C.: Theory of rational option pricing. The Bell Journal of economics and management science, 141–183 (1973)

Wang, C.W., Wu, C.W., Tzang, S.W.: Implementing option pricing models when asset returns follow an autoregressive moving average process. Int. Rev. Econ. Financ. **24**(2012), 8–25 (2012)

Yung, H.H., Zhang, H.: An empirical investigation of the GARCH option pricing model: Hedging performance. Journal of Futures Markets: Futures, Options, and Other Derivative Products **23**(12), 1191–1207 (2003)

Portfolio Performance Evaluation with Leptokurtic Asset Returns

Chin-Wen Wu[1], Chou-Wen Wang[2]([✉]), and Yang-Cheng Chen[3]

[1] Nanhua University, Chiayi County, Taiwan
[2] National Sun Yat-Sen University, Kaohsiung, Taiwan
chouwenwang@gmail.com
[3] National Kaohsiung First University of Science and Technology, Kaohsiung, Taiwan

Abstract. When asset returns follow multivariate Lévy processes, this paper derives a theoretically sound portfolio performance measure (PPM) that takes into account idiosyncratic and common jump risks. We demonstrate that the PPM can reduce to the Generalized Sharpe Ratio introduced by Zakamouline and Koekebakker (2009), resolving the Sharpe ratio paradox presented in Hodges (1998). With the data of iShares MSCI Germany Index fund, SPDR USA S&P 500 and the iShares MSCI Canada Index Fund over the period from January 1, 2001 to September 30, 2010, we attain that the optimal asset allocation obtained by maximizing the PPM can catch more detailed information of financial shock so that fund managers are able to adjust optimal investment strategy to enhance the investment performance during the period of the financial extreme risk.

Keywords: Portfolio Performance Measure · Multivariate Lévy process · Idiosyncratic Jump Risk · Common Jump Risk

1 Introduction

Portfolio selection proposed by Markowitz (1952) is one of the most important issues in finance. Relying upon the assumption that asset returns follow a normal distribution, Markowitz formulate the portfolio problem as a choice of the mean and variance of a portfolio of assets. However, in practice the asset returns usually peak around the mean and fatter tails; that is, the returns are non-normally distributed. The fast-growing concerns of investors for extreme risks lead to eager need for portfolio performance evaluation beyond the mean-variance framework.

Non-normal property can be generated by heavy-tailed distributions with higher-order moments. Concerning for higher-order moments can be traced back to Kendall and Hill (1953), Mandelbrot (1963a and 1963b), Cootner (1964) and Fama (1965) discovering the presence of significant skewness and excess kurtosis in empirical asset return distributions. In this sense, there has been substantial growth in recent years of the use of copula in finance; see for example the recent book by Cherubini et al. (2004). Further empirical evidence has been provided by Patton (2004), Jondeau and Rockinger (2006) and Harvey et al. (2004). Such models, however, can leave two issues unresolved

© Springer Nature Switzerland AG 2023
F. Fui-Hoon Nah and K. Siau (Eds.): HCII 2023, LNCS 14039, pp. 225–241, 2023.
https://doi.org/10.1007/978-3-031-36049-7_18

(Adcock, 2010). First, certain choices of the functional form of the utility function are prohibited because the expected values do not exist. Second, the use of copula is restricted to a small number of dimensions, which means that large-scale portfolio selection require simplifying assumptions to be made.

The past three decades has noticed the development of a large body of theory concerning multivariate probability distributions. Most studies about multivariate asset models are based on Brownian motions due to their simple structure. Jonathan and Ingersoll (1986) provide a theoretical basis of multinormal case for the financial using. To allow for fat-tailed phenomenon, elliptically symmetric distributions are also used as a tractable model for asset returns and hence for the development of various aspects of portfolio theory (Fang et al., 1990; Liu, 1994; Landsman, 2006; Landsman and Nešlehová, 2008). Self evidently, elliptically symmetric distributions cannot handle asset returns with nonzero skewness. One way to incorporates both kurtosis and skewness for the probability distribution of assets returns is based on the multivariate extended skew-Student-t distribution (Adcock, 2010). Another way is to model the asset returns as multivariate Lévy processes.

Barndorff-Nielsen (2001), Cont and Tankov (2004), Luciano and Schoutens (2006) and Eberlein and Madan (2009) provide a multivariate time changed Brownian motion by a common subordinator. As noted in Luciano and Schoutens (2007), however, the common subordinator puts a strict restriction on the joint process, which in turn leads to the lack of independence and limited range of dependence structure. Semeraro (2008) and Luciano and Semeraro (2007) propose a similar model with idiosyncratic and systematic subordinators to capture idiosyncratic and systematic jump shocks simultaneously. In this line, Luciano and Semeraro (2010a) consider correlated Brownian motions with idiosyncratic and systematic subordinators which possess four desired features: the existence of characteristic functions in closed form, the ability to capture a wide range of dependence structure, the ability to model the idiosyncratic and common jump shocks, and the advantage of calibrating marginal and joint parameters separately. Consequently, this paper employs the multivariate Lévy processes of Luciano and Semeraro (2010b) as a rational and tractable model for asset returns.

The most popular performance measure is the Sharpe ratio (SR), a reward-to-risk ratio commonly used measure of portfolio performance. However, because it is based on the mean-variance theory, it can lead to unacceptable paradoxes (Hodges, 1998) when the asset returns are non-normally distributed. To take into account higher moments of distribution, some researchers replace the standard deviation in the Sharpe ratio by an alternative risk measure on the basis of the downside deviation (Sortino and Price, 1994) and the value at risk (Dowd, 2000; Favre and Galeano, 2002; Rachev et al., 2007). Other researchers replace the Sharpe ratio with alternative measures of reward and risk such as Stutzer index Stutzer (2000), the Omega ratio (Shadwick and Keating, 2002), and the Kappa measure (Kaplan and Knowles, 2004). As proposed by Zakamouline and Koekebakker (2009), however, most of the alternative performance measures lack a solid theoretical foundation, taking into account only downside risk but ignoring the upside return potential. In addition, asset returns presents a significant common jump component (Lo and Wang, 2000; Luciano and Semeraro, 2010a). Consequently, the aim of this paper is to present a theoretically sound portfolio performance measure under which the asset returns adhere to the multivariate Lévy processes to takes into account higher moments of the distribution of returns as well as the idiosyncratic and common jump shocks.

In this paper we start by introducing a definition of a portfolio performance measure (PPM) related to expected utility provided by portfolio values. The higher the PPI of a portfolio, the higher level of expected utility the portfolio provides. We indicate that when the number of asset is one and initial wealth is zero, the PPM justifies the notion of the Generalized Sharpe ratio (GSR) of Hodges (1998) and Zakamouline and Koekebakker (2009). In addition, when asset returns follows a multivariate normal distribution, the PPM reduces to the portfolio Sharpe ratio and the Sharpe ratio when the number of asset is one. Lastly, we show how the PPM can mitigate the shortcomings of the Sharpe ratio in resolving Sharpe ratio paradox presented by Hodges (1998).

In empirical study, we employ the iShares MSCI Germany Index Fund (EWG), SPDR USA S&P 500 (SPY) and the iShares MSCI Canada Index Fund (EWC) as the proxy of risky assets and the three-month USA Treasury bill (T-Bill) as the proxy of risk-free asset. The data spans the period from January 1, 2001 to December 31, 2010 (2450 observations in total). We first fit the three asset returns to the multivariate Lévy model through the maximum likelihood estimation (MLE) according to a rolling window out-of-sample scheme by fixing the window size at five years. According to the empirical results, we find that the optimal asset allocation obtained by maximizing the PPM can catch more detailed information of financial shock so that fund managers are able to adjust optimal investment strategies to enhance investment performance with lower risk.

The remainder of this paper is organized as follows: In Sect. 2, we illustrate the portfolio performance measure based on the expected constant absolute risk aversion (CARA) utility function. Section 3 explains the construction of multivariate Lévy process for asset returns and its application to the PPM; it also offers the relationship between the PPM and GSR. Section 4 empirically tests the goodness of fit of multivariate Lévy model and illustrates how the PPM can resolve the Sharpe ratio paradox presented by Hodges (1998) and then provides the application of PPM in optimal asset allocation. The last section draws some conclusions about our findings.

2 Portfolio Performance Measure

In this paper, all of the results are obtained by considering the optimal asset allocation problem in the standard expected utility theory framework. Specifically, we consider an investor allocating his or her wealth between $n - 1$ risky assets and the nth risk-free asset. The risky asset returns over a small time interval are defined as follows:

$$R_j(t) = \log(A_j(t)) - \log(A_j(t - 1)), j = 1, ..., n - 1, \qquad (1)$$

where $A_j(t)$ is the jth asset price at time t. The return on the risk-free asset over the same time interval equals r_f. We further assume that the investor has an initial wealth of w and invests a_j in the jth asset. Thus, the investor's wealth $\tilde{W}$ from time t - 1 to time t is of the form:

$$\tilde{W} = a_n(1 + r_f) + \left[\sum_{j=1}^{n-1} a_j(1 + R_j(t))\right] = w + a_n r_f + \sum_{j=1}^{n-1} a_j R_j(t) \qquad (2)$$

where $a_n = w - \sum_{j=1}^{n-1} a_j$ is the investment amount in the risk-free asset. In this study, we attempt to find the optimal investment amount a_j^*. Note that when the initial wealth value is equal to one, the investment amount for each asset becomes its investment weight. Given the portfolio value in Eq. (2), in the Definition 1, the investor's expected utility is

$$E\left(U\left(\tilde{W}\right)\right) = E\left(U\left(w + a_n r_f + \sum_{j=1}^{n-1} a_j R_j(t)\right)\right), \tag{3}$$

for given utility function $U(\cdot)$. We suppose that $U(\cdot)$ is increasing, concave, and everywhere differentiable function. Consequently, the investor's objective is to choose a_j to maximize the expected utility.

One of the traditional utility functions most often used in financial economics is the CARA utility. Cass and Stiglitz (1970) show that the portfolio weight in risky asset i relative to risky asset j does not depend on the initial wealth if all investors have the same CARA utility, which makes this utility suitable for financial applications. In addition, Zakamouline and Koekebakker (2009) demonstrate that the use of CARA utility in the computation of the performance measure can alleviate the weakness of the Sharpe ratio for non-Gaussian asset returns. We therefore employ the CARA utility in this study.

The CARA utility form is given by

$$U(\tilde{W}) = -\frac{1}{\rho} \exp\left(-\rho \tilde{W}\right) \tag{4}$$

where ρ is the measure of relative risk aversion. Incorporating Eq. (4) into Eq. (3) yields

$$E\left(U(\tilde{W})\right) = E\left(-\frac{1}{\rho} exp\left(-\rho\left(w + a_n r_f + \sum_{j=1}^{n-1} a_j R_j(t)\right)\right)\right)$$
$$= -\frac{1}{\rho} exp\left(-\rho a_n r_f - \rho w\right) \Phi_R(i\rho a), \tag{5}$$

where $a = [a_1, ..., a_{n-1}]'$ is $(n-1)$-by-1 vector for the investment amounts and $\Phi(\cdot)$ denotes the characteristic function (CF) of the risky assets and satisfies

$$\Phi_R(u) = E\left(exp\left(i \sum_{j=1}^{n-1} u_j R_j(t)\right)\right), \tag{6}$$

where $u = [u_1, ..., u_{n-1}]'$.

Zakamouline and Koekebakker (2009) show that for any distribution of the risky asset returns and the investor with zero wealth, the relation between GSR and maximum investor's expected utility is given by

$$GSR = \sqrt{-2log\left(-E\left(U^*\left(\tilde{W}\right)\right)\right)}, \tag{7}$$

where $E\left(U^*\left(\widetilde{\mathcal{W}}\right)\right)$ is the investor's maximum expected utility. This performance measure is strictly positively related to the maximum expected utility, which is the cornerstone of modern finance. In this line, without loss of generality, in this paper we define the portfolio performance measure (PPM) as follows:

$$PPM(a) = \sqrt{-2\log\left(-E\left(U(\tilde{\mathcal{W}})\right)\right)} =$$

$$\sqrt{2\left(\log\rho + \rho(a_n r_f + w) + \psi_R(i\rho a)\right)},$$

(8)

where $\psi_R(u) = \log(\phi_R(u))$ is the characteristic exponent. Apparently, a higher level of PPM of a portfolio leads to a higher level of expected utility the portfolio.

3 Multivariate Lévy Processes for Asset Returns

In this section, we would introduce the multivariate model for our assets return and portfolio return process. Following the work of Luciano and Semeraro (2010a), we construct the assets return using the multivariate subordination of Brownian motions, with a common and an idiosyncratic component. Intending of directly fitting the asset returns to multivariate Lévy processes (Luciano and Semeraro, 2010b), we standardize the multivariate Lévy processes to model the residuals of risky asset returns. In this paper, the residuals follow two type of multivariate Lévy processes, multivariate Variance gamma ($\alpha - VG$) and multivariate Normal inverse Gaussian ($\gamma - NIG$), which build on the idea of splitting the multivariate subordinator into idiosyncratic and systematic components.

The Characteristic Functions of Leptokurtic Asset Returns.
We model the risky asset returns as multivariate Lévy processes, that is

$$R_j(t) = c_j + \sigma_j \varepsilon_j(t) = c_j + \sigma_j\left(\varepsilon_j^I(t) + \varepsilon_j^C(t)\right), j = 1, ..., n - 1,$$

(9)

where c_j and σ_j are mean and volatility of the asset return $R_j(t)$; $\varepsilon_j(t)$ is the standardized residual including two parts, the idiosyncratic component $\varepsilon_j^I(t)$ and common component $\varepsilon_j^C(t)$ and satisfies $E\left(\varepsilon_j(t)\right) = 0$ and $V\left(\varepsilon_j(t)\right) = 1$. We generate the standardized stochastic processes by time changed technique through the different Brownian motion. More precisely, the idiosyncratic part is modeled through the time changed independent Brownian motions denoted by $W_j^I(\cdot)$, and the common part is modeled through the time changed dependent Brownian motions dependent denoted by $W_j^C(\cdot)$. Or equivalently, we have

$$\varepsilon_j^I(t) = W_j^I(X_j) = \theta_j X_j + k_j B_j^I(X_j), j = 1, ..., n - 1,$$

(10)

$$\varepsilon_j^C(t) = W_j^C(\alpha_j Z) = \theta_j(\alpha_j Z) + k_j B_j^C(\alpha_j Z), j = 1, ..., n - 1,$$

(11)

where θ_j is the drift term; k_j is the diffusion term; X_j, independent of W_j^I, is the idiosyncratic subordinator (or non-systematic part); Z, independent of W_j^C, is the common subordinator (or systematic part); α_j is the sensitivity parameter to the common component;

and the Brownian motions are given by

$$W_j^l(t) = \theta_j t + k_j B_j^l(t),\ l = I\, orC, j = 1, \ldots, n-1, \tag{12}$$

where $B_j^I(t), j = 1, \ldots, n-1$, are independent Brownian motion and $B_j^C(t), j = 1, \ldots, n-1$, are dependent Brownian motion and satisfies $E\left(B_{\uparrow}^C(t)B_j^C(t)\right) = \rho_{\uparrow,j}t$. Following the similar derivation of Luciano and Semeraro (2010a), the characteristic function for the jth standardized stochastic process $\varepsilon_j(t)$ is of the form:

$$\Phi_{\varepsilon_j}(u_j) = \exp(iu_j\mu_j)L_{X_j}\left(\Psi_{W_j^I}(u_j)\right)L_Z\left(\alpha_j\Psi_{W_j^C}(u_j)\right), \tag{13}$$

where

$$\Psi_{W_j^l}(u_j) = \ln\left(E\left[\exp\left(u_jW_j^l(1)\right)\right]\right) = iu_j\theta_j - \frac{1}{2}k_j^2u_j^2, l = I, orC, \tag{14}$$

is the characteristic exponent of $W_j^l(1)$ for $l = I$ or C; and

$$L_{X_j}(u_j) = E\left[\exp\left(u_jX_j\right)\right], L_Z(u_j) = E\left[\exp\left(u_jZ\right)\right], \tag{15}$$

are the moment generating functions (MGF) for X_j and Z, respectively. According to Eq. (13), we can obtain the CF of the jth risky asset $R_j(t)$, called $\Phi_{R_j(t)}(u_j)$, as follows:

$$\begin{aligned}
\Phi_{R_j}(u_j) &= E\left(\exp\left(iu_jR_j(t)\right)\right) = \exp\left(iu_jc_j\right)\Phi_{\varepsilon_j}\left(\sigma_ju_j\right) \\
&= \exp\left(iu_j(c_j + \mu_j)\right)L_{X_j}\left(\Psi_{W_j^I}(\sigma_ju_j)\right)L_Z\left(\alpha_j\Psi_{W_j^C}(\sigma_ju_j)\right).
\end{aligned} \tag{16}$$

Because the PPM is mainly determined by the CF of the risky assets, we derive the CF of the multivariate asset returns, together with the correlation coefficient between the jth asset return and $\uparrow$ th asset return in Theorem 1.

Theorem 1. The characteristic function for the multivariate asset returns is of the form:

$$\begin{aligned}
\Phi_R(u) &= \exp\left(\sum_{j=1}^{n-1}(iu_j(c_j + \mu_j))\right)\left\{\prod_{j=1}^{n-1}\left[L_{X_j}\left(iu_j\theta_j - \frac{1}{2}\left(k_j^2u_j^2\right)\right)\right]\right\} \\
&\quad L_Z\left(i\hat{u}'\theta - \frac{1}{2}\left(\hat{u}'\Sigma\hat{u}\right)\right),
\end{aligned} \tag{17}$$

where

$$\hat{u} = \begin{bmatrix} \sigma_1 u_1 \alpha_1^{1/2} \\ \sigma_2 u_2 \alpha_2^{1/2} \\ \vdots \\ \sigma_{n-1} u_{n-1} \alpha_{n-1}^{1/2} \end{bmatrix}, \theta = \begin{bmatrix} \theta_1 \\ \theta_2 \\ \vdots \\ \theta_{n-1} \end{bmatrix}, \Sigma = \left[\rho_{\ell,j}k_\ell k_j\right]_{n-1 \times n-1}, \ell, j = 1, 2, \ldots, n-1. \tag{18}$$

The dependence structure is given by

$$Corr(R_j(t), R_\ell(t)) = \frac{\sigma_j \sigma_\ell \left[(\theta_j \theta_\ell)(\alpha_j \alpha_\ell) Var(Z) + \rho_{j\ell}(k_j k_\ell)(\alpha_j \alpha_\ell)^{\frac{1}{2}} E(Z) \right]}{\sigma_j \sigma_\ell} \qquad (19)$$

$$= (\theta_j \theta_l)(\alpha_j \alpha_l) Var(Z) + \rho_j l(k_j k_l)(\alpha_j \alpha_l)^{(1/2)} E(Z).$$

In view of Eq. (16), the first advantage of the multivariate Lévy model is that each marginal distribution has its own parameters. As a result, instead of calibrating the whole parameters of the multivariate Lévy processes simultaneously, we can directly calibrate the marginal parameters separately. Then, applying Eq. (19), we can estimate the correlation matrix $[\rho_{j\uparrow}]$ and the common shock coefficient α_j.

In this subsection we focus on two kinds of multivariate Lévy models, multivariate Variance gamma ($\alpha - VG$) and multivariate Normal inverse Gaussian ($\gamma - NIG$). We also provide the only continuous multivariate Lévy model, multivariate normal model for comparison. We first model the residuals as $\alpha - VG$ model. As showed by Luciano and Semeraro (2010b), we have

$$X_j = \Gamma\left(\frac{b}{\alpha_j} - J_c, \frac{b}{\alpha_j}\right). \qquad (20)$$

$$Z = \Gamma(J_c, b), \qquad (21)$$

where $\Gamma(\cdot)$ denote as the gamma distribution; $\frac{b}{\alpha_j} - J_c$ and J_c are the location parameters for gamma distribution; $\frac{b}{\alpha_j}$ and b are the shape parameters for gamma distribution; and α_j is the parameter as the bridge between the non-systematic jump and systematic jump. The MGF of idiosyncratic and common subordinations are

$$L_{X_j}(u_j) = \left[1 - \left(\frac{\alpha_j}{b}\right)(iu_j)\right]^{-\left(\left(\frac{b}{\alpha_j}\right) - J_c\right)}. \qquad (22)$$

$$L_Z(u_j) = \left[1 - \left(\frac{1}{b}\right)(iu_j)\right]^{-J_c}. \qquad (23)$$

We therefore obtain the CF of asset return margin as follows:

$$\Phi_{R_j(t)}(u_j) = \exp\left(iu_j\left(\mu_j^*\right)\right)\left[1 - \left(\alpha_j^*\right)^{-1}\left(i\theta_j^* u_j - \left(\frac{1}{2}\right)\left(k_j^*\right)^2 u_j^2\right)\right]^{-\left(\alpha_j^*\right)}, \qquad (24)$$

where $\theta_j^* = \sigma_j\theta_j$, $k_j^* = \sigma_j\sqrt{1 - \left(\frac{1}{\alpha_j}\right)\theta_j^2}$, $\mu_j^* = c_j - \sigma_j\theta_j$ and $\alpha_j^* = \frac{1}{\alpha_j}$ to ensure that $E(\varepsilon_j(t)) = 0$ and $V(\varepsilon_j(t)) = 1$. Substituting Eqs. (22) and (23) into Eq. (17) yields the CF of the risky asset returns:

$$\Phi_{R(t)}(u) = \exp\left(\sum_{j=1}^{n-1}\left(iu_j\mu_j^*\right)\right)\left(\prod_{j=1}^{n-1}\left[L_{X_j}\left(iu_j\theta_j^* - \frac{1}{2}\left(k_j^{*2}u_j^2\right)\right)\right]\right)L_Z\left(i\hat{u}^{*\prime}\theta^* - \frac{1}{2}\left(\hat{u}^{*\prime}\Sigma^*\hat{u}^*\right)\right)$$

$$= \exp\left(\sum_{j=1}^{n-1}\left(iu_j\mu_j^*\right)\right)\left(\prod_{j=1}^{n-1}\left[1 - \frac{1}{\alpha_j^*}\left(i\theta_j^*u_j - \frac{1}{2}k_j^*u_j^2\right)\right]^{-\left(\alpha_j^*-J_C\right)}\right)\left(1 + \hat{u}^{*\prime}\theta^* + \frac{1}{2}i\hat{u}^{*\prime}\Sigma^*\hat{u}^*\right)^{-J_C},$$

$$(25)$$

where

$$\hat{u}^* = \begin{bmatrix} \sigma_1 u_1\left(\alpha_1^*\right)^{-1/2} \\ \sigma_2 u_2\left(\alpha_2^*\right)^{-1/2} \\ \vdots \\ \sigma_{n-1}u_{n-1}\left(\alpha_{n-1}^*\right)^{-1/2} \end{bmatrix}, \theta^* = \begin{bmatrix} \theta_1^* \\ \theta_2^* \\ \vdots \\ \theta_{n-1}^* \end{bmatrix}, \Sigma^* = \left[\rho_{j\ell}k_\ell^*k_j^*\right]_{n-1\times n-1}, \quad (26)$$

$$\ell, j = 1, 2, \ldots, n-1.$$

Substituting Eq. (25) into Eq. (8) with $u = i\rho a$, we can obtain the PPM according to the $\alpha - VG$ model.

For the $\gamma - NIG$ model, Luciano and Semeraro (2010a) define the idiosyncratic and common subordinations as follows:

$$X_j = IG\left(1 - J_c\gamma_j, \frac{1}{\gamma_j}\right), \quad (27)$$

$$Z = IG(J_c, 1), \quad (28)$$

where $IG(\cdot)$ denote as the Inverse Gaussian distribution; $1 - J_c\gamma_j$ and J_c are the location parameters for Inverse Gaussian distribution, respectively; γ_j is the measure of the common jump component. The MGF of idiosyncratic and common subordinations are

$$L_{X_j}\left(u_j\right) = \exp\left[-(1 - J_c\gamma_j)\left(\frac{b}{\gamma_j}\right)\left(\left(1 - iu_j\left(\frac{\gamma_j}{b}\right)\right)^{\frac{1}{2}} - 1\right)\right]. \quad (29)$$

$$L_Z\left(u_j\right) = \exp\left[-(J_c)b\left(\left(1 - iu_j\left(\frac{1}{b}\right)\right)^{\frac{1}{2}} - 1\right)\right]. \quad (30)$$

As a result, we can build the CF of the risky asset return as follows:

$$\Phi_{R_j(t)}\left(u_j\right) = \exp\left(iu_j\left(\mu_j^*\right)\right)\exp\left[-\left(\gamma_j^*\right)\left(\left(1 - \left(i\theta_j^*u_j - \left(\frac{1}{2}\right)\left(\gamma_j^*\right)^2u_j^2\right)\left(\frac{1}{\gamma_j^*}\right)\right)^{\frac{1}{2}} - 1\right)\right].$$

$$(31)$$

where$\theta_j^* = \sigma_j\theta_j$, $k_j^* = \sigma_j\sqrt{1 - \left(\frac{1}{\alpha_j}\right)\theta_j^2}$, $\mu_j^* = c_j - \sigma_j\theta_j$ and $\alpha_j^* = \frac{1}{\gamma_j}$ to ensure that $E(\varepsilon_j(t)) = 0$ and $V(\varepsilon_j(t)) = 1$. Similarly, substituting Eqs. (29) and (30) into Eq. (17), we obtain the CF of the risky asset returns as follows

$$\Phi_R(u) = \exp\left(i\sum_{j=1}^{n-1} u_j\mu_j^* - J_c\left(\sqrt{1 + i\hat{u}^{*\prime}\theta - \frac{1}{2}\hat{u}^{*\prime}\Sigma^*\hat{u}^*} - 1\right)\right) \times$$

$$\exp\left(-\sum_{j=1}^{n-1}(\gamma_j^* - J_c)\left(\left(1 + \frac{1}{\gamma_j^*}\left(\theta_j^* u_j + \frac{1}{2}ik_j^{*2}u_j^2\right)\right)^{\frac{1}{2}} - 1\right)\right). \tag{32}$$

where

$$\hat{u}^* = \begin{bmatrix} \sigma_1 u_1 (\gamma_1^*)^{-1/2} \\ \sigma_2 u_2 (\gamma_2^*)^{-1/2} \\ \vdots \\ \sigma_{n-1} u_{n-1} (\gamma_{n-1}^*)^{-1/2} \end{bmatrix}, \theta^* = \begin{bmatrix} \theta_1^* \\ \theta_2^* \\ \vdots \\ \theta_{n-1}^* \end{bmatrix}, \Sigma^* = \left[\rho_{j\ell} k_\ell^* k_j^*\right]_{n-1 \times n-1}, \ell, j = 1, 2, \ldots, n- \tag{33}$$

Substituting Eq. (32) into Eq. (8) with $u = i\rho a$, we can obtain the PPM according to the $\gamma - NIG$ model.

For parameter calibration, following the similar derivation in Theorem 1, we also derive the dependence structure for the $\alpha - VG$ model and $\gamma - NIG$ model in Corollary 2.

Corollary 2. Under the $\alpha - VG$ model, the correlation coefficient between the jth asset return and ℓ^{th} asset return (i) is given by.

$$Corr(R_j(t), R_\ell(t)) = (\theta_j\theta_\ell)(\alpha_j\alpha_\ell)Var(Z) + \rho_{j\ell}^{Corr}(k_jk_\ell)(\alpha_j\alpha_\ell)^{\frac{1}{2}}E(Z)$$
$$= \theta_j\theta_\ell J_C\left(\alpha_j^*\alpha_\ell^*\right)^{-1} + \rho_{j\ell}k_jk_\ell J_C\left(\alpha_j^*\alpha_\ell^*\right)^{-\frac{1}{2}}, \tag{35}$$

Similarly, the correlation coefficient between the jth asset return and ℓ^{th} asset return (i) for the $\gamma - NIG$ model is given by

$$Corr(R_j(t), R_\ell(t)) = (\theta_j\theta_\ell)(\alpha_j\alpha_\ell)Var(Z) + \rho_{j\ell}^{Corr}(k_jk_\ell)(\alpha_j\alpha_\ell)^{\frac{1}{2}}E(Z)$$
$$= \theta_j\theta_\ell J_C\left(\gamma_j^*\gamma_\ell^*\right)^{-2} + \rho_{j\ell}k_jk_\ell J_C\left(\gamma_j^*\gamma_\ell^*\right)^{-1}. \tag{36}$$

When the asset returns follow the multivariate normal distribution with zero initial wealth, we have from Eqs. (5) and (34)

$$E(U(\tilde{W})) = -\frac{1}{\rho}\exp(-\rho a'R_f)\Phi_R(i\rho a) = -\frac{1}{\rho}\exp(-\rho a'R_f)\exp\left(-\rho a'c + \frac{1}{2}\rho^2 a'\Sigma_N a\right)$$

$$= -\frac{1}{\rho}\exp\left(-\rho a'(c - R_f) + \frac{1}{2}\rho^2 a'\Sigma_N a\right) \tag{37}$$

where $R_f = [r_f, ..., r_f]'$ is a is $(n\text{-}1)$-by-1 vector. Differentiating Eq. (37) with respect to a and setting it equal to zero, we get $a^* = \frac{1}{\rho}\Sigma_N^{-1}(c - R_f)$. Substituting it into Eq. (37) yields

$$E\left(U^*\left(\tilde{\mathcal{W}}\right)\right) = -\frac{1}{\rho}\exp\left(-\frac{1}{2}(c - R_f)'\Sigma_N^{-1}(c - R_f)\right). \tag{38}$$

When ρ equals one, Eq. (8) can be rewritten as follows:

$$PPM(a) = \sqrt{-2\log\left(-E\left(U\left(\tilde{\mathcal{W}}\right)\right)\right)} = (c - R_f)'\Sigma_N^{-1}(c - R_f), \tag{39}$$

Consequently, when ρ equals one, the initial wealth is zero, and the asset returns are multivariate normally distributed, the PPM becomes a portfolio Sharpe ratio (PSR), a reward-to-risk ratio. In addition, when the number of risky asset is one, the PSR reduces to the Sharpe ratio.

4 Empirical Analysis

In this selection, we introduce the data that we use and do some analysis. First, we present the descriptive statistics for the all sample. The sample observations for each risky asset are from January 1, 2001 to September 30, 2010, each with over 2450 daily return observations. Second, we estimate the parameters estimations. In this study, we have two steps for the parameters calibration. The first one is the estimation result for the full sample; other is the estimation result for the weekly moving estimation period. Third, we present the model selection analysis for these two cases based on the maximum log-likelihood function, AIC and BIC criteria. Last, we do the parameters analysis for the MN, $\alpha - VG$, and $\gamma - NIG$ models.

5 The Data Description

The data we use in this paper are iShares MSCI Germany Index Fund (EWG), SPDR USA S&P 500 (SPY), and the iShares MSCI Canada Index Fund (EWC). All of the above daily indices, spanning from Jan. 1, 2001 to Sep. 30, 2010, are obtained from the Yahoo Finance. Table 1 presents the descriptive statistics for the raw data.

The Descriptive statistics present that all risky assets we used exhibit non-zero skewness and positive excess kurtosis. Based on the Jarque–Bera (JB) test statistics, the null hypothesis is significantly rejected, which means that the empirical distribution of the return series do not follow the normality assumption.

Table 1. Descriptive Statistics

Statistics	EWG returns	SPY returns	EWC returns
Mean	9.02×10^{-5}	1.81×10^{-5}	3.05×10^{-4}
Median	8.30×10^{-5}	6.81×10^{-4}	1.06×10^{-5}
Maximum	0.1803	0.1356	0.1166
Minimum	-0.1197	-0.1036	-0.1165
Std. Dev	0.0186	0.0139	0.0166
Skewness	0.1213	0.05698	-0.4561
Excess Kurtosis	11.0149	13.3712	8.4786
JB test	6563.76 *	10981.78 *	3148.98 *
Observation	2450	2450	2450

6 Parameter Calibration

In this study, we adopt maximum likelihood estimation (MLE) to estimate parameter for all three models. Because each marginal distribution has its own parameters in the multivariate Lévy framework, we estimate the parameters using the following procedure. First, we estimate the marginal parameters using maximum likelihood estimation. Then applying the Eqs. (35) and (36), we can use the sample correlation matrix to determine the correlation matrix of the dependent Brownian motions and the common jumps parameter J_C for the systematic subordinator $Z(t)$ by minimizing the sum of square error estimation (MSE). We first present the results of the full samples from January 1, 2001 to September 30, 2010.

In this section we follow some criterion to find out the best model for the assets return. We compare the values of log-likelihood function (LLF), Akaike information criterion (AIC) and Bayesian information criterion (BIC). The greater the absolute value of each model selection criteria, the better the model is. We present the model selection results for full sample data from January 1, 2001 to September 30, 2010 in Table 2. Obviously, the $\gamma - NIG$ model is the best goodness-of-fit model for three asset return series.

Using the rolling window scheme, we also obtain the corresponding parameters, LLF, AIC, and BIC by fixing the window size at 5-year data. The rolling window procedure is conducted by moving the in-sample (or the estimated sample) by omitting the returns for the oldest seven calendar days and adding new ones for the latest seven calendar days. Consequently, there are 290 in-sample results. For example, the first window is January 1, 2001 to December 31, 2005. The second window is January 8, 2001 to January 7, 2006. We present the weekly re-estimated results in Table 3. Table 3 exhibits that the $\alpha - VG$ and $\gamma - NIG$ models are the best model according to the LLF, AIC, and BIC, which means that the multivariate Lévy model can best describe the empirical stylized regularities. In addition, the $\gamma - NIG$ model fits the asset return better than the $\alpha - VG$ model does. We therefore fit the $\gamma - NIG$ model to the asset returns in the following analysis.

Table 2. The LLF, AIC and BIC of Full Sample

Asset	Model	LLF	AIC	BIC
EWG	MN	6293.3	6291.3	6285.5
	$\alpha - VG$	6540.1	6536.1	6524.5
	$\gamma - NIG$	**6551.7**	**6547.7**	**6536.1**
SPY	MN	7001.4	6999.4	6993.6
	$\alpha - VG$	7354.5	7350.5	7338.9
	$\gamma - NIG$	**7361.1**	**7357.1**	**7345.5**
EWC	MN	6564.3	6562.3	6556.5
	$\alpha - VG$	6788.2	6784.8	6773.2
	$\gamma - NIG$	**6801.5**	**6797.5**	**6785.9**

Table 3. The Counts of Goodness-of-Fit for the MN, $\alpha - VG$ and $\gamma - NIG$ Models

Asset	Model	LLF	AIC	BIC
EWG	MN	0	0	0
	$\alpha - VG$	86	104	104
	$\gamma - NIG$	**204**	**186**	**186**
SPY	MN	0	0	0
	$\alpha - VG$	**169**	**167**	**167**
	$\gamma - NIG$	121	123	123
EWC	MN	0	0	0
	$\alpha - VG$	3	3	3
	$\gamma - NIG$	**287**	**287**	**287**
Total periods		290	290	290

7 Application of the PPM

In this section we first illustrate how the PPM can resolve the Sharpe ratio paradox presented in Hodges (1998). With the calibrated parameters of iShares MSCI Germany Index fund, SPDR USA S&P 500 and the iShares MSCI Canada Index Fund over the 290 weekly data sets, we obtain the optimal asset allocation obtained by maximizing the PPM.

We first employ the example provided by Hodges(1998) and Zakamouline and Koekebakker (2009) to test whether the PPM is capable of making portfolio decision more appropriately. We list the numerical example and its statistics descriptive in Table 4 and Table 5. Using the moment match technique, we obtain the parameters in Table 6 for the MN and $\gamma - NIG$ models. Table 8 shows the optimal allocation according to the MN model (or PSR) and the $\gamma - NIG$ model. Intuitively, the asset B is more valuable than asset A. The investor should invest all the money in asset B. However, the optimal asset allocation according to PSR incorrectly invests half of the wealth in asset A, while the optimal asset allocation according to $\gamma - NIG$ model provides the correct investment strategy. In addition, Table 8 also exhibits that the higher proportion investing in asset B, the higher PPM (higher expected utility) is. Therefore, the PPM is a better performance measure than the PSR (Table 7).

Table 4. The Probability Distribution of the Asset Returns

Probability	0.01	0.04	0.25	0.4	0.25	0.04	0.01
Return on asset A	-25	-15	-5	5	15	25	35
Return on asset B	-25	-15	-5	5	15	25	45

Source: Hodges(1998) andZakamouline and Koekebakker (2009)

Table 5. The Moments of the Asset Returns

Asset	Mean	Std	Skew	Kurt
A	0.05	0.100	0.000	3.4
B	0.051	0.103	0.305	4.487

Table 6. The Calibrated Parameters

Normal	Asset A	$c = 0.05$ $\sigma = 0.1$
	Asset B	$c = 0.051$ $\sigma = 0.103$
$\gamma - NIG$	Asset A	$c = 0.05 \gamma^* = 0.1333 J_C = 0.1333$ $\sigma = 0.01 \theta = 0$
	Asset B	$c = 0.051 \gamma^* = 0.4541 J_C = 0.1333$ $\sigma = 0.0107 \theta = 0.2241$

Table 7. The value of PPM and portfolio decision

	MN	$\gamma - NIG$
PPM value	1.4494	1.4742
Allocation in Asset A	0.4937	0
Allocation in Asset B	0.5063	1
Allocation in Risk-Free Rate	0	0
Rate of Return	0.0505	0.051

Note: The risk free rate equals 1%

Table 8. The PTM values for Distinct Weights

(a_1, a_2, a_3)	PTM
(1,0,0)	1.4490
(0.8,0.2,0)	1.4546
(0.6,0.4,0)	1.4598
(0.4,0.6,0)	1.4648
(0.2,0.8,0)	1.4696
(0,1,0)	1.4742

Note: a_1 and a_2 are the weights of Asset A. The a_3 is the weight of risk-free asset with risk-free rate equal to 0.01

To apply the PPM in optimal portfolio selection, our objective is to maximize the PPM (expected utility) subject to the total weight equal to one, namely,

$$Max : PPM (a_1, a_2..., a_{n-1}), \tag{40}$$

$$s.t. \sum_{j=1}^{n} a_j = 1, 0 < a_j \leq 1, j = 1, 2, \ldots n - 1. \tag{41}$$

The first three assets are the risky assets, including the iShares MSCI Germany Index Fund (EWG), SPDR USA S&P 500 (SPY), and the iShares MSCI Canada Index Fund (EWC). The proxy of the risk-free asset is the three-month USA Treasury Bill (T-Bill) rates. All the risky assets are denominated in US dollar. Given the weekly calibrated parameters of the MN and $\gamma - NIG$ model, we obtain optimal weights for each data set according to Eqs. (40) and (41), and then invest the one risk-free asset and three risky assets at adjacent seven calendar day according to the optimal weights. We present the results of the two investment strategies in Table 9–11.

Form Table 9 to Table 11, we demonstrate that the $\gamma - NIG$ strategy provides a better performance than the MN strategy, according to the portfolio value at each time. The standard deviation, skewness, and excess kurtosis of the portfolio value according to $\gamma - NIG$ strategy are also lower than those of the MN strategy, while the mean of the $\gamma - NIG$ strategy is higher than the MN strategy. The $\gamma - NIG$ strategy according to Eqs. (40) and (41) not only has better performance than the MN strategy but also avoid the financial meltdown such as the subprime crisis.

Table 9. The Portfolio Value

Performance Measure	Present value at 2005/12/31	Future value at 2010/09/30	Value at 2008/10/04
MN	100	120.61	108.15
$\gamma - NIG$	100	**171.81**	**138.18**

Note: The Present value and Future value are denominated in US Dollars

Table 10. The Counts of Value of the $\gamma - NIG$ Strategy Larger than that of MN Strategy

Results	Number of periods	Odd ratio
$\gamma - NIG > MN$	227	78%

Note: There are total 290 periods

Table 11. The First Four Moment for the Portfolio Values

Model	Portfolio Mean	Standard Deviation	Portfolio Skewness	Portfolio Kurtosis
MN	6.48×10^{-4}	0.025	-1.297	9.878
$\gamma - NIG$	1.90×10^{-3}	**0.016**	**-0.480**	**9.752**

8 Conclusion

Large negative returns have been demonstrated to occur more frequently than predicted under the assumption of normality. An un-appropriate asset model may therefore lead to a serious underestimation of the true risk involved in holding such assets. In our study we construct a multivariate Lévy asset return model that consider the system jump (risk) and non-system jump (risk) simultaneously. Then, we use this model to derivate the portfolio performance measure.

There are some contributions in this paper. The first main contribution is to construct a portfolio performance measure based on accurate asset models which can catch much more detailed information of the financial crisis. As pointed by Luciano and Schoutens (2006), with a reasonable volatility parameter, say around 20 percent, the probability of a significant move, say more than 5 percent, is completely unrealistic for a Brownian motion with continuous sample paths because asset prices are in reality driven by jumps. As a result, it would be better to have a more flexible multivariate distribution, in order to take into account skewness, excess kurtosis, non-systematic jump and systematic jump. Next, we demonstrate that the Generalized Sharpe Ratio introduced by Zakamouline and Koekebakker (2009) is a special case of the PPM, resolving the Sharpe ratio paradox presented in Hodges (1998). With the data of iShares MSCI Germany Index fund, SPDR USA S&P 500 and the iShares MSCI Canada Index Fund over the period from January 1, 2001 to September 30, 2010, the empirical results proves that the proposed $\gamma - NIG$ model can catch more detailed information of financial shock. Consequently, the optimal asset allocation obtained by maximizing the PPM not only has better performance than the MN strategy case but is capable of allowing fund managers to adjust optimal investment strategy to enhance the investment performance during the period of the financial crisis.

References

Abdelmalek, W., Hamida, S.B., Abid, F.: Selecting the best forecasting-implied volatility model using genetic programming. Advances in Decision Sciences (2009)

Adcock, C.J.: Asset pricing and portfolio selection based on the multivariate extended Skew-Student-t distribution. Annuity Operation Research **176**, 221–234 (2010)

Barndorff-Nielsen, O.E., Pedersen, J., Sato, K.I.: Multivariate subordination self- decomposability and stability. Advance Application Probability. **33**, 160–187 (2001)

Cass, D., Stiglitz, J.: The Structure of investor preferences and asset returns, and separability in portfolio allocation. Journal of Economic Theory **2**, 122–160 (1970)

Cont, R., Tankov, P.: Financial modelling with jump processes. Chapman and hall-CRC financial mathematics series (2004)

Cherubini, U., Luciano, E., Vecchiato, W.: Copula Methods in Finance. Wiley (2004)

Dowd, K.: Adjusting for risk: An improved Sharpe ratio. Int. Rev. Econ. Financ. **9**, 209–222 (2000)

Eberlein, E., Madan, D.B.: On Correlating Lévy Processes. Working paper (2009)

Favre, L., Galeano, J.: An analysis of hedge fund performance using loess fit regression. Journal of Alternative Investments, pp. 8–24. Spring (2002)

Fang, H., Lai, T.: Co-kurtosis and capital asset pricing. Financ. Rev. **32**, 293–307 (1997)

Harvey, C., Lietchty, J., Lietchty, M., Müller, P.: Portfolio Selection with Higher Moments, Working Paper, p. 51. Duke University (2004)

Hodges, S.: A generalization of the sharpe ratio and its applications to valuation bounds and risk measures, Working Paper. Financial Options Research Centre, University of Warwick (1998)

Ingersoll, J.E., Jr.: Theory of financial decision making. The handbook Yale University Ch4 Section 4.7, p. 64 (1986)

Jondeau, E., Rockinger, M.: Conditional Volatility, Skewness and Kurtosis: Existence, Persistence and Comovements. J. Econ. Dyn. Control **27**(10), 1699–1737 (2003)

Kaplan, P.D., Knowles, J.A.: Kappa: a generalized downside risk-adjusted performance measure, Working paper (2004)

Kendall, M., Hill, B.: The analysis of economic time-series – Part I: Prices. J. R. Stat. Soc. Ser. A **116**(1), 11–34 (1953)

Landsman, Z.: On the generalization of Stein's lemma for elliptical class of distributions. Statist. Probab. Lett. **76**, 1012–1016 (2006)

Landsman, Z., Nešlehová, J.: Stein's lemma for elliptical random vectors. J. Multivar. Anal. **99**, 912–927 (2008)

Liu, J.S.: Siegel's formula via Stein's identities. Statist. Probab. Lett. **21**, 247–251 (1994)

Lo, A.W., Wang, J.: Trading volume: Definitions, data analysis, and implications of portfolio theory. Review of Financial Studies **13**(2), 257–300 (2000)

Luciano, E., Schoutens, W.: A multivariate jump-driven financial asset Model. Quantitative, 385–402 (2006)

Luciano, E., Semeraro, P.: Extending time-changed Lévy asset models through multivariate Subordinators Working paper (2007)

Luciano, E., Semeraro, P.: A generalized normal mean variance mixture for return processes in finance. Int. J. Theoret. Appli. Fina. 415–440 (2010a)

Luciano, E., Semeraro, P.: Multivariate time changes for Lévy asset models: characterization and calibration. J. Computat. Appli. Maths. 1937–1953 (2010b)

Mandelbrot, B.: The variation of certain speculative prices. J. Bus. **36**, 394–419 (1963)

Mandelbrot, B.: New methods in statistical economics. J. Polit. Econ. **71**, 421–440 (1963)

Markowitz, H.: Portfolio selection. Journal of Finance **7**(1), 77–91 (1952)

Patton, A.: On the out-of-sample importance of skewness and asymmetric dependence for asset allocation. J. Financ. Economet. **2**(1), 130–168 (2004)

Rachev, S., Mittnik, S.: Stable Paretian Models in Finance Series. John Wiley & Sons Ltd, Chichester (2000)

Shadwick, W.F., Keating, C.: A universal performance measure. Journal Performance Measurement **6**(3), 59–84 (2002)

Sortino, F.A., Price, L.N.: Performance measurement in a downside risk framework. The Journal of Investing **3**(3), 59–64 (1994)

Stutzer, M.: A portfolio performance index. Financ. Anal. J. **56**, 3 (2000)

Zakamouline, V., Koekebakker, S.: Portfolio performance evaluation with generalized sharpe ratios: beyond the mean and variance. J. Bank. Finance **33**, 1242–1254 (2009)

Distinguishing Good from Bad: Distributed-Collaborative-Representation-Based Data Fraud Detection in Federated Learning

Zongxiang Zhang[1], Chenghong Zhang[1]([⊠]), Gang Chen[2], Shuaiyong Xiao[3], and Lihua Huang[1]

[1] School of Management, Fudan University, Shanghai 200433, People's Republic of China
zongxiangzhang21@m.fudan.edu.cn
[2] School of Management, Zhejiang University, Hangzhou 310058, People's Republic of China
[3] School of Economics and Management, Tongji University, Shanghai 200092, People's Republic of China

Abstract. Breaking down data silos and promoting data circulation and cooperation is an important topic in the digital age. As data security and privacy protection have received widespread attention, the traditional cooperation model based on data centralization has been challenged. Federated learning provides technical solutions to solve this problem, but the characteristics of multi-party cooperation and data invisibility make it face the risk of data fraud. Malicious participants can manipulate data individually or in collusion to illegally obtain data or influence federated learning model. This paper proposes a novel data fraud detection method based on distributed collaborative representation and realizes the effective detection of federated learning data fraud through collaborative clustering, adaptive representation and dynamic weighting. The method proposed in this paper overcomes weakness in the existing methods that detect data fraud mechanically and statically, which cannot be organically combined with the training objectives and process. It realizes the dynamically continuous anti-collusion soft constraint detection while ensuring fraud detection and contribution evaluation are relatively independent. Our research is of great significance for federated learning to deal with the risk of data fraud and better apply to real-world scenarios.

Keywords: Federated learning · Anomaly detection · Adversarial machine learning

1 Introduction

With the rapid development and application of Internet of Things (IoT), Big Data, cloud computing, and Artificial Intelligence (AI), human production and business activity generates great amount of valuable data, widely distributed in various organizations and groups. To find natural and social law behind the data and promote human production and business activity, it is important to find methods for effective aggregation and analyzation of distributed data. However, data privacy concerns also require data

© Springer Nature Switzerland AG 2023
F. Fui-Hoon Nah and K. Siau (Eds.): HCII 2023, LNCS 14039, pp. 242–255, 2023.
https://doi.org/10.1007/978-3-031-36049-7_19

circulation and cooperation can be carried out only under the premise of adequate protection of data privacy. Under this circumstance, data circulation and cooperation under the requirements of data privacy protection urgently needs a new technical framework at the application execution level. Federated learning, proposed by Google in 2016, has begun to receive widespread attention [1]. Federated learning utilizes data distributed in participants to train a high-quality machine learning models shared by all participants with the constraints of limited communications and security. With further research, federated learning's core target focuses on how to construct machine learning model with encryption, transmission, and aggregation optimization of participant's model parameters and approach effect of directly centralizing and training raw data [2]. The core character of federated learning is "data is immutable and available but not visible while model is dynamic". It not only realizes the collaborative value mining of multiple participant's data, but also meets the requirements of data security and privacy protection. Consequently, federated learning has been regarded as a crucial technical solution for data circulation and cooperation under the premise of data privacy protection.

Although federated learning has unique advantages in distributed machine learning modeling, it still faces many potential dangers and threats in real-world applications. Comparing with other general distributed computing frameworks, federated learning participants have more autonomous controlling capabilities since participants keep raw data local and only upload encrypted optimized model parameters. Thus, some malicious participants may deliberately manipulate their data or model, which is called "federated learning data fraud". Formally, federated learning data fraud is that some of the participants manipulate their raw data, intermediate calculation results, or model parameters to fulfil purposes like decreasing federated learning model effect [3, 4], retrieving other participants' data or models illegally [5], accessing federated learning model speculatively without useful contribution [6], or increasing training cost [7]. For example, malicious participants could randomly flip specific samples' labels and thus train biased models [8]. Malicious participants could also directly send biased model updates to central server by adding random noises or generating poisoned updates according to optimization. Such fraud activities have extremely negative effect on normal federated learning process. First, poisoned raw data and model updates lead optimization of the aggregated federated learning model to the wrong direction, decreasing model performance and hindering accomplishing established federated learning tasks. Second, malicious participants can harm other normal participants and destroy fairness and stability of the federation. Third, circulation of poisoned data and models in federation produces irreversible and persistent negative effects on the credibility and explanability of the federation in the long run. Thus, it is of vital importance to establish a fraud detection mechanism for federated learning to guarantee normal progress of federated learning and rights and interests of all legal participants. First, such mechanism can find and stop potential or ongoing fraud activity and guarantee that the federated learning model is trained in the correct optimization direction. Second, detecting and stopping fraud activity protect data and model security of normal participants and maintains fairness and stability of the federation. Third, existence of fraud detection mechanism deters the participants who may commit fraud activity and reduces occurrence of fraud from the source, which helps maintain a stable and trust-worthy federated learning group. To sum up, fraud detection is of great significance to theoretical completeness and real-world applications of federated learning.

As mentioned above, data fraud problem has become a nonnegligible challenge for federated learning. Many researches have spent effort on establishing an effective fraud detection mechanism. There are mainly four kinds of fraud detection mechanism [9]. Distance-based detection mechanism is the most common mechanism. It first computes distance of participants' update information like gradients and then keep those within a certain range while excludes participants whose distances are too far (potential fraud). Commonly used distance metrics include Euclidean Distance, cosine similarity, KL-divergence, and so on. The other frequently used category of mechanisms is to judge whether the participant commit fraud activity by testing its model performance on a public preserved validation dataset. The key to this kind of mechanism is establishing a reasonable performance evaluation index. Previous methods have tried sum of gradient's L2-norm and difference between loss function, difference of cross entropy, and accuracy of a pre-trained model like autoencoder. Except for these two widely used mechanisms, precious studies also focus on statistic-based and optimization-based mechanisms. The former kind of mechanisms mainly utilizes statistics such as mean, median, coordinate-wise median to distinguish normal participants' training information from malicious ones. The latter dynamically adjusts weight of participants by adding regularization terms of their update information into the federated learning loss function. In general, federated learning fraud detection has formed a basic method system and proposed some significant methods.

Although existing studies have explored a series of fraud detection methods, these methods are imperfect in some aspects and hard to deal with complicated challenges in real scenarios. First, data is usually non i.i.d in federated learning. Existing methods are mainly based on rigid indicators such as centroid distance or statistical indicators such as median to make judgments, which are easily disturbed by sample imbalance or data heterogeneity [10], and cannot effectively identify fraud data. Second, federated learning is a distributed multi-round dynamic machine learning framework. Thus, different malicious parties can commit fraud activity at different training rounds. One-time static judgment cannot clear fraud data and it needs to combine with the training process to form a dynamic and continuous detection process. Third, federated learning is a cooperative machine learning model involving multiple participants. Existing studies have showed that malicious participants can collude to achieve collaborative fraud activity [11, 12]. So, it is incomplete to detect fraud activity in isolation without considering interactions between participants. Finally, existing methods don't well distinguish contribution evaluation [13] between fraud detection and the relationship between them and simply consider that small contribution equals to fraud activity while large contribution means no harm. To sum up, previous fraud detection methods cannot effectively handle complicate real-world challenges. It is necessary and urgent to establish a comprehensively collaborative fraud detection method to solve the above problems for federated learning.

To solve challenges mentioned above, our research proposed a novel fraud detection method for federated learning based on distributed collaborative representation, called Distributed-collaborative-representation-based Data Fraud Detection (Dcr-DFD). Our method can achieve the following goals. First, our method overcomes non i.i.d disturbance of federated learning through dynamic clustering, solving the problem that previous methods cannot effectively distinguish between low-quality data and data fraud. Second, through dynamic combination of fraud detection and training process, our research method realizes detecting all participants in all time rounds instead of previous one-time static detection. Third, the proposed methos carries out internal and external collaborative fraud detection of participants through dynamic clustering and group weighting and overcomes shortcomings of existing methods that can only detect each participant in isolation. Simultaneously, the intermediate calculation results of dynamic clustering and group weighting, such as intra-cluster distance, can also be used as indicators to distinguish contribution evaluation and fraud detection, which prevent confusing contribution evaluation with fraud detection in previous methods.

To better solve fraud detection challenges of federated learning, based on the idea of distributed collaborative representation, our proposed method achieves whole-process all-participant fraud detection goal guided by maximization of training task effect. This method mainly includes three main parts: dynamic clustering, learning-based intra-class weighting, and inter-class weighting. First, we cluster model update information of all participants, use information of other participants in the same cluster to represent each participant, and calculate KL divergence between actual value and representation as fraud detection weight. Second, we construct intra-class and inner-class weight for each participant. Third, our method aggregates update information combining with fraud detection weight, intra-class weight, and inter-class weight in the form of coefficients and regularization items, to realize synergy of fraud detection and federated learning tasks.

In the era of digital economy, federated learning provides solutions for data circulation cooperation under privacy protection. However, data fraud brings practical challenges to application of federated learning. Some studies have tried to propose solutions but cannot completely solve this problem. This research proposed a fraud detection method based on distributed collaborative representation for federated learning, which effectively promoted fraud detection research of federated learning. This research mainly has the following four contributions. First, we proposed a novel mechanism based on existing four types of detection mechanisms. Since federated learning is a distributed framework, our mechanism utilizes this character to dynamically cluster participants to do soft constraints, avoiding disturbance of data quality heterogeneity or imbalance. Second, our proposed method dynamically combines fraud detection and federated learning training process. This improvement changes the situation that detection and training is separated in previous methods and guarantees that detection mechanism serves for maximizing federated learning training effect instead of merely mechanically clearing fraud data. Third, through dynamic clustering and intra-group and inter-group weights, our method considers not only unilateral influence of each participant but also collaborative effect among participants globally, which can cope with more complex application scenarios. Finally, our detection mechanism provides basis for contribution evaluation and benefit distribution of federated learning. The dynamic clustering gap and intra-group and inter-group weights can be used as the calculation basis for contribution evaluation and benefit distribution, which profoundly illustrates the importance of fraud detection for federated learning. In general, our method makes targeted improvements on short-comings of fraud detection for federated learning, which is of great significance for enriching federated learning theories and promoting federated learning to applications.

2 Literature Review

2.1 Federated Learning

Federated learning is a privacy-preserving distributed machine learning based on feder-ated optimization [14]. In a federated learning system, a group of participants (hardware such as remote sensors and mobile phones or data owners like hospitals and financial institutions) collaborate to train a global machine learning model while keeping each one's data local to preserve data privacy [1, 2, 15–19]. During a federated learning training process, each participant trains a local model using its own data, generates and encrypts the model optimization output like gradient or loss, and sends it to a central coordinator. The coordinator harmonizes participants' model output based on specific strategy like FedAvg [20] and FedOpt [21], generates a global update, and sends it back to all participants. Each participant uses the global update to update its local model. As described above, data circulation in federated learning only exchanges encrypted model update and no raw data is exposed. Thus, federated learning realizes data circulation and data privacy preserving simultaneously.

Having noticed this peculiar advantage of federated learning in privacy-preserved machine learning, a growing number of studies have developed federated learning meth-ods for data collaboration applications in various areas such as finance [22], healthcare [23, 24], vehicle transportation [25, 26], and recommendation system [27]. Yang et al. [22] proposed FFD, a federated learning framework for credit card fraud detection. Chen et al. [23] used federated learning in wearable healthcare of Parkinson's disease auxiliary diagnosis application. Saputra et al. [25] proposed a federated energy demand learning (FEDL) to predict energy demand in vehicle electronic network. Wu et al. [27] com-bined federated learning with the recently popular neural network model, Graph Neural Network (GNN), to do recommendations and got competitive results with GNN-based recommendation methods. Federated learning has been widely applied in real world scenario and proved to have competitive performance and great value.

With its vital practical utility, federated learning has had some meaningful attempts in real-world applications, but it still needs to solve the data fraud problem to be comprehen-sively applied in real-world scenarios. Since federated learning requires all participants to keep their data local, this special property, on the one hand, preserve data privacy. On the other hand, some potential malicious parties can commit fraud activity during federated learning training process. Dishonest participants could use poison data to cheat the federated learning system or do free-riding in training process [6, 28]. Hostile partic-ipants use manipulated data to decrease training efficiency [29]. In some cases, hostile participants can even collude to commit more sophisticated and imperceptible fraud activity to compromise federated learning models [3, 12]. Thus, to give full play to the advantages of federated learning's privacy preserving property, while avoiding harmful parties to use this feature to carry out data fraud activity, federated learning desperately needs a fair and objective data fraud detection mechanism in real world applications.

2.2 Data Fraud Detection in Federated Learning

Data fraud is manipulating machine learning model's data to achieve illegal purposes, including misleading model optimization direction, obtaining data illegally, and extorting training results [8, 30, 31]. In federated learning, multiple participants work together to complete model training and share training outcomes. Malicious participants have motivations to obtain data models of normal participants or interfere and mislead federated training through data fraud activity. Besides, distributed training and data privacy protection also make data fraud more difficult to detect and control in federated learning than other machine learning models. Based on different goals, objects, and strategies, harmful participants can commit fraud activity in a series of complex and undefendable ways [32]. To tackle with data fraud problem in federated learning, existing studies have explored various detection methods from multiple perspectives [9, 33]. Generally, these methods can be divided into three categories: anomaly update detection, robust federated training, and backdoored model restoration [33]. Anomaly update detection mainly focuses on specific calculation logic and judgment rules of data fraud judgment criterions. These criterions use participants' update information such as gradient information or local representations to calculate and determine whether participants commit data fraud. Robust federated learning trains the model and mitigates backdoor attacks simultaneously. Backdoored model restoration tries to find malicious models and repair the backdoored global model after training.

Anomaly update detection is the category that most research focus on and can be further divided into four categories: distance-based detection, performance-based detection, statistics-based detection, and optimization-based detection [9]. Distance-based detection calculates and compares distances between participants' update information. Participants whose update information is far away from others are more likely to commit fraud activity. Most research calculates Euclidian distance between participants' update information and uses them to detect possible malicious participants by ranking [34, 35] or combining with other models like graph networks [6]. Fung et al. [28] use cosine similarity to distinguish benign and malicious participants. Research of statistics-based detection mainly exploits how to use statistical characteristics of uploaded updates, such as the median or mean, to detect abnormal updates and malicious participants. Yin et al. [36] compute coordinate-wise median and the coordinate-wise trimmed mean of participants' updates to find abnormal updates. Xie et al. [37] proposed three aggregation rules: geometric median, marginal median, and "mean around median" as index of data fraud detection. Performance-based detection evaluates participants' update information over a clean dataset maintained by the coordinator. Performance indexes like accuracy or entropy indicate whether participants commit data fraud activity. Xie et al. [38] proposed Zeno, which uses loss function value and L2-norm of gradients as performance indexes. Park et al. [39] use cross entropy to evaluate participants' update information and find malicious participants. Li et al. [40] use participants' update information to train an autoencoder and design an index using its accuracy. Participants with low indexes are judges as malicious participants. Optimization-based judge criterion refers to optimizing modified objective function of the original federated learning task to detect and find malicious participants. Li et al. [41] add a regularization term to the objective loss function, such that weights of malicious participants' update information are forced to decrease. As for robust federated learning and backdoored model restoration, previous research mainly studies how to reduce impact of data fraud on federated learning tasks. Sun et al. [42] put forward a defense strategy that clips model weights and injects noise for mitigating malicious participants' updates on the global model. Wu et al. [43] proposed a post-training defense strategy against backdoor attacks in federated learning.

2.3   Data Fraud Detection in Federated Learning

With the development of federated learning, data fraud detection of federated learning has entered a new stage. Although it has similarities with previous data fraud detection in federated learning system, the detection mechanism has significantly changed. Peculiar properties and special functions of federated learning have raised unprecedented novel demands for data fraud detection in federated learning. Hence, data fraud detection of federated learning faces some research gaps worth further effort to tackle.

First, non-i.i.d property of data is a common phenomenon in federated learning. Previous methods are mainly used rigid indicators such as centroid distance or statistical indicators such as median to detect, which are easily disturbed by sample imbalance or data heterogeneity. This may confuse non-i.i.d data with fraud data. The first research gap is how to dynamically assess data quality during the training process under the constrains.

Second, as a distributed multi-round dynamic machine learning framework, federated learning can face fraud activities committed by different malicious parties at different training rounds. One-time static detection cannot clear fraud data and negative impact caused by fraud data from the long run. This situation induces the second research gap, that is, how to combine data fraud detection with the training process to form a dynamic and continuous detection process.

Third, federated learning is a cooperative machine learning model involving multiple participants. Malicious participants can make collusions to achieve collaborative fraud activity. So, it is incomplete to detect fraud activity in isolation without considering interactions between participants. Consequently, how to construct a mechanism which is effective in collaborative fraud activity is a research gap for data fraud detection in federated learning.

Finally, existing methods don't well distinguish contribution evaluation between fraud detection and the relationship between them. It is imperfect to simply consider that small contribution equals to fraud activity while large contribution means no harm. The final research gap is how to construct fraud-contribution independency to make data fraud detection of federated learning fairer and more reasonable. Table 1 compares recently popular federated learning algorithm with data fraud detection mechanisms and Dcr-DFD in four different dimensions.

Table 1. Comparison of Dcr-DFD with Existing Relevant Methods.

	Constrain softness	Dynamic continuity	Anti-collusion	Fraud-contributionindependency
FedAvg [20]	×	×	×	×
Krum [34]	×	×	×	×
FABA [35]	×	×	×	×
Sniper [6]	×	✓	×	×
FoolsGold [28]	×	✓	×	×
Zeno [38]	✓	✓	×	×
SageFlow [39]	✓	✓	×	×
Median [36]	✓	✓	×	×
Mean [36]	✓	✓	×	×
Dcr-DFD	✓	✓	✓	✓

3 Proposed Method

3.1 Framework Overview

To solve data fraud detection problem in federated learning, we propose a novel method, Distributed-collaborative-representation-based Data Fraud Detection (Dcr-DFD). Figure 1 shows general framework of our proposed method. This method includes three modules, Frobenius-norm-based participants update information clustering, adaptive statistical participant update information evaluation, and inter-cluster & intra-cluster participant update information aggregation. In this section, we briefly introduce the framework and each parts of our proposed method for data fraud detection of federated learning.

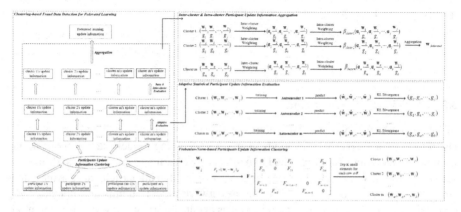

Fig. 1. Framework of Dcr-DFD

3.2 Problem Formulation

Considering in a federated learning system, there are m participants and some of them are malicious. Let G as global aggregation function and as local models correspond to i-th federated learning participants. Then, our problem can be formulated as:

$$\min_{\mathbf{w}} G(F_1(\mathbf{w}), F_2(\mathbf{w}) \cdots F_m(\mathbf{w}))$$

To protect participants' privacy, each participant's raw data never are not allowed leaves leakage from its local device. The participants only transmit encrypted update information like representation and gradient to central server during federated learning process. Table 2 summarizes the key notations used for model description in this section.

Table 2. Notation List.

Notation	Description
m	Number of participants in the federated learning system
T	Total training rounds
$\mathbf{w}^t$	Global model parameter at round t
G	Global aggregation function
$F_i(\mathbf{w})$	Local bjective function of participant i
$\nabla F_i(\mathbf{w}^t)$	model parameter gradient of participant i at round t
$\tilde{\nabla} F_{ij}(\mathbf{w}^t)$	Encoded model parameter gradient of participant i by participant j at round t
$\tilde{\nabla} F_i(\mathbf{w}^t)$	Encoded model parameter gradient of participant i at round t
K	Number of participants in each cluster
$cluster_i^t$	The cluster of participant i at round t
$clusternorm_{ij}^t$	The distance between participant i and j at round t
$\{\alpha_i\}_{i \in [m]}$	Participants' inner-cluster weight
$\{\beta_i\}_{i \in [m]}$	Participants' intra-cluster weight
g_i^t	Participant i's data fraud detection weight

3.3 Local Model Construction

In the local model construction, participants use their own local machine learning models to train parameters. Various kinds of machine learning models can be used in our framework, including but not limited to neural networks, linear regression, and support vector machines. Without losing generality, we assume that all participants use the same structure neural network model. The neural network model outputs and uploads hidden representations of participants' raw data as update information, which is a common practice in previous federated learning studies [44, 45].

3.4 Encrypted Mechanism

Secure parameters transmission between participants and central server is necessary for federated learning, including additive masking [46], differential privacy [47], and trusted execution environment (TEE) [48]. Our research tries to solve challenge of fraud activity caused by malicious participant's data or model manipulation, not data leakage or adversarial attack caused by unsecure encryption or communication. So, we mainly focus on constructing fraud detection mechanism and simplify encryption and communication process in our algorithm framework. Previous research about federated learning algorithm design have taken similar settings like simplifying encryption process [42] or assuming information transmitted safely in the federated learning system [39, 40, 49]. Thus, such simplified encryption setting is reasonable and supposed not to affect correctness and generality of our experiments and analysis. Still, for the sake of completeness, we describe a classic encryption process based on Yang et al. [2] under our algorithm setting as following.

The federated learning system first uses encryption-based user ID alignment algorithm [50, 51] to ensure common entity consistency of all participants. Since this procedure is encrypted, none of the participants' raw data is exposed. After common entity is ensured, the federation trains a global machine learning model collaboratively based on the following encryption process:

1. Central server creates and sends encryption keys to all participants;
2. Each participant trains their local model and generates local update information. After added additional mask, local update information is encrypted and sent to central server;
3. Central server receives, decrypts and aggregates all participants' update information to train the global model. Then, central server generates global update information and sends it back to all participants;
4. Each participant receives, decrypts and uses global update information to do next round training.

3.5 Distributed Collaborative Representation

Distributed collaborative representation is the key part of our algorithm. First, after each participant uploads its training representation, central coordinator calculates Frobenius norm of participant's training representations pairwise as a measure of similarity. Then, central coordinator takes each participant as the center and select m other participants that are most like it to form a cluster centered on this participant. Besides, central coordinator also maintains an autoencoder for each cluster. In each cluster, participants use the corresponding autoencoder to represent other participants. Then central coordinator takes weighted average of all other participants' representations of a participant as predicted fusion representation of that participant and calculates KL divergence between predicted representation and true representation as participant's fraud detection weight.

3.6 Federated Representation

In the federated level, central coordinator combines fraud detection with model optimization process of federated learning to detect participants dynamically and simultaneously as well as to guarantee effectively finishing federated learning task. In each aggregation round, after each participant computes representations locally and sends them to central coordinator, central coordinator first processes representations as mentioned in 4.4. Then it applies within-cluster and between-cluster weights to each participant's representations along with fraud detection weight. Finally, it uses the global federated learning model to derive the global gradients, and sends them back to update local models. This process will be executed iteratively until the model converges.

3.7 Algorithm Process

In this part, we describe detailed algorithms progress of our proposed Dcr-DFD method.

Algorithm 1: Dcr-DFD for Fraud Detection in FL

1 **Input:** $\mathbf{w}^t$, $\{\alpha_i\}_{i\in[m]}$, $\{\beta_i\}_{i\in[m]}$

2 **for** $t = 0, \cdots, T\text{-}1$ do

3   Central coordinator sends $\mathbf{w}^t$ to each participant.

4   **for** $i \in [m]$ do:

5   Participant i uses $\mathbf{w}^t$ and its local data to compute local gradient $\nabla F_i(\mathbf{w}^t)$

6   **end**

7   **for** $i \in [m]$ do: //dynamic clustering

8   **for** $j \in [m], j \neq i$ do:

              Calculate $clusternorm_{ij}^t = \| \nabla F_i(\mathbf{w}^t) - \nabla F_j(\mathbf{w}^t) \|_F$

9          $cluster_i^t = \{\text{Top } k \text{ participants that have smallest } clusternorm \text{ with } i\}$

10   **end**

11   **for** $i \in [m]$ do: //fraud detection

12          $\nabla F_j(\mathbf{w}^t) \xrightarrow[\text{represent}]{\text{Autoencoder } i} \hat{\nabla} F_{ij}(\mathbf{w}^t)$

13          $\hat{\nabla} F_i(\mathbf{w}^t) = \dfrac{1}{K} \sum\limits_{j \in cluster_i^t} \hat{\nabla} F_{ij}(\mathbf{w}^t)$

14          $g_i^t = D(\Delta_i^t \| \tilde{\Delta}_i^t)$

15          $\alpha_i \longleftarrow \alpha_i - \eta \nabla F_i(\alpha_i)$

16          $\beta_i \longleftarrow \beta_i - \eta \nabla F_i(\beta_i)$

17   **end**

18      $\mathbf{w}^{t+1} \longleftarrow G(\mathbf{w}^t, \{\beta_i \alpha_i \dfrac{\nabla F_i(\mathbf{w}_i^t)}{g_i^t}\}_{k\in[m]})$

19 **return** $\{\alpha_i\}_{i\in[m]}$, $\{\beta_i\}_{i\in[m]}$, $\{g_i^t\}_{i\in[m], t\in\{0,1,\cdots T-1\}}$, $\mathbf{w}^T$

Acknowledgments. This work was supported in part by the National Natural Science Foundation of China (grant numbers 72271059, 71971067). Chenghong Zhang is the corresponding author.

References

1. Konečný, J., McMahan, H.B., Yu, F.X., Richtárik, P., Suresh, A.T., Bacon, D.: Federated learning: Strategies for improving communication efficiency, arXiv preprint arXiv:1610.05492 (2016)

2. Yang, Q., Liu, Y., Chen, T., Tong, Y.: Federated machine learning: concept and applications. ACM Trans. Intell. Sys. Technol. (TIST) **10**(2), 1–19 (2019)
3. Fang, M., Cao, X., Jia, J., Gong, N.Z.: Local model poisoning attacks to byzantine-robust federated learning. In: Proceedings of the 29th USENIX Conference on Security Symposium, pp. 1623–1640 (2020)
4. Xie, C., Huang, K., Chen, P.-Y., Li, B.: Dba: distributed backdoor attacks against federated learning. In: International conference on learning representations (2020)
5. Hitaj, B., Ateniese, G., Perez-Cruz, F.: Deep models under the GAN: information leakage from collaborative deep learning. In: Proceedings of the 2017 ACM SIGSAC conference on computer and communications security, pp. 603–618 (2017)
6. Cao, D., Chang, S., Lin, Z., Liu, G., Sun, D.: Understanding distributed poisoning attack in federated learning. In: 2019 IEEE 25th International Conference on Parallel and Distributed Systems (ICPADS), (IEEE, 2019), pp. 233–239 (2019)
7. Mahsereci, M., Balles, L., Lassner, C., Hennig, P.: Early stopping without a validation set, arXiv preprint arXiv:1703.09580 (2017)
8. Biggio, B., Nelson, B., Laskov, P.: Poisoning attacks against support vector machines. In: Proceedings of the 29 th International Conference on Machine Learning (2012)
9. Hu, S., Lu, J., Wan, W., Zhang, L.Y.: Challenges and approaches for mitigating byzantine attacks in federated learning, arXiv preprint arXiv:2112.14468 (2021)
10. Zhao, Y., Li, M., Lai, L., Suda, N., Civin, D., Chandra, V.: Federated learning with non-iid data, arXiv preprint arXiv:1806.00582 (2018)
11. Mahloujifar, S., Mahmoody, M., Mohammed, A.: Multi-party poisoning through generalized p-tampering, arXiv preprint arXiv:1809.03474, 5 (2018)
12. Xie, C., Koyejo, O., Gupta, I.: Fall of empires: breaking byzantine-tolerant sgd by inner product manipulation. In: Uncertainty in Artificial Intelligence, (PMLR, 2020), pp. 261–270 (2020)
13. Liu, Z., Chen, Y., Yu, H., Liu, Y., Cui, L.: Gtg-shapley: Efficient and accurate participant contribution evaluation in federated learning. ACM Trans. Intelli. Sys. Technol. (TIST) **13**(4), 1–21 (2022)
14. Konečný, J., McMahan, H.B., Ramage, D., Richtárik, P.: Federated optimization: distributed machine learning for on-device intelligence, arXiv preprint arXiv:1610.02527 (2016)
15. Hard, A., et al.: Federated learning for mobile keyboard prediction. arXiv preprint arXiv: 1811.03604 (2018)
16. Li, L., Fan, Y., Tse, M., Lin, K.-Y.: A review of applications in federated learning. Computers & Industrial Engineering, 106854 (2020)
17. Li, T., Sahu, A.K., Talwalkar, A., Smith, V.: Federated learning: challenges, methods, and future directions. IEEE Signal Process. Mag. **37**(3), 50–60 (2020)
18. Bonawitz, K. et al.: Towards federated learning at scale: System design, arXiv preprint arXiv: 1902.01046 (2019)
19. Kairouz, P., et al.: Advances and open problems in federated learning, arXiv preprint arXiv: 1912.04977 (2019)
20. McMahan, B., Moore, E., Ramage, D., Hampson, S., Arcas, B.A.: Communication-efficient learning of deep networks from decentralized data. In: Artificial intelligence and statistics, (PMLR, 2017), pp. 1273–1282 (2017)
21. Asad, M., Moustafa, A., Ito, T.: FedOpt: towards communication efficiency and privacy preservation in federated learning. Appl. Sci. **10**(8), 2864 (2020)
22. Yang, W., Zhang, Y., Ye, K., Li, L., Xu, C.-Z.: Ffd: a federated learning based method for credit card fraud detection. In: International conference on big data, pp. 18–32. Springer (2019)
23. Chen, Y., Qin, X., Wang, J., Yu, C., Gao, W.: Fedhealth: a federated transfer learning framework for wearable healthcare. IEEE Intell. Syst. **35**(4), 83–93 (2020)

24. Brisimi, T.S., Chen, R., Mela, T., Olshevsky, A., Paschalidis, I.C., Shi, W.: Federated learning of predictive models from federated electronic health records. Int. J. Med. Informatics **112**, 59–67 (2018)
25. Saputra, Y.M., Hoang, D.T., Nguyen, D.N., Dutkiewicz, E., Mueck, M.D., Srikanteswara, S.: Energy demand prediction with federated learning for electric vehicle networks. In: 2019 IEEE Global Communications Conference (GLOBECOM), (IEEE, 2019), pp. 1–6 (2019)
26. Lu, Y., Huang, X., Dai, Y., Maharjan, S., Zhang, Y.: Federated learning for data privacy preservation in vehicular cyber-physical systems. IEEE Network **34**(3), 50–56 (2020)
27. Wu, C., Wu, F., Cao, Y., Huang, Y., Xie, X.: Fedgnn: federated graph neural network for privacy-preserving recommendation, arXiv preprint arXiv:2102.04925 (2021)
28. Fung, C., Yoon, C.J., Beschastnikh, I.: Mitigating sybils in federated learning poisoning, arXiv preprint arXiv:1808.04866 (2018)
29. He, X., Ling, Q., Chen, T.: Byzantine-robust stochastic gradient descent for distributed low-rank matrix completion. In: 2019 IEEE Data Science Workshop (DSW), (IEEE, 2019), pp. 322–326 (2019)
30. Jagielski, M., Oprea, A., Biggio, B., Liu, C., Nita-Rotaru, C., Li, B.: Manipulating machine learning: poisoning attacks and countermeasures for regression learning. In: 2018 IEEE symposium on security and privacy (SP), (IEEE, 2018), pp. 19–35 (2018)
31. Xiao, H., Biggio, B., Brown, G., Fumera, G., Eckert, C., Roli, F.: Is feature selection secure against training data poisoning?. In: international conference on machine learning, (PMLR, 2015), pp. 1689–1698 (2015)
32. Fung, C., Yoon, C.J., Beschastnikh, I.: The limitations of federated learning in sybil settings. In: RAID, pp. 301–316 (2020)
33. Gong, X., Chen, Y., Wang, Q., Kong, W.: Backdoor attacks and defenses in federated learning: state-of-the-art, taxonomy, and future directions. IEEE Wireless Communications (2022)
34. Blanchard, P., El Mhamdi, E.M., Guerraoui, R., Stainer, J.: Machine learning with adversaries: byzantine tolerant gradient descent. Advances in neural information processing systems 30 (2017)
35. Xia, Q., Tao, Z., Hao, Z., Li, Q.: FABA: an algorithm for fast aggregation against byzantine attacks in distributed neural networks. In: International Joint Conference on Artificial Intelligence (IJCAI), pp. 4824–4830 (2019)
36. Yin, D., Chen, Y., Kannan, R., Bartlett, P.: Byzantine-robust distributed learning: towards optimal statistical rates. In: International Conference on Machine Learning, (PMLR, 2018), pp. 5650–5659 (2018)
37. Xie, C., Koyejo, O., Gupta, I.: Generalized byzantine-tolerant sgd, arXiv preprint arXiv:1802.10116 (2018)
38. Xie, C., Koyejo, S., Gupta, I.: Zeno: distributed stochastic gradient descent with suspicion-based fault-tolerance. In: International Conference on Machine Learning, (PMLR, 2019), pp. 6893–6901 (2019)
39. Park, J., Han, D.-J., Choi, M., Moon, J.: Sageflow: robust federated learning against both stragglers and adversaries. Adv. Neural. Inf. Process. Syst. **34**, 840–851 (2021)
40. Li, S., Cheng, Y., Liu, Y., Wang, W., Chen, T.: Abnormal client behavior detection in federated learning, arXiv preprint arXiv:1910.09933 (2019)
41. Li, L., Xu, W., Chen, T., Giannakis, G.B., Ling, Q.: RSA: Byzantine-robust stochastic aggregation methods for distributed learning from heterogeneous datasets. In: Proceedings of the AAAI Conference on Artificial Intelligence, pp. 1544–1551 (2019)
42. Sun, Z., Kairouz, P., Suresh, A.T., McMahan, H.B.: Can you really backdoor federated learning?, arXiv preprint arXiv:1911.07963 (2019)
43. Wu, C., Yang, X., Zhu, S., Mitra, P.: Mitigating backdoor attacks in federated learning, arXiv preprint arXiv:2011.01767 (2020)

44. Smith, V., Chiang, C.-K., Sanjabi, M., Talwalkar, A.S.: Federated multi-task learning. Advances in neural information processing systems 30 (2017)
45. Nguyen, D.C., et al.: Federated learning for smart healthcare: a survey. ACM Computing Surveys (CSUR) **55**(3), 1–37 (2022)
46. Bonawitz, K., et al.: Practical secure aggregation for federated learning on user-held data, arXiv preprint arXiv:1611.04482 (2016)
47. Geyer, R.C., Klein, T., Nabi, M.: Differentially private federated learning: a client level perspective, arXiv preprint arXiv:1712.07557 (2017)
48. Ohrimenko, O.: et al.: Oblivious {Multi-Party} machine learning on trusted processors. In: 25th USENIX Security Symposium (USENIX Security 16), pp. 619–636 (2016)
49. Li, X., Qu, Z., Zhao, S., Tang, B., Lu, Z., Liu, Y.: Lomar: a local defense against poisoning attack on federated learning. IEEE Transactions on Dependable and Secure Computing (2021)
50. Liang, G., Chawathe, S.S.: Privacy-preserving inter-database operations. In: International Conference on Intelligence and Security Informatics, pp. 66–82. Springer (2004)
51. Scannapieco, M., Figotin, I., Bertino, E., Elmagarmid, A.K.: Privacy preserving schema and data matching. In: Proceedings of the 2007 ACM SIGMOD international conference on Management of data, pp. 653–664 (2007)

Exploring Human Behavior and Communication in Business: Case Studies and Empirical Research

Why Do People Donate for Live Streaming? Examining the S-O-R Process of Sponsors in the Live Streaming Context

Tsai-Hsin Chu[1], Wei-Hsin Chu[1], and Yen-Hsien Lee[2(✉)]

[1] Department of E-Learning Design and Management, National Chiayi University, Chiayi City, Taiwan, R.O.C.
[2] Department of Information Management, National Chiayi University, Chiayi City, Taiwan, R.O.C.
yhlee@mail.ncyu.edu.tw

Abstract. As live streaming becomes more popular, research has not fully explored how and why people donate in live streaming. Current studies examine motivations and need satisfaction to explain why people participate in live streaming. However, these studies do not provide a satisfactory answer to the question of why people continue to donate in live streaming. Studies also fail to show what stimuli perceived in a live stream can generate intense participation, such as subscription and donation. This study clarifies the behavioral mechanism that 'explains people's intense participation in live streaming. Applying the Stimulus-Organism-Response (S-O-R) model, this study attempts to answer the research question: How are donation behaviors (R) performed in response to individuals' sensemaking (O) to environmental stimuli (S)? We conducted an interpretive qualitative research to explore the sponsors' S-O-R process to clarify the above research question. Our findings indicated that sponsors interpreted the live stream as a party, and they paid more attention to the way they enjoyed the party. Sponsors donated a live stream to heat up the atmosphere and make the party fun. They left messages and attached funny videos to the pop-up window enabled by the donation to show their enthusiasm and support for the live streamer. Research implications are also discussed.

Keywords: Live streaming · Donators · S-O-R model · Interpretive research

1 Introduction

Live streaming is becoming increasingly popular and its impact is growing over time. According to a report by TwitchTracker in 2020, there are about 100,000 live stream channels on Twitch per month and they attract an average of over 2 million viewers to watch live streaming [1]. Existing studies link interest in live streaming to behavioral issues, particularly motivation, to explain why people watch a live stream. Collectively, these studies show that people watch a live streaming to satisfy needs such as social

© Springer Nature Switzerland AG 2023
F. Fui-Hoon Nah and K. Siau (Eds.): HCII 2023, LNCS 14039, pp. 259–273, 2023.
https://doi.org/10.1007/978-3-031-36049-7_20

satisfaction [2], sense of belonging [3], and identity [4]. While these studies have initially explained why people watch live streaming, we know less about what stimuli in live streaming can trigger people's motivation and, in turn, lead them to donate in live streaming.

This study uses the Stimulus-Organism-Response (S-O-R) model [5] as a theoretical lens to examine the underlying mechanism for explaining sponsors' live streaming participation behavior. The S-O-R model explains individual behavior as a response to how individuals interpret environmental stimuli [5]. Using this lens, this study argues for the need to pay attention to the information processing process from environmental stimuli to individual cognition and then to responsive behavior to explain individual participatory behavior in the live streaming context. The research question is: How can live streaming audience behavior be explained by environmental stimuli and individual cognition? We seek a satisfactory answer to why some viewers watch a live streaming and donate as sponsors. Specifically, we address the following four research questions: (1) What are the environmental stimuli (S) that sponsors selectively perceive in a live streaming context? (2) How do sponsors make sense (O) of the environmental stimuli? (3) What is the behavior (R) that sponsors choose in response to the environmental stimuli?

2   Literature Review

2.1   Research on Live Streaming

The literature on live streaming explains why people participate in live streaming by identifying their motivations [2, 4, 6]. Overall, the existing research concludes that people seek to satisfy needs by participating in a live stream. For example, Sjöblom and Hamari [2] proposed that cognitive, affective, personal integration, social integration, and tension release are motivations for people to participate in live streaming. Hilvert-Bruce, Neill, Sjöblom and Hamari [6] proposed six motivations for live streaming engagement, including social interaction, sense of community, meeting new people, entertainment, information seeking, and lack of external support in real life. They found that entertainment, information seeking, meeting new people, social interaction, and sense of community could significantly predict emotional connectedness. Entertainment, social interaction, and lack of external support were positively correlated with time spent, while social interaction and sense of community were positively correlated with subscriptions and donations [6].

Some studies examine the predictors of active and passive participation behaviors. For example, Bründl, Matt and Hess [7] found that perceived co-experience can increase the enjoyment of chatting and watching, but perceived efficacy only affected the enjoyment of chatting. Hu, Zhang and Wang [4] suggested that viewers continue to watch a live stream because they perceive the streamer as a desirable role model, or perceive the streamer and other viewers as similar to themselves. And Li, Lu, Ma and Wang [3] proposed that class identity (i.e., personal identity) and relational identity (i.e., the role of fans in the relationship with the streamer) influence viewers' gifting behavior in the live stream context. Both class identity and relational identity positively influenced the number of paid gifts given [3].

While current literature enhances our understanding of individual live stream motivations and participatory behaviors, it remains unclear about which stimuli create motivations and lead to particular actions as a response. It is also necessary to clarify the interpretations that an individual makes sense of and uses to respond to a live stream. This knowledge gap should be bridged so that live streamers can better manage their services by providing appropriate stimuli to elicit specific sensemaking and subsequently elicit the expected participatory behaviors (e.g., subscription and donation).

2.2 Stimulus-Organism-Response (S-O-R) Model

Derived from environmental psychology, the S-O-R model suggests that material and social stimuli in the external environment (i.e., stimulus, S) generate the individual's internal cognitions and affections (i.e., organism, O), which in turn elicit behavioral responses (i.e., response, R) [5]. This model has been successfully used to study individuals' behavior in e-commerce and social commerce contexts, such as website stickiness [8], online impulse buying [9, 10], online purchase intention [11, 12], virtual goods purchase intention [13], m-payment adoption [14], and discontinuous social media use intention [15].

While Mehrabian and Russell [5] propose the conceptual definition of S, O, and R, subsequent studies specify the definition of the three constructs to fit specific research concerns and settings. For example, to investigate the non-transactional behavior of live stream customers, Kang, Lu, Guo and Li [16] specified interactivity as the social stimulus, tie strength as the organism, and customer engagement behavior as the response. In the context of social games, Huang [13] defined information, system, and service quality as stimuli, satisfaction, trust, and intimacy as organism, and customer loyalty as response. In social media use, Yuan, Liu, Su and Zhang [14] identified overload as a stimulus, exhaustion and regret as an organism, and discontinuous use intention as a response. In this light, this study defined the S-O-R according to our research concerns in the context of SLSS as follows.

Technology Features as Stimuli. Stimuli refer to the material and social cues in the external environment [5]. In the physical context, it can be manifested in various formats, such as a product display, the store environment, and the availability of a salesperson [17]. In the online context, stimuli refer to the technological features of the website or APP, such as environmental conditions, symbols, artifacts, architecture, and spatial arrangement [18]. The live stream platforms such as twitch typically Provide various embedded Internet Relay Chat (IRC) channels, including streaming video, chat rooms, pop-up messages, and/or Damuku [19, 20]. With these features, viewers can have a shared experience with live streams and other viewers through viewing and chatting behaviors [7, 21].

Sensemaking as Organism. Organism refers to the individual's organic experiences, such as flow, involvement, cognitive network, and schema [13]. In studies of technology use, this concept is often constructed as technological sensemaking. Technological sensemaking is a process by which an individual constructs a frame of reference for understanding the unknown technology and taking action toward that technology. This process helps people understand the unknown, but it limits the action space as a response

[22–24]. When faced with an unknown, people look for external cues and search for similar mental scripts that built from past experiences, apply the script to understand the unknown, and act on it. This cognitive mechanism allows people to quickly construct the unknown into something they can understand so that they can perform the cognitive functions of integration, comprehension, explanation, and prediction.

Sensemaking is essentially a kind of plausibility rather than accuracy [23]. People may make different senses of the same technology due to the different external cues they observe, as well as the different context and social interactions they are in [24]. IS research have used the perspective of technological sensemaking to explain technology use. These studies concern with how users interpret technology and present the technological interpretations (also called technological frames) to explain users' enactments of technology use [25, 26].

Participatory Behavior as Response. In live streaming, the streamer broadcasts live video content (such as playing video games, eating, painting, and dancing), along with humorous banter and lighthearted conversation, to a public audience [6, 20]. Participation in a live stream is highly open, meaning that audiences are free to watch and listen to the streamer, interact with streamers and other viewers, and contribute to the streamer's income through subscriptions, donations, and virtual gifts [6, 27].

3   Research Methodology

3.1   Research Design

This study used an interpretive case study approach to investigate our research question. Theoretical sampling was the principle for selecting the research site and potential respondents. Theoretical sampling refers to a theory-driven sampling approach in which the selected case phenomenon should have the potential to explore a specific research question [28, 29]. In this study, we consider Twitch to be an appropriate research site for the following reasons. First, Twitch is a live stream platform whose technological features are dedicated to live streaming. Therefore, studying the phenomenon on Twitch allows us to focus on the effects of live streaming context compared to studying live streaming on social media platforms such as Facebook. Second, Twitch is a leading live stream platform in terms of the number of streamers and viewers [1, 20]. As of February 2020, there were 3.8 million streamers on Twitch, and in March 2020, an average of 1.44 million viewers were watching live streams online at the same time [30]. Third, Twitch is the most popular and well-known live stream platform [31]. In 2018, Twitch was reported to be the 30th most viewed website in the world, with approximately 15 million users online every day [32].

The research design of this study was iterative, with the research questions gradually converging based on the themes that emerged from the field data, as well as the dialogue between the field data and theory. This research design was consistent with the principle of theoretical sampling, which allowed the researcher to decide which groups to target, what data to collect, and where to obtain these data based on the preliminary analysis

[33], and provided the researcher with the flexibility to explore in depth the concepts that emerged from the field data [28].

This study went through three phases of investigation to gradually converge the research focus. The first phase of investigation was to have a general understanding of the characteristics, ecology, and business models about the live streaming industry. We conducted four interviews with a manager and an agent of a livestream platform. In this phase, we excluded the potential effect of technology features by setting our research context on the platform mainly designed for live streaming (e.g., Twitch, 17Live, LiveMe). In addition, we set the research target on live game broadcast to control the potential effect of different business models in other live streaming, such as talk show and talent show. For example, livestream on talent shows attracted sponsors to donate virtual gifts as revenue, while those on game broadcasts attracted subscriptions and business endorsements to generate revenue.

Then, in the second phase of investigation, we explore viewers' participation in live streaming, including why they participated in a live stream, what they did in a live stream, and why they participated in a live stream as it was. In this phase of the investigation, our field data revealed that people paid attention to different environmental stimuli when participating in a live stream. These emerging themes initiated the third phase of investigation, in which the S-O-R model was used as a theoretical lens for data collection and analysis.

3.2 Data Collection

This study used in-depth interviews as the main approach to data collection, supplemented by field observations and archival documents. During field observations, researchers participated in livestreams as viewers to deepen our understanding of the technological features, content, and typical interactions in the livestream context. For archival data, we collected reports from livestream platforms and the press.

For the interview, we recruited the potential respondents who were familiar with and had donated to a livestream on Twitch. The interviews were conducted in October 2020. We posted a pre-screening questionnaire on social media platforms and online forums in Taiwan to invite the potential respondents who not only watched a livestream but also paid to donate. Among the participants of the screening, we selected qualified respondents as interviewees with the set conditions, including those who were between 16 and 34 years old, had the habit of watching, and had donated on a livestream. As a result, 10 sponsors were interviewed. The profile of the respondents is shown in Table 1. The majority of respondents were male, between the ages of 20 and 23, and experienced participants, having participated in livestreams for at least three years. The 10 sponsors interacted with the livestreamers through pop-up messages after making donations. These sponsors made multiple donations with small amounts each time (approximately USD$3–17).

Semi-structured interviews were conducted using open-ended questions. An interview schedule was prepared to guide, but not constrain, the interview process. A series of follow-up questions were used to encourage feedback and elaboration from the interviewees in order to gain deeper opinions and insights about the livestream. The interview schedule is shown in Table 2.

Table 1. Respondent Demographics

Attribute	Numbers	Attribute	Numbers
Gender		**Participation experience (year)**	
Male	7	<1	0
Female	3	1–2	1
		3–4	5
Age (years old)		5–6	2
<16	1	7–9	1
20–21	3	**Amount donated per time ($NT, $30 NT ≈ $1 USD)**	
22–23	3	<100	4
>= 24	3	101~500	5
		500~1000	1

3.3 Data Analysis

Data analysis was carried out in two phases. The first phase of analysis began with open coding. We examined each interview transcript sentence by sentence, dividing and extracting meaningful units from the semantic paragraphs by assigning an appropriate label to represent the concept. The labels of similar concepts were then grouped into categories to form themes. At the beginning of this study, we examined participants' motivations for explaining these participatory behaviors. However, we found that people expressed similar motivations even though they exhibited different engagement and participation behaviors. Some viewers stayed at a distance to watch, while others actively participated in the action of heating up the livestream. This observation led us to focus on the link between internal interpretation and participatory behavior. Another emerging theme was the complexity of the livestream interface and features. The Twitch interface included a variety of icons and messages, such as games, streamer images, chat rooms, pop-ups, and voices. If users received different symbols as stimuli, their interpretation of the livestream would be different. In this sense, this study sought the S-O-R model as a theoretical lens for analysis.

In the second phase of analysis, we used the S-O-R model as a theoretical lens to develop the relationship among the emerging themes for theory building and insight generation. We define S as the live stream content and technological features that viewers notice during live streaming. In our field data, the identified stimuli included the live content provided by the streamer, the chat room messages from viewers, and text or video messages appearing briefly in the form of sponsored pop-ups.

For O, we initially followed [34] distinction between cognition and emotion to operationalize O. However, our field data showed that cognition and emotion were not independent concepts, but intertwined to shape understanding of a livestream and corresponding behaviors. For example, respondents noted that in some stream rooms that were perceived as negative environments, they would mimic the live streamer's provocative comments

Table 2. The interview scheme

Concepts and example questions
Live stream participation experience
• When did you begin to watch (or how long have you been watching)?   • What prompted you to start watching or accessing live streaming?   • What is your most watched or favorite live stream? What platform(s) do you use to watch this live stream? What is the content of this live stream?
Stimuli perceived
• In the live broadcast above, how did the streamer conduct the live steam? Which technology features did she or he use for live streaming?   • Which features did the live steamer use to interact with her or his fans during the live stream mentioned above?   • What is the event that stood out to you during this live steaming? Please give us some examples
Sensemaking about Live Steam participation
• Please use one or two adjectives to describe the live streamers. What event made you feel like this? Please give us an example   • Please use one or two adjectives to describe the live stream? What event made you feel like this? Please give us an example   • Please describe the fans gathered in this livestream. What was the trigger event for this feeling? Please give us an example   • In summary, what do you think it means to you to donate the live stream? Why?
Participation behaviors
• What did you do while participating in the above live streams? Please give us an example   • What made you want to sponsor this live stream?   • Please describe the most impressive sponsorship you have provided

in the chat room. They pointed out that it was normal interaction to provoke the streamer in these places, and that such behavior made them happy. However, they also said that in other peaceful live stream rooms, they could not provoke the streamers, but were happy with the calm interactions. They had different interpretations of how to feel happy and whether a particular behavior was appropriate to participate in different live streams. These observations led us to believe that people's interpretations of live environmental events were important in determining their participation actions, and we therefore focused on sensemaking as a manipulation of O.

Behavioral responses refer to approach and avoidance behaviors in the S-O-R model [5]. However, our field data showed that all respondents continuously watch live streams, but have different levels of involvement in live interactions. Therefore, we classify participation behaviors according to their level of involvement in live interactions. The emerging codes representing a high level of engagement include behaviors such as leaving messages in chat rooms, sending pop-up videos and messages, sharing live streams with friends, and donating. The code representing a lower level of engagement includes behaviors such as watching and not leaving messages.

4   Research Findings

We found that sponsors understood and participated in a livestream as if they were at a party. In contrast to the free viewers, who mostly talked about events created by others, the sponsors focused more on themselves and talked about how they participated in a livestream (Fig. 1).

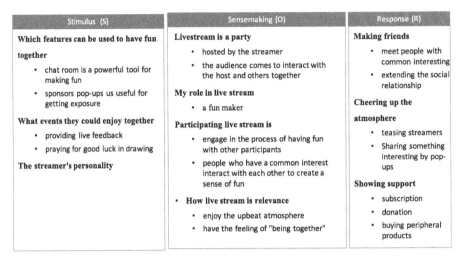

Fig. 1. The Sponsor's S-O-R: A Journey of Party Enjoyment

4.1   Stimulus(s): What the Sponsor Perceived

In addition to noting live content, sponsors pay more attention to features that are used to have fun together. Sponsors noted that chat rooms were not only used to ask questions, but were also a powerful tool for sharing a joke with people in the livestream. For example, one sponsor mentioned what he found in the chat room:

> The chat room would be flooded with something like…'broom…lend me a broom'. It was because her hair looks like a broom at that live stream… They like to make jokes about her hair, like "Sorry I spilled my drink, can I borrow that mop on your head to mop the floor? Something like that…" ~M2

In addition to chat rooms, sponsors have found the sponsorship pop-ups to be very useful for getting exposure. They found that messages in chat rooms were often quickly swiped, so streamers might not always see what was being said. This led sponsors to use the sponsorship pop-ups, which is a pop-up message that appears at the top of the screen, to get the attention of the live streamer and the audience. Two respondents said:

> …There were too many messages in the chat room that the streamer couldn't see, but it was impossible to ignore the pop-up message along with Donate. So if I

wanted to say something, I would use Donate to talk to the streamer, he would never ignore it. ~M4

DT (a pseudonym) had a big American sponsor, and he donates $3,000 (about USD$100) every time. ... One time, this sponsor was arguing with another major sponsor, and we found that two of them were arguing through the pop-ups enabled by the donation. ... It was impressive because the amount of donation was large. We, as the audience, became riotous over this big event! ~M8

Sponsors paid attention to various events they could enjoy together on the livestream, such as providing live feedback and supporting livestream activities. For example, one sponsor noted what he noticed in a live stream:

...He [the streamer] would let the audience decide the game options. For example, when he played a horror game, he asked, "I'm going to do Plan A or B, which one do you prefer?" People would donate to him and use the pop-ups to tell him their preference...and then he would do it. ~M5

Sponsors also mentioned the event where a live streamer held a drawing to win a virtual treasure in a game as a prize in his livestream. If the live streamer was unable to draw the particular virtual treasure they wanted, the audience who donated to the live streamer could continue to draw the virtual treasure. When he started drawing, the live streamer would invite the audience to pray together for him. For example, one sponsor mentioned:

There was a time when he drew a special treasure in the game "Maplestory", he failed many times, the chat room fooled with pray for wish him luck. People watched the drawing, and people donated to him to help him draw for rounds. I thought it was pretty good to see this process: we work together to pray for him. "Come on! Let's win!" Then he fails and gets depressed. We support him and cheer him up. ~M2

Sponsors also pay attention to the personality of the broadcaster and the atmosphere of the livestream. Sponsors identify the streamer's image by his or her actions during the livestream. They were also sensitive to the atmosphere they perceived from the interaction between the streamer and the audience.

The livestream was more like playing a video game with a friend. I played video games, he played video games. Then we watched the same video game together to pass the time. ~M3

4.2 Sensmaking (O): How the Sponsors Understand Live Streaming

Sponsors understood live streaming as a small party where the audience in the chat room comes to interact with the live streamer and others together. When they participated in the activities in the party, the viewers could get entertainment, fun and satisfaction. Sponsors expected live streaming to have a degree of enjoyment by allowing viewers not only to watch, but also to participate in the process of having fun with other participants. For example, one sponsor mentioned what live streaming meant to them:

It was more of a feeling that everyone was having a party. We came to a show and we had fun together. ~M2

The sponsors also explained what a live party is like, saying that a live party is where people who have a common interest interact with each other to create a sense of fun. The typical script of the party they attended was similar to a person playing a computer game with a group of people watching and making many comments. Some of the sponsors described the livestream as a party for gamers. Viewers came to the livestream of a particular game because they were interested in that game.

It was a group of game losers who got together, watched the same video, and did the same thing. Sometimes people would come up with a silly idea, they would share it, and we could laugh together. ~M3

In addition, the sponsors understood the audience as a group of people who came together in a live room because of their common interests. The sponsors mentioned that they were familiar with some of the other viewers because they usually met in livestreams on the same topic. Their para-social relationships were more like close friends than people with shared interests. They chatted like old friends who appreciated each other's jokes when they met in the chatroom. For example, one sponsor said:

It was easy to communicate. We had interesting things in common, and everyone knew the punch line of a joke we were talking about. ~M5

Sponsors made sense to a live streamer through the way he or she presented the live content and interacted with the audience. Sponsors treated the streamer as the host of the live party and themselves as the followers and sidekicks. Sponsors identified live streamers as capable game masters who could demonstrate an excellent game with strong skills. Sponsors found that the streamer engaged the audience in interaction in addition to show mastery.

We followed X (a pseudonym) because he was a skilled gamer. Although he was also funny, his image was about a game master. He shared a lot of tips and concepts on how to master the game. ~M8

...Egg's performance was so entertaining that the audience called him "Funny Egg". He said that he was the top 1 master in game streaming, but the audience teased him that he was the top 1 funny cast in game streaming...I thought he was a master in engaging the audience in the entertaining situation. ~M7

Why did sponsors want to join the livestream party? Sponsors explained that they were attracted by the upbeat atmosphere created by the streamer's discourse, actions, and the joy shared by the audience. For example, one sponsor recalled what happened at a live party:

After the donation, there came out the music with a Foxmont's image shaking. The music was an Indian song, sounding like "Du Li De Du Da Da Da Da", and the streamer song among with it. It was quite funny. We donated to see that. ~M5

The sponsors participated in a live stream because of the feeling of "being together". Compared with the entertainment they usually did alone, such as playing games, watching dramas, and reading novels, participating in a livestream gave them the feeling of doing something with a group of people. This was the main reason why they participated in the interactions of livestreaming. One sponsor explained:

> I felt that people were watching the live stream, that we were all similar and had a common concern. I thought that was different from what you did on your own. You could play games by yourself, watch dramas by yourself, read books by yourself, and do other things by yourself. But the live stream was like people coming together, even though I didn't know those people personally. I probably... I was looking for the joy of people having fun together. ~M1

Sponsors saw sponsorships as a way to support the live party. They noted that donating was like paying for drinks at a party to have fun, to get the attention of the party attendees, or to show support for the party host. For them, sponsorship allowed them to present themselves, including making comments and teasing someone. By spending a small amount of money, they could have a lot of fun with the audience and the streamer. Two sponsors explained:

> You donated a small amount, usually one US dollar, to say something or to tease someone. You did this to feel a sense of closeness and belonging. You could enjoy the happy atmosphere in the live stream, have fun with your friends, and make jokes to the streamer. What you paid was only one dollar. ~M2

> I shared a funny video clip with my donation. It was the joy of sharing! It meant that I could not be the only one to see such a good video. It was like tagging friends when you share something interesting. I want to see people's reactions to what I've shared. ~M3

4.3 Response (R): Sponsor's Participatory Behaviors

Sponsors' participatory behaviors included making friends, cheering up the atmosphere, and supporting with donations.

Making Friends. The sponsors who made friends focused on communicating and interacting with the audience, just like extending the social relationship at a party. These sponsors said that viewers in the chat room were like friends who shared the experience of teasing streamers. Over time, they became friends and may extend their interactions to play together. one sponsor said:

> We were like old friends. In other live streams, people paid more attention to talking to the streamers and maybe following the trends in the chat room. But we, the PUBG viewers, were different. We talked to each other in the chat room. ~M4

Cheering Up the Atmosphere. Some sponsors took action to lighten the mood. they used sponsored messages, chat room messages, and other features to bring excitement to the live party. For example, two of the sponsors mentioned:

I thought it was funny to tease someone with others in the chat room. ...For example, one day the streamer was demonstrating something, but he did not notice that his phone number was displayed. The next day I donated and said, "Can I ask the streamer if there is a call-in session today? It's just for fun. It's pretty funny. ~M2

We had fun together. For example, he would scream when he played a scary game. We would type "someone is behind you" or "the chair has moved" and so on. We teased him together. ~M5

Supporting with Donations. The sponsor supported the streamer through donations and subscriptions. Sponsors paid around USD$5 per month to subscribe to a live stream. sponsors also supported a live streamer by purchasing peripheral products.

I bought more and more of the live streamer's peripheral products. One reason was that the good quality pictures drawn by the artist he works with made me want to buy them. The other reason was to show my support for the live streamer. ~M3

5 Discussion

Our findings suggest that the S-O-R journey of sponsors is more like attending a party. In terms of **stimuli (S)**, sponsors pay more attention to the features that allow them to create and participate in events as stimuli. They pay more attention to the pop-ups for sending sponsored messages and attaching videos or music to the pop-ups to create fun. Sponsors also put more emphasis on their actions to create an exciting, fun atmosphere in the live stream. In terms of **organism (O)**, sponsors consider attending a livestream to be like attending a party, where attendees support the host with actions to create a fun atmosphere. They understand the live streamer as the host of the live party, inviting the audience to participate in exciting games and have fun. They also express a strong sense of seeking the joyful atmosphere created in a live stream. In the **participation (R),** sponsors are actively involved and become part of the live interaction. When some viewers type symbols such as 777 (i.e. great), 4 (i.e. yes), 54 (i.e. no), ha ha ha (i.e. laugh) to follow the trend in the chat room, sponsors use pop-up messages with sponsorship to invite streamers to interact or make fun of streamers to heat up the atmosphere. They were more likely to engage in intense interactions with streamers and others.

5.1 Research Implications

Our research findings have two theoretical implications. First, this study presents the process of stimulus, sensemaking, and response behavior of sponsors. Previous studies explain live streaming behavior in terms of need satisfaction, but lack an explanation of what stimulates viewers' behavioral arousal in the live streaming context. This study bridges the theoretical gap by exploring the mental processes of sponsors in live streaming through the S-O-R model.

Second, this study presents the information processing processes of the sponsors' participation behavior. Our findings illustrate the stimuli perceived by sponsors, unfold their interpretations of the live stream, and present their participation behaviors. Such a pragmatic analysis is particularly important in the context of live streaming. This is because the livestream platform provides many stimuli simultaneously on one screen, including game content and sound, live camera, live streamer's voice, chat room messages, pop-ups, and pop-up sponsorship boxes. Such simultaneous stimuli created a huge information load, and people could not notice all of them. Instead, to reduce information overload, they selectively pay attention to some of the external stimuli and make sense of the received stimuli to take responsive actions. Therefore, to understand how and why viewers act in live streams, we need to reconsider the entire process from stimulus to organization to response.

In practice, this study provides live streamers with ways to attract sponsorship to increase revenue. When attracting a pool of viewers, live streamers can manipulate the stimuli to drive a particular sensemaking to convert free views into sponsorships. The live streamer makes viewers aware of the benefits of using sponsorship pop-up messages. This is because the sponsors show uniqueness in treating sponsorship messages as a cheerful tool to heat up the atmosphere of a party. It makes the sponsors to create a sense of engagement on joyful interactions and activities that allow people to have fun together.

6 Conclusion

This study explains sponsor behavior by exploring their S-O-R process in the context of live streaming. The research results revealed that sponsors pay more attention to the affordance of technology features that they can use to achieve social presence and a sense of togetherness when participating in a livestream. For the sponsors, a livestream is like a party where they can participate in the activities to enjoy the atmosphere of being with people. We also present the process participation behaviors of sponsors, based on their perceived stimuli and interpretations.

References

1. Twitchtrack Homepage: https://twitchtracker.com/statistics. Last accessed 1 Apr 2021
2. Sjöblom, M., Hamari, J.: Why do people watch others play video games? An empirical study on the motivations of Twitch users. Comput. Hum. Behav. **75**, 985–996 (2017)
3. Li, R., Lu, Y., Ma, J., Wang, W.: Examining gifting behavior on live streaming platforms: an identity-based motivation model. Inf. Manag. **58**, 103–406 (2020)
4. Hu, M., Zhang, M., Wang, Y.: Why do audiences choose to keep watching on live video streaming platforms? An explanation of dual identification framework. Comput. Hum. Behav. **75**, 594–606 (2017)
5. Mehrabian, A., Russell, J.A.: The basic emotional impact of environments. Percept. Mot. Skills **38**, 283–301 (1974)
6. Hilvert-Bruce, Z., Neill, J.T., Sjöblom, M., Hamari, J.: Social motivations of live-streaming viewer engagement on Twitch. Comput. Hum. Behav. **84**, 58–67 (2018)
7. Bründl, S., Matt, C., Hess, T.: Consumer use of social live streaming services: the influence of co-experience and effectance on enjoyment. In: the 25th European Conference on Information Systems (ECIS 2017)

8. Friedrich, T., Schlauderer, S., Overhage, S.: The impact of social commerce feature richness on website stickiness through cognitive and affective factors: an experimental study. Electron. Commer. Res. Appl. **36**, 100861 (2019)
9. Floh, A., Madlberger, M.: The role of atmospheric cues in online impulse-buying behavior. Electron. Commer. Res. Appl. **12**, 425–439 (2013)
10. Chen, C.C., Yao, J.Y.: What drives impulse buying behaviors in a mobile auction? The perspective of the stimulus-organism-response model. Telematics Inform. **35**, 1249–1262 (2018)
11. Kaur, S., Lal, A.K., Bedi, S.S.: Do vendor cues influence purchase intention of online shoppers? An empirical study using S-O-R framework. J. Internet Comm. **16**, 343–363 (2017)
12. Kuhn, S.W., Petzer, D.J.: Fostering purchase intentions toward online retailer websites in an emerging market: an S-O-R perspective. J. Internet Comm. **17**, 255–282 (2018)
13. Huang, E.: Online experiences and virtual goods purchase intention. Internet Res. **22**, 252–274 (2012)
14. Yuan, S., Liu, L., Su, B., Zhang, H.: Determining the antecedents of mobile payment loyalty: cognitive and affective perspectives. Electron. Commer. Res. Appl. **41**, 100–971 (2020)
15. Cao, X., Sun, J.: Exploring the effect of overload on the discontinuous intention of social media users: an S-O-R perspective. Comput. Hum. Behav. **81**, 10–18 (2018)
16. Kang, K., Lu, J., Guo, L., Li, W.: The dynamic effect of interactivity on customer engagement behavior through tie strength: evidence from live streaming commerce platforms. Int. J. Inf. Manage. **56**, 102251 (2021)
17. Jacoby, J.: Stimulus-organism-response reconsidered: an evolutionary step in modeling (consumer) behavior. J. Consum. Psychol. **12**, 51–57 (2002)
18. Eroglu, S.A., Machleit, K.A., Davis, L.M.: Empirical testing of a model of online store atmospherics and shopper responses. Psychol. Mark. **20**, 139–150 (2003)
19. Tang, J.C., Venolia, G., Inkpen, K.M.: Meerkat and Periscope: I Stream, You Stream, Apps Stream for live streams. In: Conference on Human Factors in Computing Systems (2016)
20. Hamilton, W., Garretson, O., Kerne, A.: Streaming on Twitch: fostering participatory communities of play within live mixed media. In: CHI 2014: CHI Conference on Human Factors in Computing Systems. Association for Computing Machinery, New York, NY (2014)
21. Lim, S., Cha, S.Y., Park, C., Lee, I., Kim, J.: Getting closer and experiencing together: antecedents and consequences of psychological distance in social media-enhanced real-time streaming video. Comput. Hum. Behav. **28**(4), 1365–1378 (2012)
22. Orlikowski, W.J., Gash, D.C.: Technology frames: making sense of information technology in organizations. ACM Trans. Inf. Syst. **12**, 174–207 (1994)
23. Weick, K.E.: Sensemaking in Organizations. Sage, Thousand Oaks, CA (1995)
24. Hsu, C.: Frame misalignment: interpreting the implementation of information systems security certification in an organization. Eur. J. Inf. Syst. **18**, 140–150 (2009)
25. Faulker, P., Runde, J.: On the identity of technological objects and user innovation in functions. Acad. Manag. Rev. **34**, 442–462 (2009)
26. Chu, T.H., Robey, D.: Explaining changes in learning and work practice following the adoption of online learning: a human agency perspective. Eur. J. Inf. Syst. **17**, 79–98 (2008)
27. Scheibe, K., Fietkiewicz, K.J., Stock, W.G.: Information behavior on social live streaming service. J. Inf. Sci. Theor. Pract. **4**, 6–20 (2016)
28. Mason, J.: Qualitative Researching. Sage, London (2002)
29. Strauss, A., Corbin, J.M.: Basics of Qualitative Research: Grounded Theory Procedures and Techniques. Sage (1990)
30. BusinessofApps: https://www.businessofapps.com/data/twitch-statistics/. Last accessed 21 Mar 2021
31. Twitch: Thanks to you, we reached a record 1 trillion Minutes Watched in 2020. Twitter (2021)

32. Johnson, M.R., Woodcock, J.: "And today's top donator Is": how live streamers on Twitch.Tv monetize and gamify their broadcasts. Social Media + Society **5**, 4–10 (2019)
33. Strauss, A., Corbin, J.: Grounded theory methodology: an overview. In: Denzin, N.K., Lincoln, Y.S. (eds.) Handbook of Qualitative Research, pp. 273–285. Sage, Thousand Oaks, CA (1994)
34. Vergura, D.T., Zerbini, C., Luceri, B.: Consumers' attitude and purchase intention towards organic personal care products. An application of the S-O-R model. Sinergie Italian J. Manag. **38**, 121–137 (2020)

The Socioemotional Wealth Effect on Intra-family Succession in Family Businesses

Pi Hui Chung$^{(\boxtimes)}$

Fu Jen Catholic University, New Taipei City 242062, Taiwan
152396@mail.fju.edu.tw

Abstract. CEO succession brings temporary internal disruption and opportunities to adapt firms' strategies and fast-changing environments. To better understand CEO succession in family businesses, this study draws on socioemotional wealth (SEW) perspectives to illustrate the effect of SEW on intra-family succession. The present study argues that the presence of higher family SEW (family ownership) is positively related to intra-family succession in a family firm, and this relationship is stronger when the high family involvement in board and longer family candidate's TMT tenure in a focal family firm, respectively. By using a logistic regression model with data on Taiwanese publicly listed firms from 2001 to 2013, this study finds that family firms prefer adopting intra-family succession to preserve family SEW, such a tendency is stronger when the family involvement in board is higher and when the family candidate's TMT tenure is longer, respectively. This study contributes to the CEO succession and family business literature by emphasizing the importance of governance and managerial arrangements.

Keywords: Socioemotional Wealth (SEW) Perspective · Intra-Family Succession · Family Involvement in Board · Family Candidate's TMT Tenure

1 Introduction

CEO succession, one of the most important research topics in the family business, can lead to substantial changes in the structure, volatility, strategy, and outcome of firms (e.g., [1−3]) and determine the organizational path for several years or decades [4]. Given the importance of CEO succession, numerous studies have generated many insights and managerial implications of CEO succession by examining the antecedent of CEO selection [5], the origin of successors [6], the impact on firms' strategic decisions and performance [7, 8]. But there has little attempt to include the affective endowment to understand its influence on CEO selection. To fill this research gap, this study attempts to propose and test a model that introduces the affective endowment on CEO succession.

The effect of non-economic and affective utilities—what this study refers to as socioemotional wealth (SEW)—plays a major role in family firms' strategies and outcomes [9]. This unique feature dominates family owners and influences everything family firms do. To better understand CEO succession in family businesses, this study draws on SEW perspectives to illustrate the effect of SEW on intra-family succession. Unlike

© Springer Nature Switzerland AG 2023
F. Fui-Hoon Nah and K. Siau (Eds.): HCII 2023, LNCS 14039, pp. 274–284, 2023.
https://doi.org/10.1007/978-3-031-36049-7_21

non-family counterparts, family firms use SEW as their primary reference point to balance the concern of economic and non-economic goals, such as financial performance versus the preservation of affective endowment related to their firms. This study further investigates how the influence of SEW on intra-family succession is conditioned by the family involvement in board and the candidate's top management team (TMT) tenure in the focal family firm.

The present study utilizes a sample of Taiwanese publicly listed firms in traditional industries from 2001 to 2013 to test hypotheses. The corporate governance and financial information were obtained from Taiwan Economic Journal (TEJ) database and the information on CEO succession was captured by firms' annual reports. The final sample of this study includes 455 CEO successions events in 248 firms. By using a logistic regression model with data on Taiwanese publicly listed firms from 2001 to 2013, this study finds that family firms prefer adopting intra-family succession to preserve family SEW, such tendency is stronger when the family involvement in board is higher and when the successor's TMT tenure is longer, respectively. This study contributes to the CEO succession and family business literature by emphasizing the importance of governance and managerial arrangements.

2 Literature Review

CEO successions are essential for firms to face temporary internal disruption and realign firms with fast-changing environments. Prior works pay more attention to the fit between top managers and the corporate environment that is conducive to surviving and prospering. The empirical findings suggest that outsiders may outperform insiders in certain contingencies, such as turbulent or competitive environment [6, 10], and simpler internal power structures [8]. However, CEO succession in family and nonfamily firms may differ because family firms' primary concern is whether families can influence and control firms by taking the CEO position. Adopting the traditional distinction between insider and outsider cannot differentiate family and nonfamily firms.

Family firms are more concerned with whether the successor should be a member of owning family [11], this has been recognized as one of the most critical and challenging choices for family firms [7, 12]. Gomez-Mejia et al. [13] proposed the socioemotional wealth perspective to emphasize the importance of the affective endowment of family firms. The notion of SEW is that the potential gain and loss in socioemotional wealth as family firms' primary reference point. Therefore, families may make decisions that are helpful to pursue socioemotional wealth but at the expense of other principals [13, 14]. This study extends the concept of SEW to CEO succession and investigates how affective endowment influence intra-family succession in family businesses.

2.1 Antecedent of Intra-family Succession

Despite abundant literature demonstrating the influences of CEO succession in family firms, little is known about the antecedents of CEO succession [4, 15, 16]. Furthermore, the mixed results and a lack of integrating affective endowment show the requirement for further investigation.

SEW is one of the theoretical perspectives to incorporate family firms' affective needs into consideration. The core of SEW emphasizes the non-economic motives in family firms, for example, the pursuit of the family's affective needs, the perpetuation of the family dynasty through family control and influence, sustaining social capital and status, and continuation of family values through the firm [13, 17]. Indeed, family firms' non-economic motives reflect the perceptions, values, attitudes, and intentions of the controlling family [9, 18, 19]. The loss of the family SEW implies lost intimacy, reduced status, and failure to meet the family's expectations [13]. Consequently, family firms tend to adopt strategies that are helpful to preserve SEW. Grounded in the behavioral agency model (BAM), which suggests decision makers' risk preferences vary with different governance contexts or situations, SEW perspective views the preservation of SEW or affective endowment as family firms' primary reference point rather than economic concerns for guiding their CEO succession since family members receive greater utility from preserving SEW even at the expense of financial or economic well-being [13, 14].

CEO succession, one of the critical strategic decisions, is often influenced by SEW to select the next leaders in family firms. To preserve and maintain the current SEW endowment, this study proposes that family firms have the motivation to choose successors in line with the social norm and family tradition, the intra-family succession thus would be a primary option for family firms. Indeed, adopting intra-family succession ensures the continuity of the family dynasty and conforms to the established social norm and family traditions. Therefore, if family firms have higher SEW endowments, the possibility of adopting intra-family succession would be higher as families may maximize their continuity. Family ownership was widely recognized as a suitable proxy for SEW endowment. If a family with higher ownership of the firm represents a higher SEW endowment that the family tends to preserve [14, 20], the owning family would prefer the decision to adopt intra-family succession to reinforce family control and dynasty.

H1: The presence of higher family SEW (family ownership) is positively related to intra-family succession in a family firm.

2.2 The Moderating Effect of Family Involvement in Board

The major tasks of a corporate board of directors are hiring and monitoring CEOs. The decisions of CEO succession are under the purview of the board. As for the pre-succession phase of CEO succession, boards play an essential role in establishing rules that are conducive to providing a sense of order in governance decisions, communicating the corporate mission, and guiding political dynamics into appropriate practices and executive turnover [5]. Meanwhile, CEO succession is deemed a vital opportunity to realign the firms with the controlling interests of the board of directors and with fast-changing environments.

The board of directors is essential for CEO succession by establishing a succession plan, selecting, and fostering adequate candidates in the pre-succession phase [5], evaluating and hiring the next CEO, and monitoring new CEOs' performance in the post-succession stage [21]. Family members are often arranged to be on the board to maintain the dominant control of family businesses. Family involvement in board is considered to be the degree of overlap between the family and the firm; different levels of family involvement in board are related to diverse behaviors on CEO selection [22, 23]. And yet, the role of the board has been less discussed in the literature on the CEO succession process [24]. Understanding the notion of family involvement in board would improve our knowledge of how families utilize their power to influence CEO succession.

Family involvement in board is viewed as active control of the firm via the family board of directors' decision-making in CEO selection. To ensure the control power of firms, the family board of directors usually selects a successor who identifies with business goals as well as family goals and values. To meet the criteria for following business and family goals, selecting a successor from owning family members could reduce information asymmetry and reach the goal of preserving family control and socioemotional wealth. Before choosing a successor, the board of directors often assesses the qualification of potential candidates, such as their commitment, willingness, and behavior in accordance with the firm's culture [25–27]. The board of directors usually lacks comprehensive information to assess potential successors' qualifications, but information on family candidates could be obtained from the family board of directors. Hence, the board of directors has more information to assess family successors' qualifications. Moreover, higher family involvement in board represents the substantial portion of the family wealth tied to the firm, the family board of directors thus tends to select a family successor to preserve their control and socioemotional wealth of the firm. This study thus hypothesizes.

H2: The possibility of adopting intra-family succession would be higher if jointly considering higher family SEW and higher family involvement in board in a family firm.

2.3   The Moderating Effect of Family Candidates' TMT Tenure

TMT plays a pivotal role in developing candidates' managerial capabilities and preparing them for a higher position. Thus, candidates' TMT tenure becomes one of the important factors influencing CEO succession. Indeed, tenure is associated with domain familiarity [28, 29]. A candidate with longer TMT tenure has developed familiarity with the firm and the greater familiarity is helpful to overcome the post-succession challenge [30]. Furthermore, a family candidate with longer TMT tenure could understand the mixed motives and desire the owning family has as family members would like to preserve and protect their interest and socioemotional wealth in business [31], intra-family succession thus would be preferred by family firms.

Next, longer TMT tenure also enhances the commitment to current strategies and practices [32], which is one of the key factors in determining the success of CEO succession in family firms. Organizational members usually become committed to their past actions over time [33], and a committed candidate shows a willingness to be a leader in the family firms and consequently facilitate a higher level of satisfaction with the succession process [25]. Family candidates with longer TMT tenure escalate their commitment to focal family firms, their commitment comes from not only identification and involvement with firms but also the emotional links with business [34, 35]. Furthermore, board of directors in family firms usually prefer family candidates who have a strong affective commitment and expect them to continue existing strategies, rather than introduce strategic change that might disrupt internal and external relationships [36]. This study thus hypothesizes.

H3: The possibility of adopting intra-family succession would be higher if jointly considering higher family SEW and longer TMT tenure of the family candidate in a family firm.

3   Method

3.1   Sample

The sample for this study was drawn from the population of publicly traded family firms listed on the Taiwan Economic Journal (TEJ) database between 2001 and 2013. Many firms in Taiwan's traditional industry are run by family members, therefore Taiwan's sample is considered appropriate for testing proposed hypotheses. Firstly, this study identified all firms which are owned by families from the TEJ database. Next, the present study identified 705 CEO successions during the period of 2001 to 2013 from the TEJ database and firms' annual reports. Observations with missing information would be removed, and the final sample of this study includes 455 CEO successions in 248 firms.

3.2   Dependent Variable

This study classified CEO successions into intra-family succession and non-intra-family succession, the variable *intra-family succession* was inserted as a dummy variable and coded 1 if a new CEO was the family member and 0 otherwise. A new CEO's family membership was checked and confirmed by the firms' annual reports and the TEJ database.

3.3   Independent Variable

In accordance with recent studies, this study captures *family SEW* by using family ownership of the firm [37–39], whereas the values of this variable were continuous shares of equity held by the family.

3.4   Moderators

Family involvement in board was included in this study to proxy the influence of family and was measured by the percentage of family members on the board [40, 41]. *A candidate's TMT tenure* is the number of years in which the candidate is employed by the focal family firm [42].

3.5   Control Variables

To rule out the potential effects, this study includes several control variables from firm, board, and individual levels that are known in the literature to affect the adoption of intra-family succession. Firm-related control variables include firm age, debt ratio, and prior performance. Firm age reflects that older firms are more likely to have well-established succession processes for selecting new CEOs. This study measures the firm age from the founding year to the time of succession [43]. Next, a firm's financial structure and prior performance influence CEO selection, thus debt ratio and prior performance (ROA_{t-1}) in the previous year were included in the analysis [44, 45].

Control variables from the board and individual levels include institutional ownership, CEO education, and CEO interlock. Institutional ownership may influence a board's decisions on CEO selection. Institutional ownership has figured out the ratio of shares held by banks, investment firms, and pension funds. CEO education, which relates to CEO selection, was measured by a categorical variable with three levels (master or Ph.D. degree = 3, undergraduate degree = 2, and no college degree = 0). To capture a CEO's information network extensiveness, CEO interlock was adopted as a proxy to control the effect on intra-family succession.

The year dummies were also included in this study to capture the year effect, with 2001 being used as a control. The industry dummy was also considered to control for the types of industry and was coded with a value of 1 if a firm in the manufacturing sector and 0 otherwise.

4 Results

Table 1 presents the descriptive statistics and bivariate correlations for all variables of our model. All correlation coefficients of the main variables are modest. To reduce the likelihood of multicollinearity, this study mean-centered all independent and moderating variables that do not belong to dummy variables. We also checked the variance inflation factors (VIFs) and found that the highest VIF among those in our model was 1.2, well below the general cutoff value of 10 [46]. Thus, multicollinearity does not pose a serious concern. All these models yield significant explanatory power (*Pseudo R2 = 0.1093–0.3784, p < 0.001*).

Table 2 shows the result of the logistic regression on the CEO selection in family firms. Model 1 presents the baseline model with only the control variables. Model 2 shows the main effects of the socioemotional wealth of CEO succession in family firms. Model 3 and Model 4 test the moderating effects of family involvement in board and the successor's TMT tenure in the focal family firm. Model 5 is the full model.

Model 2 examines the direct influence of CEO selection, and the results show that socioemotional wealth has a significant positive effect on intra-family succession ($\beta = 0.0273, p = 0.000$). In model 3, a positive interaction effect between socioemotional wealth and family involvement in board is associated with intra-family succession ($\beta = 0.1095, p = 0.0190$). Model 4 shows the results of the interaction effect between the effect of socioemotional wealth and the successor's TMT tenure, which receives support ($\beta = 0.0045, p = 0.001$). Model 5 shows the full model.

Table 1. Descriptive Statistics and Correlation Matrix.

		1	2	3	4	5	6	7	8	9	10
1	Firm age	1									
2	Debt ratio	0.0434	1								
3	Prior performance (ROA$_{t\_1}$)	0.0871	−0.1134*	1							
4	CEO interlock	0.2063*	−0.0901	0.057	1						
5	Institutional ownership	0.0331	0.0706	0.1007*	0.1290*	1					
6	CEO Education	−0.0794	−0.0359	0.0126	0.0977*	0.02	1				
7	Intra-family succession	0.0965*	0.0204	0.1139*	0.2233*	−0.0941*	−0.0437	1			
8	SEW (family ownership)	−0.1137*	0.0457	0.047	−0.0238	0.4380*	−0.0614	0.0835	1		
9	Family involvement in board (FIB)	0.1771*	0.0010	0.1728*	0.09	−0.1672*	−0.0270	0.4180*	−0.0397	1	
10	TMT tenure	0.1272*	−0.1079*	0.1580*	0.2085*	−0.0782	−0.1163*	0.4270*	0.0301	0.2221*	1

Note: 1. N = 455.

2. *represents statistical significance ($p < 0.05$).

3. Year effects (year dummy 1–12) were omitted in Table 1.

5 Discussion

5.1 Theoretical Contributions

This study makes several contributions to the literature on CEO succession and family business by exploring the relationships among SEW, family involvement in board, candidate's TMT tenure, and intra-family succession.

First, this study sheds light on the effect of SEW on intra-family succession. Most CEO succession research introduces SEW perspective to illustrate the influence on family firms' strategies and outcomes, but SEW has been ignored as a potential determinant of CEO selection. The findings show that SEW has a positive effect on intra-family succession. Given the results of this study, introducing SEW as an antecedent could be adequately included and assessed in the CEO succession.

This study also complements family business literature for a better understanding of the antecedents of CEO succession by introducing the contextual effects. The results show that the possibility of adopting intra-family succession would increase if taking high family involvement in board and a longer candidate's TMT tenure into consideration. The findings of this study imply the importance of contextual effects on CEO succession with the governance aspect of the succession arrangement.

Table 2. Logit Regression.

Intra-family succession	Model 1	Model 2	Model 3	Model 4	Model 5
Constant	−1.1800	−2.2064*	−3.1394*	−3.2115**	−4.0241**
Firm age	0.0019	0.0077	−0.0071	0.0027	−0.0123
Debt ratio	0.9611	1.0097	1.0370	1.5332†	1.4230
Prior performance (ROA$_{t\_1}$)	0.03006*	0.0284*	0.0093	0.0188	0.0011
CEO interlock	0.1431***	0.1540***	0.1670***	0.1317***	0.1384**
Institutional ownership	−0.0154**	−0.0243***	−0.0138†	−0.0248**	−0.0142†
CEO Education	−0.2591†	-0.2271	−0.2180	−0.0071	0.0040
Industry dummy	0.570235†	0.6802*	0.2912	0.5514	0.2200
Year dummy 1	−0.8803	−1.0303	−2.0148*	−0.8561	−2.0747†
Year dummy 2	25.9673	28.8251	29.8991	39.3072	44.7901
Year dummy 3	−0.8934	−1.0047	−1.2738	−.4422†	−1.9471*
Year dummy 4	−0.2919	−0.3632	−0.5352	−0.4836	−0.8385
Year dummy 5	−0.3468	−0.4988	−0.6670	−0.6804	−1.0016
Year dummy 6	−0.3371	−0.3514	−0.6088	−0.3552	−0.7590
Year dummy 7	−1.0001	−0.9481	−1.1380	−1.7067	−2.1390
Year dummy 8	0.1183	0.0667	−0.0958	−0.0897	−0.3033
Year dummy 9	−0.4011	−0.3899	−0.3689	−1.3995	−1.2626
Year dummy 10	−0.1939	−0.2483	−0.3302	−0.3698	−0.5117
Year dummy 11	−1.1880	−1.3747†	−1.2848	−1.5124†	−1.5036
Year dummy 12	−0.9222	−0.9885	−0.9154	−0.7777	−0.8961
H1: SEW (family ownership)		0.0273***	0.0208*	0.0281**	0.0219*
Family involvement in board (FIB)			5.5502***		5.4166***
H2: SEW X FIB			0.1095*		0.0954†
TMT tenure				0.1339***	0.1381***
H3: SEW X TMT tenure				0.0045**	0.0043*
LR chi2(19)	53.45***	67.41***	137.99***	130.49***	185.1***
Pseudo R^2	0.1093	0.1378	0.2821	0.2668	0.3784

Note: 1. N = 455.
2. †p < 0.1; *p < 0.05; **p < 0.01; ***p < 0.001.

5.2 Practical Implications

This study has several practical implications for family firms. First, this study argues that SEW exists in the intra-family succession in family firms, showing that SEW takes place in the family business among family members to protect family interests and wealth. Second, this study follows Zahra [47] to recognize the complex relationships existing in family firms and thus consider governance and candidate-related factors that may influence CEO succession. Finally, the findings of this study show that a candidate with longer TMT tenure in focal family firms increases the possibility of adopting intra-family succession. Long tenure gives candidates time to learn tacit knowledge and develop familiarity with the firm [48], possibly reducing the misfits between successors and focal family firms and overcoming the challenge of post-succession [30].

5.3 Limitations

This study, however, has some limitations that highlight promising avenues for future research. One of the limitations is the measurement of SEW. This study views SEW as an aggregated construct and uses family ownership as a proxy to capture SEW. However, this measurement is unable to show the multidimensional nature of SEW, which may lead to mixed predictions on CEO succession [49]. To advance the SEW literature, future studies can introduce different dimensions of SEW and examine the heterogenous effects of SEW on CEO selection.

A limitation of this study is that by focusing on internal influences, such as family and governance factors, the external environment could be included, such as market dynamism, market competition, and the change in digital technology. Future research can develop a more comprehensive model that includes internal and external factors of CEO succession in family firms to balance and compare influences from both sides.

Finally, the data are used from the unique context of Taiwan family firms. The generalizability of this study is limited because the country-specific effects cannot be ruled out. To increase the generalizability of findings, future research could redefine and extend the argument of this study to other countries or regions.

References

1. Huson, M.R., Malatesta, P.H., Parrino, R.: Managerial succession and firm performance. J. Financ. Econ. **74**(2), 237–275 (2004)
2. Denis, D.J., Denis, D.K.: Performance changes following top management dismissals. J. Finance **50**(4), 1029–1057 (1995)
3. Weisbach, M.S.: CEO turnover and the firm's investment decisions. J. Financ. Econ. **37**(2), 159–188 (1995)
4. Lee, K.S., Lim, G.H., Lim, W.S.: Family business succession: appropriation risk and choice of successor". Acad. Manage. Rev. **28**(4), 657–666 (2003)
5. Zhang, Y., Rajagopalan, N.: When the known devil is better than an unknown god: an empirical study of the antecedents and consequences of relay CEO successions. Acad. Manage. J. **47**(4), 483–500 (2004)
6. Zhang, Y., Rajagopalan, N.: Explaining new CEO origin: firm versus industry antecedents. Acad. Manage. J. **46**(3), 327–338 (2003)
7. Chang, S.-J., Shim, J.: When does transitioning from family to professional management improve firm performance? Strateg. Manag. J. **36**(9), 1297–1316 (2015)

8. Shen, W., Cannella, A.A.: Revisiting the performance consequences of CEO succession: the impacts of successor type, postsuccession senior executive turnover, and departing CEO tenure". Acad. Manage. J. **45**(4), 717–733 (2002)

9. Gomez-Mejia, L.R., Cruz, C., Berrone, P., De Castro, J.: The bind that ties: socioemotional wealth preservation in family firms. Acad. Manag. Ann. **5**(1), 653–707 (2011)

10. Virany, B., Tushman, M.L., Romanelli, E.: Executive succession and organization outcomes in turbulent environments: an organization learning approach. Organ. Sci. **3**(1), 72–91 (1992)

11. Minichilli, A., Nordqvist, M., Corbetta, G., Amore, M.D.: CEO succession mechanisms, organizational context, and performance: a socio-emotional wealth perspective on family-controlled firms. J. Manag. Stud. **51**(7), 153–1179 (2014)

12. Miller, D., Steier, L., Le Breton-Miller, I.: Lost in time: Intergenerational succession, change, and failure in family business. J. Bus. Ventur. **18**(4), 513–531 (2003)

13. Gómez-Mejía, L.R., Haynes, K.T., Núñez-Nickel, M., Jacobson, K.J.L., Moyano-Fuentes, J.: Socioemotional wealth and business risks in family-controlled firms: evidence from Spanish olive oil mills. Adm. Sci. Q. **52**(1), 106–137 (2007)

14. Berrone, P., Cruz, C., Gomez-Mejia, L.R.: Socioemotional wealth in family firms: theoretical dimensions, assessment approaches, and agenda for future research. Fam. Bus. Rev. **25**(3), 258–279 (2012)

15. Amore, M.D., Minichilli, A., Corbetta, G.: How do managerial successions shape corporate financial policies in family firms? J. Corp. Finance **17**(4), 1016–1027 (2011)

16. Miller, D., Le Breton-Miller, I., Scholnick, B.: Stewardship vs. stagnation: an empirical comparison of small family and non-family businesses. J. Manag. Stud.**45**(1), 51–78 (2008)

17. Zellweger, T.M., Kellermanns, F.W., Chrisman, J.J., Chua, J.H.: Family control and family firm valuation by family CEOs: the importance of intentions for transgenerational control. Organ. Sci. **23**(3), 851–868 (2012)

18. Argote, L., Greve, H.R.: 'A behavioral theory of the firm': 40 years and counting: Introduction and impact. Organ. Sci. **18**(3), 337–349 (2007)

19. Cyert, R.M., March, J.G.: A behavioral theory of the firm. Prentice Hall/Pearson Education, Upper Saddle River (1963)

20. Zellweger, T.M., Dehlen, T.: Value is in the eye of the owner: affect infusion and socioemotional wealth among family firm owners. Fam. Bus. Rev. **25**(3), 280–297 (2012)

21. Biggs, E.L.: CEO succession planning: an emerging challenge for boards of directors. Acad. Manag. Perspect. **18**(1), 105–107 (2004)

22. Minichilli, A., Corbetta, G., MacMillan, I.C.: Top management teams in family-controlled companies: 'Familiness', 'faultlines', and their impact on financial performance. J. Manag. Stud. **47**(2), 205–222 (2010)

23. Nordqvist, M., Sharma, P., Chirico, F.: Family firm heterogeneity and governance: a configuration approach. J. Small Bus. Manag. **52**(2), 192–209 (2014)

24. Berns, K.V.D., Klarner, P.: A review of the CEO succession literature and a future research program. Acad. Manag. Perspect. **31**(2), 83–108 (2017)

25. Sharma, P., Irving, P.G.: Four bases of family business successor commitment: antecedents and consequences. Entrep. Theory Pract. **29**(1), 13–33 (2005)

26. Zellweger, T., Sieger, P., Halter, F.: Should I stay or should I go? Career choice intentions of students with family business background. J. Bus. Ventur. **26**(5), 521–536 (2011)

27. Jaskiewicz, P., Uhlenbruck, K., Balkin, D.B., Reay, T.: Is nepotism good or bad? Types of nepotism and implications for knowledge management. Fam. Bus. Rev. **26**(2), 121–139 (2013)

28. Sitkin, S.B., Pablo, A.L.: Reconceptualizing the determinants of risk behavior. Acad. Manage. Rev. **17**(1), 9–38 (1992)

29. Haleblian, J., Finkelstein, S.: Top management team size, CEO dominance, and firm performance: the moderating roles of environmental turbulence and discretion. Acad. Manage. J. **36**(4), 844–863 (1993)

30. Georgakakis, D., Ruigrok, W.: CEO succession origin and firm performance: a multilevel study. J. Manag. Stud. **54**(1), 58–87 (2017)
31. Gedajlovic, E., Carney, M., Chrisman, J.J., Kellermanns, F.W.: The adolescence of family firm research: taking stock and planning for the future. J. Manag. **38**(4), 1010–1037 (2012)
32. Katz, R.: The effects of group longevity on project communication and performance. Adm. Sci. Q. **27**(1), 81–104 (1982)
33. Salancik, G.: Commitment and the control of organizational behavior and belief. In: New Directions in Organizational Behavior, St. Clair Press, Chicago, pp. 1–54 (1977)
34. Allen, N.J., Meyer, J.P.: Affective, continuance, and normative commitment to the organization: an examination of construct validity. J. Vocat. Behav. **49**(3), 252–276 (1996)
35. Meyer, J.P., Allen, N.J., Smith, C.A.: Commitment to organizations and occupations: extension and test of a three-component conceptualization. J. Appl. Psychol. **78**, 538–551 (1993)
36. Wennberg, K., Wiklund, J., Hellerstedt, K., Nordqvist, M.: Implications of intra-family and external ownership transfer of family firms: short-term and long-term performance differences. Strateg. Entrep. J. **5**(4), 352–372 (2011)
37. Chrisman, J.J., Patel, P.C.: Variations in R&D investments of family and nonfamily firms: behavioral agency and myopic loss aversion perspectives. Acad. Manage. J. **55**(4), 976–997 (2012)
38. Patel, P.C., Chrisman, J.J.: Risk abatement as a strategy for R&D investments in family firms. Strateg. Manag. J. **35**(4), 617–627 (2014)
39. "In the Horns of the Dilemma: Socioemotional Wealth, Financial Wealth, and Acquisitions in Family Firms - Luis R. Gomez-Mejia, Pankaj C. Patel, Thomas M. Zellweger, 2018." https://journals.sagepub.com/https://doi.org/10.1177/0149206315614375. Accessed 10 Feb 2023
40. Zahra, S.A.: International expansion of U.S. manufacturing family businesses: the effect of ownership and involvement. J. Bus. Ventur. **18**(4), 495–512 (2003)
41. Revilla, A.J., Pérez-Luño, A., Nieto, M.J.: Does family involvement in management reduce the risk of business failure? The moderating role of entrepreneurial orientation. Fam. Bus. Rev. **29**(4), 365–379 (2016)
42. Wangrow, D.B., Schepker, D.J., Barker, V.L.: When does CEO succession lead to strategic change? The mediating role of top management team replacement. J. Gen. Manag. 03063070221126267 (2022)
43. Quigley, T.J., Hambrick, D.C., Misangyi, V.F., Rizzi, G.A.: CEO selection as risk-taking: a new vantage on the debate about the consequences of insiders versus outsiders. Strateg. Manag. J. **40**(9), 1453–1470 (2019)
44. Arnoldi, J., Villadsen, A.R.: Political ties of listed Chinese companies, performance effects, and moderating institutional factors. Manag. Organ. Rev. **11**(2), 217–236 (2015)
45. Luo, J., Wan, D., Cai, D., Liu, H.: Multiple large shareholder structure and governance: the role of shareholder numbers, contest for control, and formal institutions in Chinese family firms. Manag. Organ. Rev. **9**(2), 265–294 (2013)
46. Kutner, M., Nachtsheim, C., Neter, J.: Applied Linear Regression Model, 4th edn.. McGraw-Hill Irwin (2004)
47. Zahra, S.A.: Entrepreneurial risk taking in family firms. Fam. Bus. Rev. **18**(1), 23–40 (2005)
48. Simsek, Z.: CEO tenure and organizational performance: an intervening model. Strateg. Manag. J. **28**(6), 653–662 (2007)
49. Huang, X., Chen, L., Xu, E., Lu, F., Tam, K.-C.: Shadow of the prince: parent-incumbents' coercive control over child-successors in family organizations. Adm. Sci. Q. **65**(3), 710–750 (2020)

An Experimental Study of the Relationship Between Static Advertisements Effectiveness and Personality Traits: Using Big-Five, Eye-Tracking, and Interviews

Semira Maria Evangelou$^{(\boxtimes)}$ ⓘ and Michalis Xenos ⓘ

Department of Computer Engineering and Informatics, University of Patras, Patras, Greece
{evangelou,xenos}@ceid.upatras.gr

Abstract. The study aimed to explore the influence of advertising on purchasing behavior in an online store. 43 participants (17 female) were given 1,000 euros to navigate the store and make purchases. Eye-tracking data, questionnaires, and semi-structured interviews were collected and analyzed to comprehend the factors that affect the effectiveness of advertising. Results indicated that 32 participants were classified as goal-oriented users and the semi-structured interviews revealed that color and offers were crucial factors in attracting participants' attention to advertisements. The findings also showed a correlation between participants' Big-Five personality traits and their response to advertising annoyance, with those lower in Openness to Experience having higher scores of annoyance towards advertising. Additionally, a positive correlation between Extraversion scores and a favorable attitude towards advertising was identified, and the existence of banner blindness was confirmed. The results suggest that internet users are predominantly goal-oriented, and that advertising plays a significant role in shaping purchasing behavior in online stores. Future studies should aim to increase the sample size to gain a more comprehensive understanding of the correlations with personality traits.

Keywords: Online advertising · Eye-tracking · Big-Five model

1 Introduction

Advertising began several years ago when entrepreneurs sought a successful way to promote their products or services. Today, advertising has become a vast industry that is present in many aspects of our daily lives. Advertisements (ads) can be found on TV, radio, in newspapers or magazines, on the internet, on social media, on mobile devices, and even in emails. The term "digital marketing" first emerged in the 1990s. The first banner advertisements appeared in 1993 on Hotwired, the first web magazine to embed ads on its website [1]. These ads, known as "banner ads," were sold based on the number of "impressions" (i.e., people who viewed the ad), a model that was later adopted by many traditional media outlets for brand advertising. In 1994, Hotwired launched the

© Springer Nature Switzerland AG 2023
F. Fui-Hoon Nah and K. Siau (Eds.): HCII 2023, LNCS 14039, pp. 285–299, 2023.
https://doi.org/10.1007/978-3-031-36049-7_22

first banner ad, titled "Have you ever clicked your mouse right HERE? YOU WILL," as part of AT&T's "You Will" campaign. This marked the start of a transformation in web content and advertising that continues to shape the industry today.

Digital advertising is a major topic in the marketing literature and considers how consumers respond to various aspects of digital advertisements [2]. Online advertising can be defined as any kind of marketing message that is mediated with the help of the Internet. Many consumers find online advertising disruptive and have increasingly turned to ad blocking for a variety of reasons.

Using the internet, social media, mobile applications (apps), and other digital communication technologies has become part of billions of people's daily lives. In modern society, many people spend hours connected to the internet and exposed daily to a variety of online ads. Indeed, the development of technology is directly related to the advertising industry and is constantly influencing it. In particular, the internet has been a milestone in the history of advertising.

Today, banners are typically presented as either static images or multimedia objects and are often used to encourage users to click through to a website. The goal of banners is to grab the viewer's attention, and they often use movement or short videos to make them more visually appealing. As a result, banner advertisements are commonly found on high-traffic websites. Ad networks serve as advertising agencies by finding available advertising space on the internet and matching it with the needs of advertisers. The technology for this whole process relates to one central ad server [3]. According to surveys, on one hand men tend to be less patient when it comes to dealing with pop-up banner advertisements and often seek ways to avoid them. On the other hand women tend to be more patient and make better use of them. However, both groups agree that banners often have negative aspects [4].

1.1 Factors that Contribute to Advertising Effectiveness

Media professional and researchers remain greatly interested in examining the influence of media context on advertising effectiveness [5]. Media context is an important factor for determining advertising effectiveness and it is considered that some media contexts are more suitable for different types of advertising than others. Today one of the most important and fastest-growing media options is through the Internet, making Internet advertising predominant. The following factors appear mostly in research regarding the advertising effectiveness:

- Attitudes toward the medium
- Uses of the medium
- Involvement while using the medium
- Mood states affecting media usage
- Interactivity of the medium
- Recall
- Attitude toward the advertisements

1.2   Color in Advertising

In today's competitive businesses world, the use of advertising is prominent to attract potential clients [6]. Billions are spent to build, reach and influence target consumers to buy products or services. Creating awareness is one of the most important advertising goals and companies are using a variety of constructive elements in advertising to communicate and influence buyer behavior.

One of the most powerful advertising tools is color. Color advertisements are essential for establishing the brand image and are more likely to grab users' attention than those without color. Over the past years, many well-known brands have been associated with colors. In essence, color in a brand is so important that it may become its own identity. Vodafone, Coca-Cola, Starbucks, and Facebook are such examples, as their brands are directly associated with a specific color. Vodafone as well as Coca-Cola were directly linked to red, Starbucks to green, and Facebook to blue. In general color affects the psychology and behaviors of consumers. Colors based on their wavelength have been categorized into cool and warm colors [7]. Colors can have different meanings in terms of personality traits, salient qualities, and emotions. Blue and green, which are considered "cool" colors, tend to evoke different emotions than red and yellow, which are considered "warm" colors. For example, red is often associated with emotions such as passion, excitement, and danger, but in a business context, it can also be associated with power, energy, speed, and courage. Blue is often associated with high quality, trustworthiness, and dependability, while purple is associated with progressiveness and affordability. Yellow is often associated with happiness and gray is often associated with dependability and high quality. However, it's important to note that colors can also elicit negative meanings and reactions, such as fear, anger, intolerance, and exaggeration. In general, warm colors tend to elicit uplifting effects such as increased blood pressure and frequent eye blinking, although these effects are not always pleasurable. Research suggests that cool color interiors are more appealing to customers and tend to elicit more positive emotions than warm color interiors. A study also showed that users tend to prefer cool colors, specifically blue versions, in websites. In this study, the colors used were chosen based on the literature on effective color usage.

1.3   Banner Blindness

The increase in advertising saturation is high, thus the term "banner blindness" has arisen [8]. The phenomenon of banner blindness occurs when internet users actively ignore banner ads while exposed to them [9]. Therefore, advertisers need to find solutions to measure and improve the effectiveness of online advertising. Another definition describes this term as the situation in which web users tend to ignore online page elements that perceive them as ads. Banner blindness is based on selective attention, the tendency of people to direct their attention in elements of their environment that are related to their aims.

2 The Big-Five Model & Measurements of Advertising Effectiveness

2.1 The Big-Five Model

The Big-Five Model of personality is a model in trait psychology which describes individual differences in human personality and determines the features of human thinking, feeling, and behavior [10]. This model suggests that all people, regardless of age, culture, or gender share the same basic traits, but differ in the degree of their manifestation. More specifically, it conceptualizes personality traits in five basic dimensions of personality: Extraversion (E), Neuroticism (N), Agreeableness (A), Conscientiousness (C), and Openness to experience (O). Common acronyms for the Big-Five are OCEAN, NEOAC, or CANOE. Each of the Big-Five dimensions is a combination of several distinct but closely related traits, which describe a continuum between two extreme poles.

- **Extraversion:** The characteristic of extroversion is sometimes related primarily to sociability and evaluates the amount and intensity of interpersonal interactions and ambition. People who show a high degree of extroversion are sociable, talkative, gregarious, impetuous, ambitious, assertive, expressive, exhibitionist, and energetic.
- **Agreeableness:** The characteristic of agreeableness refers to the quality of the individual's interpersonal relationships. It is considered that people who score high in this dimension have strong empathy; they are cooperative, kind, forgiving, compassionate, willing, kind, trustworthy, and tolerant.
- **Consciousness:** Consciousness reflects dependability. People with a high degree of conscientiousness are usually hardworking, organized, disciplined, responsible, meticulous, ambitious, and reliable.
- **Neuroticism:** Neuroticism is also called emotional Stability and people with high are usually anxious, embarrassed, insecure, angry, and worried.
- **Openness to Experience:** The last dimension has been called Openness to Experience (also known as Intellect). People with a high rating of this feature are usually original, imaginative, cultured, artistically sensitive, curious, intelligent, and broad-minded. Lists the dominant characteristics of people with high score in each dimension of personality.

2.2 Visual Attention and Recall

It is essential to understand how and when internet users allocate their attention to commercial stimuli and what factors influence their attentional strategies and patterns [11]. There are two primary measurements to evaluate advertising effectiveness: visual attention and recall [12], as both affect brand and product awareness. Researchers can use eye-tracking techniques to investigate how the elements of online banner advertisements influence customers' attention.

Not only is it a gate, but it also serves as a key coordinating mechanism that helps to maintain information processing and other goals over time [13]. Studies have shown that visual attention to advertising has systematic downstream effects on brand memory, indicating its predictive validity [14]. However, it is not only attention that plays a significant role, as people do not always remember everything they notice. Sometimes, an advertisement's message is cognitively processed and remains in the consumer's memory, but at other times it does not [15]. One of the prevalent approaches to measuring the amount of attention related to advertising is to use self-reported memory measures by asking questions such as "Did you remember this advertisement?" or "How well do you remember the advertisement?".

2.3 Eye-Tracking

Eye-movement tracking is a technique that is being used to study usability issues in Human-Computer Interaction (HCI) contexts [16]. It is important for the reader before reading the analysis using eye-tracking data to be introduced to the basics of eye-movement technology and be informed about several key aspects of practical guidance. Eye tracking is a method of recording a person's eye movements so that the researcher knows how the user behaves during browsing related to what they saw. Researchers by using eye-tracking techniques can answer questions like where the user looked, for how long and can know the sequence in which their eyes change from one location to another.

Eye-tracking provides such measures. Eye movements are tightly coupled with visual attention, making them eminent indicators of the visual attention process, which is now easy to assess with modern eye-tracking equipment. Eye movements are behavioral measures of the unobservable visual attention process of prime interest. "The eyes don't lie. If you want to know what people are paying attention to, follow what they are looking at".

An eye-tracking device can measure:

- Fixation frequency (i.e., number of eye fixations on target stimuli)
- Fixation duration (i.e., total duration of eye fixation on target stimuli)
- Scan path (fixation sequence)
- Location of the first fixation
- Time of the first fixation

3 Research Questions and Methodology

3.1 Research Questions

The goals of the evaluation include the following research questions:

- RQ1: Is the Banner blindness phenomenon confirmed in this study?
- RQ2: Are there correlations between the participants' Big-Five characteristics scores and the percentage of the advertisements that were noticed?
- RQ3: Are there any differences in the participants' Big-Five characteristics scores related to whether they are goal-oriented?
- RQ4: Are there correlations between participants' Big-Five personality traits and the percentage of the advertisements that were recalled?
- RQ5: Are there any correlations in participants' Big-Five characteristics scores related to their attitude towards the advertising?
- RQ6: Were the participants goal-oriented users (regardless of their Big-Five characteristics)?

3.2 Participants and Procedure

The research was conducted at the Human-Computer Interaction Laboratory, located within the Computer Engineering Department of the University of Patras. Participants participated in the experiment for a duration of 20 min. Data collection was performed over two phases during the academic year 2020–2021. A preliminary sample of participants took part in the study in 2020, and these results indicated positive outcomes [17]. Due to the COVID-19 lockdown, the study was temporarily suspended for three months but resumed in June 2021, with all necessary safety measures in place.

A total of 43 individuals, including 17 females, participated in the study. They were exposed to a simulated online store created for the purpose of the study. Participants were informed that the store was not real and were given a hypothetical budget of 1,000 euros to make purchases. The scenario presented to them was as follows: "You have been given 1,000 euros to purchase anything you desire from this online store. You may choose to spend the entire amount or opt not to make any purchases. You have all the time you need to complete your purchases."

3.3 Experiment Protocol, Metrics, and Instruments

Before arriving in the laboratory, participants received an email containing details on the experiment, possible dates they could participate, the 50-item IPIP FFM, and instructions. The 50-item Big-Five questionnaire had to be completed before their arrival in the laboratory. Moreover, the participants were told that during the experiment they would be going to undergo overall three questionnaires including the Big-Five one. An additional and important reason for sending this email was also to inform the participants about the ID that they had to complete in all questionnaires. There was clearly explained that the IDs would not be associated with their names. The reason for the existence of the unique IDs was the correlation of the three questionnaires and the interview with the recorded eye-tracking data. The first field in every questionnaire asked the participants to enter their ID. For this study, a form was created which included the 50 questions. When participants arrived at the laboratory, they signed a consent form to declare their agreement to take part in the study. The consent form was in Greek and is shown in Appendices. After that, they were given the second which included few demographic and marketing questions. The most important goal of it was to determine the purchase goals of each participant. They were asked whether they are interested in buying any product from the available product categories in the online shop.

At this stage, participants were advised to take a comfortable position and were requested to keep their heads as stable as possible during their browsing in the online shop. Then participants were left free to interact with the store and "spend" their 1,000 euros. Eye-tracking data were collected during their navigation to examine how the participants viewed the banners and the products.

A photograph from a participant that took part in the experiment during the eye-tracking calibration is shown in Fig. 1.

The eye-tracking data also facilitate the interview that followed (the evaluator had access to these data, helping her to ask the right questions during the interviews). When the participants had completed their purchases, a third was given to them to examine

Fig. 1. Participant during the eye-tracking calibration

whether, how many and at which extend was possible for the participants to recall the advertisements they were exposed to. The interview took place mostly to identify the reasons why the participants noticed some advertisements and ignored others. And finally, a semi-structured interview with the participants followed and was the last stage of the experiment. An objective of this interview was also to identify discrepancies between participants' behavior towards advertising and understand their general attitude towards advertising. Each participant was interviewed for approximately 10 min. The entire interview process was recorded with the consent of each participant. The questions were about the general experience of the participants in the online store focusing on the advertisements. Indicatively some of the questions were "Did you notice the advertisements that were in the online store?", "Why did you look at this particular advertisement?" or "Why did you ignore this particular advertisement?". The questions were tailored to each participant, according to the data collected from the eye-tracker. Figure 2 depicts a brief visual representation of the experimental protocol. Overall, three questionnaires and an online store were created for the purposes of this experiment. All the text in the questionnaires and in the online store was in Greek, except for the Big-Five question-naire which was in English. In addition, all interviews were conducted in Greek so that participants could express themselves in their mother tongue.

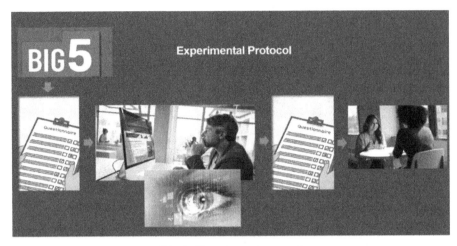

Fig. 2. Experimental protocol

4 Results

4.1 Gender and Age

Overall, forty-three participants (17 women) participated in the experiment. Their ages ranges from [18–60].

4.2 Interests in Buying Products from Each Category

The second questionnaire aimed to determine the purchase goals of each participant. Participants were asked if they were interested in buying any products from the available categories in the online shop. They were allowed to select more than one option. In total, 22 participants were interested in buying books (24.7%), 20 in smartphones and tablets (22.5%), 17 in sports and fitness products (19.1%), 11 in gaming products (12.4%), 10 in TVs (11.2%), and 9 in wearables, drones, and high-tech products (10.1%).

4.3 Daily Internet Usage and Frequency of Online Purchasing

The study examines the daily internet usage habits of the participants. The results show that most participants (82.8%) reported spending between 3 to 10 h online daily, with 41.9% stating they spend 3–5 h daily and 41.9% stating they spend 6–10 h daily. Additionally, a small percentage of participants, 7%, stated that they are always connected to the internet.

Furthermore, the data related to the frequency of purchasing products online indicates that most participants (46.5%), reported buying products online once a month, while 23.3% reported buying them once every six months. Additionally, 14% of participants reported buying products online once a week and 9.3% reported buying them twice a week. Lastly, 7% of participants selected the option "other" without providing further details.

4.4 Personality Dimensions of the Participants

Participants' scores for each dimension were determined by their responses to a questionnaire, with scores ranging from 0 to 100. Table 1 presents the descriptive statistics for the 43 participants of the experiment for each of the Big-Five dimensions of personality.

Table 1. Descriptive statistics for the 43 participants of the experiment in Big-five

Dimension	N	Min	Max	Mean	Std. Dev.
Extraversion (E)	43	25.0	85.0	60.930	15.0101
Agreeableness (A)	43	35.0	100.0	74.302	14.4371
Consciousness (C)	43	22.5	97.5	64.767	17.1288
Neuroticism (N)	43	7.5	70.0	36.919	16.4646
Openness to Experience (O)	43	30.0	70.0	50.988	9.9274

4.5 Eye-Tracking Insights

Insights from eye-tracking recordings revealed that the longer users pay attention to a banner advertisement, the better their recognition performance. In Fig. 3, the left image shows a gaze plot that the participant clearly looked at the advertisement, and subsequently recalled it. Conversely, the right image shows an example of a participant who completely ignored the advertisement on the visited page.

4.6 Banner Blindness Phenomenon

The study aimed to investigate whether the phenomenon of banner blindness occurs (see RQ1). To do this, we analyzed the interactions of 43 participants with a variety of clickable banner advertisements that were displayed on different pages of an online store. Eye-tracking recordings were used to determine how many participants viewed each advertisement. Participants were given the freedom to interact with the online shop and make their own decisions without any restrictions or guidance. As a result, some participants did not visit all the product category pages and some visited some product category pages more than once. However, in this analysis, it was assumed that browsing on a page was continuous, regardless of how many times a participant visited it.

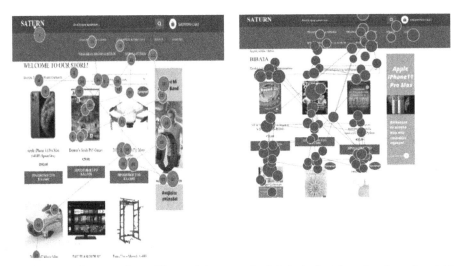

Fig. 3. Gaze plot examples – The participant on the left looked at the ad and, on the right, completely ignored the ad

It is determined that a participant has noticed an advertisement if they have focused on it for at least two consecutive fixations and a total fixation duration of at least 500 ms. Advertisement recall is assumed to occur when participants can remember and recall the advertising message from their memory. This is determined by the participants' responses in the third questionnaire, where they indicate if they remember the advertisement. The eye-tracking recordings were examined to confirm whether a participant had seen an advertisement or not. In the third questionnaire, participants were asked to indicate the extent to which they can recall the advertisement.

The study identified the following categories to correlate advertisement observation and recall:

- True Positive (TP): Participants viewed the advertisement and remembered it.
- True Negative (TN): Participants viewed the advertisement but did not remember it.
- False Positive (FP): Participants did not view the advertisement, but they reported remembering it in the third questionnaire.
- False Negative (FN): Participants did not view the advertisement and did not remember it.

When a participant was unable to fully recall an advertisement, stating that they "remember seeing the advertisement, but did not remember the product shown" or "only remember the product but could not remember details of what it was like", it is considered not to belong to the True Positive (TP) or True Negative (TN) category. TP refers to participants who viewed the advertisement and remembered it in the third questionnaire, while TN refers to participants who viewed the advertisement but did not remember any information about it. The study also examined the factors that contribute to the viewing of an advertisement through an interview process. Total the number of participants who accessed the page where the advertisement was and viewed the number of people who viewed the advertisement. In the Table 2 below, the "Total" column represents the number of individuals who accessed the page where the advertisement was located, while the "Viewed" column indicates the number of people who actually viewed the advertisement.

Table 2. Overall data about how participants interact with the advertisements

	Total	Viewed	TP	TN	FP	FN
Ad1	43	22	12	6	1	6
Ad2	27	14	7	5	1	7
Ad3	31	11	4	4	0	13
Ad4	29	11	8	2	2	9
Ad5	25	12	2	6	3	7
Ad6	32	9	3	3	5	12
Ad7	28	13	2	8	2	11
Total	**215**	**92**	**38**	**34**	**14**	**65**

The study found that a high number of participants completely ignored the advertisement, as indicated by the high False Negative (FN) rate, for Ad3, Ad4, Ad5, and Ad7. The high True Negative (TN) rate also suggests that many participants were unable to recall the advertisement, despite viewing it according to their eye-tracking recordings. Additionally, the low rate of True Positive (TP) indicates that only a small number of participants noticed and remembered the advertisements. Overall, there were 215 advertising exposures, and a significant portion of these exposures went unnoticed and unremembered by the participants. To further confirm the occurrence of the banner blindness phenomenon, the study also utilized semi-structured interviews with the participants to gain a deeper understanding of their experiences related to advertisements in the online store.

Twenty-one participants completely ignored the advertisements, as reported in their interview responses, where they questioned the purpose of the experiment and whether they had done something wrong for not noticing the advertisements. This observation was supported by the eye-tracking recordings, which showed that these participants had not looked at the advertisements or had fewer gaze points on them. Considering the high rate of False Negative (FN), the rate of True Negative (TN), and the findings from the semi-structured interview, the study confirms the occurrence of banner blindness phenomenon, as most participants failed to recall the advertisements even when they had viewed them.

4.7 Percentage of the Advertisements Noticed Based on Big-Five

To investigate the RQ2 about the correlations between participants' Big-Five personality traits and the percentage of advertisements they were noticed, a Kendall's tau-b correlation was used to test whether there is an association between two variables. Kendall's tau-b correlation is considered a nonparametric alternative to the Pearson's product-moment correlation when the data has failed one or more of the assumptions of this test, which is applicable in our case.

So, the test was run to determine the relationship between participants' **Extraversion** (E) scores and the percentage of the ads that were seen. The correlation between the two variables, with 43 participants, is rs $= -0.178$. From our 'two-tailed' prediction of the relationship, it would be necessary to accept the null hypothesis that no association exists (p $= 0.110$). The same result was found in all other scores, with the results for **Agreeableness** (A) scores being rs $= -0.062$ and no association (p $= 0.576$), for **Consciousness** (C) rs $= -0.078$ and no association (p $= 0.480$), for **Neuroticism** (N) rs $= 0.11$ and no association (p $= 0.924$), and for **Openness to Experience** (O) rs $= 0.12$ and no association (p $= 0.41$). These results suggest that people with similar trait scores do not necessarily share a similar percentage of ads that were seen for all the examined Big-Five personality traits.

4.8 Participants Being Goal-Oriented Based on Big-Five

To investigate the RQ3 about correlations between the participants' Big-Five characteristics scores of the participants related to whether they are goal-oriented, an independent t-test was conducted since in terms of the Shapiro-Wilk test, we can assume that the data are approximately normally distributed for each group. This study did not show that the groups means were statistically significant.

4.9 Percentage of the Advertisements that were Recalled

To investigate the RQ4 about the correlations between participants' Big-Five personality traits and the percentage of advertisements that were recalled, a Kendall's tau-b correlation was run.

So, the test was run to determine the relationship between participants' **Extraversion** (E) scores and the percentage of the ads recalled. The correlation between the two variables, with 43 participants, is rs $= -0.091$. From our 'two-tailed' prediction of the relationship, it would be necessary to accept the null hypothesis that no association exists (p $= 0.436$). The same result was found in all other scores, with the results for **Agreeableness** (A) scores being rs $= -0.156$ and no association (p $= 0.178$), for **Consciousness** (C) rs $= -0.011$ and no association (p $= 0.922$), for **Neuroticism** (N) rs $= 0.071$ and no association (p $= 0.540$), and for **Openness to Experience** (O) rs $= 0.090$ and no association (p $= 0.442$). These results suggest that people with similar trait scores do not necessarily share a similar percentage of ads that were noticed for all the examined Big-Five personality traits.

4.10 Attitude Towards the Advertising

To investigate RQ5 about the correlations in participants' Big-Five characteristics scores related to their attitude towards the advertising, the participants were asked to rate their level of annoyance with advertising on a Likert scale of 1 to 5. A Kendall's tau-b correlation was run to determine any correlation between participants' Big-Five characteristics scores related to their attitude towards the advertising.

Based on the results of the study, participants who had the lowest scores in the **Openness to Experience** dimension were more likely to have scores ranked higher in the response to advertising annoyance ($rs = -0.27$, $p < 0.05$). Additionally, there also was a positive correlation between participants' **Extraversion** scores of the participants and their attitude towards the advertising, which was statistically significant ($rs = -0.23$, $p < 0.05$).

4.11 Overall Percentage of Participants Being Goal-Oriented

Finally, examining where the participants were goal-oriented (regardless of their individual Big-Five traits), we found that 74% of them were goal-oriented. Results indicated that 32 of the 43 participants were goal-oriented as they made purchases consistent with the product categories they expressed interest in during the second questionnaire. Participants were probed regarding their intention to buy products from available categories in the online store. Conversely, 11 participants were classified as non-goal-oriented as they made purchases that deviated from their initially declared preferences.

The high proportion of goal-oriented participants provides evidence for the notion that internet users exhibit a strong inclination towards their set goals.

4.12 Interview Findings

At the conclusion of the experiment, each participant was quizzed about their experience with the online store and their views on advertising. The interviews provided valuable insight into the banner blindness phenomenon and participants' attitudes towards advertising. Out of the participants, 21 completely ignored the advertisements and were surprised to learn of their existence during the interview. Most participants stated that they only pay attention to advertisements if the products advertised align with their interests and needs. Those who did take notice of the advertisements were asked why they did so, with reasons ranging from matching interests, recognizable brand, eye-catching colors, and attractive offers. Despite this, most participants reported being highly annoyed by advertisements while browsing and some expressed a desire to never be exposed to advertisements again.

Regarding the participants who had a limited memory of the advertisements, the semi-structured interview delved into what exactly they remembered and why. The results showed that factors such as vibrant colors and compelling offers played a significant role in attracting participants' attention. Some participants remembered seeing a certain color or always looking at offers. Additionally, the participants were asked if they made any unexpected purchases. As mentioned in the previous section, based on the eye-tracking data, 11 participants that were categorized as not being goal-oriented online shoppers as they made purchases that did not align with their initial plans. However, during the interview, 9 of these participants explained that they wanted to spend the hypothetical money given to them fully, leading to their deviation from their original plans, since they felt compelled to spend all the "free money".

5 Conclusion, Limitations and Future Work

In this study, a simulated online store was developed with six product categories and various clickable banner advertisements. Overall, 43 participants, including 17 women, took part in the study. They were given 1,000 euros to buy products from the online store and were free to navigate it. To understand the factors affecting the effectiveness of advertising, eye-tracking data, questionnaires, and semi-structured interviews were used and analyzed. Three questionnaires were used, including a pre-arrival questionnaire with 50 questions on the participants' personalities based on the Big-Five personality trait model. Upon arrival, a second questionnaire was given to gather demographic and marketing information and to determine the participants' initial purchase plans. After browsing the online store, a final questionnaire was given to assess advertisement recall. A semi-structured interview was also conducted to gain a better understanding of the participants' experience with the advertisements.

Eye-tracking data was used to determine if a participant noticed an advertisement and if they remembered it. Advertisement recall was defined as the participant remembering the advertisement message. Four categories emerged from the correlation of advertisement observation and recall: True Positive (TP), True Negative (TN), False Positive (FP), and False Negative (FN). Participants who only remembered some information about the advertisement were not included in any of these categories. The semi-structured interview revealed that colors and offers were important factors in attracting participants' attention to advertisements. Most participants reported that they were more likely to remember advertisements when they depicted products of interest. Additionally, the results of the eye-tracking data and questionnaires were compared to determine if the participants were goal-oriented users.

The results of this study provide valuable insight into the banner blindness phenomenon and the attitudes of internet users towards advertising. Out of the 43 participants, 32 were found to be goal-oriented, making purchases that aligned with the product categories they expressed interest in. The results of the eye-tracking data showed that a high proportion of participants focused solely on the content of the page and ignored the advertisements, confirming the existence of the banner blindness phenomenon. Additionally, the interviews with participants provided further insights into their attitudes towards advertising. 21 participants reported completely ignoring the advertisements and being highly annoyed by them while browsing. The study found that factors such as bright colors and attractive offers played a significant role in attracting participants' attention. These findings contribute to the existing literature on banner blindness and the impact of advertising on online shopping behavior.

The findings indicate that there was a positive correlation between the Extraversion scores of the participants and their attitude towards the advertising, which was statistically significant, while those with lower scores in Openness to Experience were more likely to have a negative attitude towards advertising, as evidenced by the higher scores in annoyance towards advertising. These findings provide a deeper understanding of how personality traits can influence the perception and interaction with online advertisements.

The study has a clear limitation in the number of participants. Despite an effort to identify correlations between participants' personality traits and various measurements, the small sample size resulted in limited representation of the dimensions of the Big-Five, potentially limiting the demonstration of correlations. A future goal is to expand the sample size, allowing for a more comprehensive examination of the correlations with the Big-Five dimensions. Additionally, the study was conducted during the COVID-19 lockdown, making it difficult to find participants as university courses were not in session.

References

1. Kingsnorth, S.: Digital Marketing Strategy: An Integrated Approach to Online Marketing. Kogan Page Publishers (2022)
2. Stephen, A.T.: The role of digital and social media marketing in consumer behavior. Curr. Opin. Psychol. **10**, 17–21 (2016)
3. North, M., Ficorilli, M.: Click me: an examination of the impact size, color, and design has on banner advertisements generating clicks. J. Financ. Serv. Market. **22**(3), 99–108 (2017)
4. Krushali, S., Jojo, N., Manivannan, A.S.: Cognitive marketing and purchase decision with reference to pop up and banner advertisements. J. Soc. Sci. Res. **4**(12), 718–735 (2018)
5. Kwon, E.S., King, K.W., Nyilasy, G., Reid, L.N.: Impact of media context on advertising memory: a meta-analysis of advertising effectiveness. J. Advert. Res. **59**(1), 99–128 (2019)
6. Manickam, S.A.: Do advertising tools create awareness, provide information, and enhance knowledge? An exploratory study. J. Promot. Manag. **20**(3), 291–310 (2014)
7. Crowley, A.E.: The two-dimensional impact of color on shopping. Mark. Lett. **4**(1), 59–69 (1993)
8. Benway, J.P.: Banner blindness: The irony of attention grabbing on the World Wide Web. In: Proceedings of the Human Factors and Ergonomics Society Annual Meeting (vol. 42, No. 5, pp. 463–467). Sage CA: Los Angeles, CA: SAGE Publications (1998, October).
9. Owens, J.W., Chaparro, B.S., Palmer, E.M.: Text advertising blindness: the new banner blindness? J. Usability Stud. **6**(3), 172–197 (2011)
10. Novikova, I.A.: Big Five (the five-factor model and the five-factor theory). Encyclop. Cross-cult. Psychol. **1**, 136–138 (2013)
11. Li, Q., Huang, Z.J., Christianson, K.: Visual attention toward tourism photographs with text: an eye-tracking study. Tour. Manage. **54**, 243–258 (2016)
12. Kong, S., Huang, Z., Scott, N., Zhang, Z.A., Shen, Z.: Web advertisement effectiveness evaluation: attention and memory. J. Vacat. Mark. **25**(1), 130–146 (2019)
13. Wedel, M., Pieters, R.: A review of eye-tracking research in marketing. Rev. Mark. Res. **4**, 123–147 (2017)
14. Davenport, T.H., Beck, J.C.: The attention economy. Ubiquity 2001(May), 1–es (2001)
15. Lee, J., Ahn, J.H.: Attention to banner ads and their effectiveness: An eye-tracking approach. Int. J. Electron. Commer. **17**(1), 119–137 (2012)
16. Poole, A., Ball, L.J.: Eye tracking in HCI and usability research. In Encyclopedia of Human Computer Interaction, pp. 211–219. IGI global (2006)
17. Evangelou, S.M., Xenos, M.: Banner advertisement effectiveness using Big-5 personality traits, advertisement recall, and visual attention. In: 24th Pan-Hellenic Conference on Informatics, pp. 256–259 (2020, November)

User Study on a Multi-view Environment to Identify Differences Between Biological Taxonomies

Manuel Figueroa-Montero and Lilliana Sancho-Chavarría[✉]

School of Computing, Costa Rica Institute of Technology, Cartago, Costa Rica
mfigueroacr@gmail.com,
lsancho@tec.ac.cr
https://www.tec.ac.cr/escuelas/escuela-ingenieria-computacion

Abstract. The classification of living organisms is essential to understand biodiversity and, therefore, to support its conservation. The taxonomic system classifies living organisms into hierarchically organized groups. Classification is not always a simple task, as even today taxonomists face the challenge that data are frequently revised, and species may be reclassified into other groups. Visualization of hierarchies, particularly biological taxonomies, has been a relevant area of research to support this challenge. Previous user studies determined that expert taxonomists preferred the edge-drawing method to visualize changes between two versions of a taxonomy, using the tool called Diaforá, which presents a single view. In this paper we propose a visualization that consists of using the coordinated multiple view (CMV) technique and focuses on displaying only the differences between two alternative versions of a biological taxonomy. The hypothesis is that the proposed visualization would provide a better cognitive approach to users, facilitating the identification of taxonomic changes. The work consisted first in designing other views and, through a user study, selecting one of them to be integrated into Diaforá to form a CMV system. Then, a second user study was carried out to contrast the original version of Diaforá with its extended version. Participants performed a series of tasks and we obtained measures of effectiveness, efficiency, trust rate and overall engagement rate. Results indicate that multiple coordinated views (CMV) improved users' performance in identifying differences.

Keywords: Information visualization · Human-computer interaction · Hierarchy comparison · Coordinated multiple views

1 Introduction

The identification of changes between alternative versions of a biological taxonomy has been an important research topic in information visualization [15,16,20,24,30,31]. In this work we address the issue of enhancing the user overall performance when interacting with a multi-view visualization environment designed to recognize differences of alternative versions of a biological taxonomy.

© Springer Nature Switzerland AG 2023
F. Fui-Hoon Nah and K. Siau (Eds.): HCII 2023, LNCS 14039, pp. 300–319, 2023.
https://doi.org/10.1007/978-3-031-36049-7_23

Biological taxonomies are hierarchical structures where living organisms on Earth are classified into groups according to their morphological characteristics. Accurate species classification records are necessary for understanding biodiversity and therefore to contribute conservation efforts [23]. Taxonomy is a dynamic field of study. The constant revision, discovery and reclassification of species bring on continuous changes to the classification; therefore, taxonomists often face the challenge of handling large taxonomies that need to be compared in order to correct formerly recorded data.

Previous works investigated visual methods for comparing hierarchies [15,16]. The *edge drawing* method was applied and tested on large taxonomic datasets with Diaforá, which is a *single view* interactive tool that automatically infers and visualizes differences when comparing alternative versions of a biological taxonomy [30]. Although users (taxonomists) showed a high level of satisfaction with the edge drawing method, scalability limitations were present since the method requires the use of scrolling, which forces the user to recall a part of the hierarchy when the hierarchies being compared are very large.

In this work, we investigated whether the use of coordinated multiple views (CMV) [7] can improve user performance when comparing alternative versions of a biological taxonomy. This work aims to increase achievement in the Diaforá system by improving its visualization method in two ways: through the design of a multiple-view approach that would contribute to faster identification of differences, as well as the application of a visual data reduction approach that displays only taxa (i.e., nodes) affected by changes. The proposed environment is evaluated through user studies consisting of two stages. The first stage aims to select the *other view* that would be integrated into an extended CMV version of Diaforá. The second stage contrasts the single view approach of the original version of Diaforá tool against the version that integrates the coordinated views.

This paper is structured as follows. Section 2 presents related work on hierarchy comparison and its application in the comparison of biological taxonomies. Section 3 describes the rationale of the study and the hypothesis. In Sect. 4 we describe the study design and methodology. Section 5 presents and discusses the experiment results. Section 6 summarizes our conclusions and future work.

2   Related Work

The hierarchical categorization system proposed by the scientist and naturalist Carl Nilsson Linnaeus in 1735 [21] is still prevalent today and widely used by taxonomists, museums, herbaria and other institutions that preserve information about living organisms. Proper identification and classification of living beings are of vital importance for nature conservation efforts [23,27]. The task of maintaining these classifications is complex and involves the effort of analyzing and comparing different versions of the taxonomic hierarchies. Versions arise mainly because of the different classification criteria that experts use to organize species and because the information is scattered throughout different sources, which requires data integration efforts [3].

2.1   Hierarchy Comparison

Comparison of complex entities is usual in the field of visual analytic [14]. Visual comparison difficulty level increases in relation to the size and complexity of the data set to be analyzed, because it is limited by the user's cognitive and perceptual capacity. Even though designers have many ways to visually represent data, representations are perceived by the users in different ways [26]. Visual perception has an important role in how the information presented is analyzed and in the relative performance of getting the right answer when someone is doing visual comparison tasks; for example, the identification of a maximum delta between data series.

The problem of comparing hierarchical data structures has been a notable area of research. Holten et al. [17] presented hierarchical edge bundling, a visualization technique that allows the comparison of two hierarchies simultaneously. Li et al. [19] propose a visualization technique that allows comparison of the topological structure of multiple encrypted trees in a simplified graphical representation similar to a barcode. Dong et al. [9] describe a technique called PansyTree that merges multiple hierarchies in a single representation. Bongshin et al. [18], present the *CandidTree* visualization system that shows the differences between two hierarchical structures. Graham et al. [16] present an overview of various techniques that allow visual comparisons of trees.

Topics from the perceptual psychology field are used to evaluate the performance of a visual comparison representation [26]:

– **Co-location:** involves visual features like length, orientation, and motion, within the same space are very useful to represent distinct features among data sets. Comparison between two regions implicates the effort to "remember" part of the visualization in order go on with the comparison. This has a limit given by the cognitive capacity of people.
– **Symmetry**: humans are specially sensitive to symmetry. That feature can be useful when representing data sets that have a minor difference but when you put a visualization next to the other it is easy to find the differences in images that are otherwise mirror images.
– **Movement:** wired as a primitive element of human vision, this can be used to extract information like statistics and structure. It also has a limited span of attention from the user that rapidly decays when several elements are moving simultaneously.

2.2   Focused and Contextual Views

To avoid overcrowded and cluttered visualizations, strategies such as overview+detail, zooming, and focus+context are used [8]. The use of interactive lenses is widely extended to provide alternative visual representations for selected regions of interest [32]. According to Björk et al. [6] the basic idea with focus and context visualizations is to enable users to have the object of primary interest presented in detail on the screen while giving an overview or context of the other data at the same time. Focus and context techniques are useful when users require both a detailed visualization and an overview of the data simultaneously.

2.3 Multi-view Visualizations

A multiple view system uses two or more distinct visualizations to support the investigation of a single conceptual entity [34]. With this approach, it is possible to provide a complementary visualization to enhance the efficiency and efficacy of a single-view system [7]. Coordinated Multiple Views (CMV) are used in several user interfaces were comparing and contrasting different aspects of the data is needed. CMVs are widely used to explore and analyze large datasets. By providing consistency and different viewpoints and user interactions, CMVs can become a valuable tool for users to achieve a better understanding of their data [28].

2.4 Taxonomy Comparison

Biological taxonomies [21] are structures in which known living organisms are organized and classified. These hierarchical structures can have a significant number of taxa (i.e., tree nodes) based on the number of individuals recorded in each of the groups. According to Ball-Damerow et al. [4], the number of digitized biological datasets is increasing and becoming openly accessible for use and consultation by any researcher who needs them. Efforts such as *GBIF, BOLDSystems, SpeciesLink,* and Catalogue of Life *COL* are among the most widely used biodiversity databases that require constant expert review and cleaning due to the dynamism of the field. For instance, Catalogue of Life [29] is a global initiative that collects data from over 200 expert taxonomic databases; the data exposed through a COL annual checklist contains a peer-reviewed taxonomic classification of a vast number of organisms and by 2020 COL was tracking 95% of species known to science [29]. The dynamism of the data comes, for example, from new discoveries of species to be added to the records, from experts decisions to split a previously named group into several species groups or, on the contrary, to merge several species groups into a single group, given a revision of their characteristics.

Comparison of taxonomies is key to accomplish the data cleaning, visual comparison can especially support this task. Four methods for the visualization of relationships that aid comparison between hierarchies have been distinguished: *edge drawing, matrix representation, animation* and *agglomeration* [16]. Results from a user study with expert taxonomists that evaluates those four methods for the comparison of two alternative biological taxonomies indicated that the participant users preferred the *edge drawing* method while performing taxonomy comparisons [31]. Based on these results, further work produced Diaforá, an interactive prototype tool that supports taxonomy curation by the visual comparison of two alternative versions of a biological taxonomy [30]. The tool automatically identifies and presents visual keys for the identification of the types of changes that commonly occur, such as *splits, merges,* and taxa *rename;* for example, experts might split a group of species into several groups after discovering that not all recorded species in the group have the same morphological traits, and therefore should be grouped separately. Figure 1 presents Diaforá when comparing two alternative versions of the *Bryozoa* phylum.

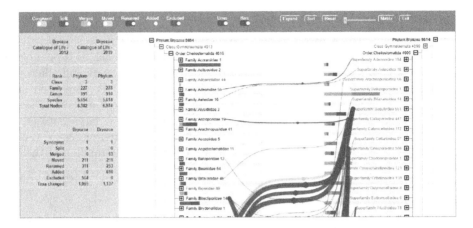

Fig. 1. Diaforá: A visualization tool for the comparison of biological taxonomies [30].

2.5   Information Visualization Evaluation

Literature indicates that heuristics are particularly useful when evaluating information visualization interfaces [33]. Heuristics are focused on how information visualization techniques enable people to get a better understanding of a dataset [36]. Heuristics evaluations are a useful method to discover possible usability problems. Forsell and Johansson [13] propose a set of ten heuristics used to evaluate information visualization techniques.

Evaluation is conducted through user studies. Although there is still controversy on the amount of users required for these studies, some experts argue that between 12 and 20 users is a reliable number [35] whereas other experts indicate that 5 users is enough for a user study to test usability [25]. In our experience we noticed that the more specialized the experts required to conduct a user study, the more difficult it will be to find a larger number of participants.

3   Problem Definition

The *edge drawing* method has been preferred by taxonomists when comparing biological taxonomies for the identification of similarities and differences [31]. However, this type of visualization has scalability issues due to the space occupied by the lines that connect the nodes between the different hierarchies, as well as by the chance that related nodes are placed out of the visual focus of the screen. The visualization in Diaforá is presented in a single view.

This work focuses on extending the single view in Diaforá with another view to form a *coordinated multiple view (CMV)* environment, and evaluating whether users have better performance with this new visualization environment.

4   Study Design

Let us recapitulate that the edge-drawing method –evaluated with the Diaforá visualization environment– was preferred by taxonomists in a previous user study. Diaforá presents a single view. To evaluate the effectiveness of using a multiple-view environment versus a single-view environment, it is necessary to first design and select *another* view. The design of the *other* view was done taking into account criteria that will be explained later, and the selection was done through a user study. Once the *other* view was selected, it was integrated into a new extended CMV version of Diaforá. The final evaluation takes place through a final user study. Therefore, the study involves two stages: one to evaluate and select the *other* view and one to evaluate the CMV proposal.

Instruments with questions were defined for the user studies of each stage. Questions were based on the common tasks [31] involved when identifying differences between alternative taxonomies. For the design of the instruments we had the advice of an expert taxonomist, who revised the questions and did a pilot test of the instruments online.

The datasets were carefully selected from lists in the Catalogue of Life. We selected versions from different years (2012 and 2019) which included more than 9,000 species which qualified for the study according to the expert taxonomist's criteria. Also, in order to minimize bias, we selected data with a high probability of being unknown to the potential participants (most participants were familiar with plant species, therefore the selected datasets belong to Animalia).

For the user studies we employed *asynchronous remote usability testing* which would make it easier for test subjects to participate, considering that some experts may be overseas on different time zones and availability spans, and also considering sanitary measures due to COVID-19 restrictions. Literature reports a comparison between synchronous and asynchronous remote testing and authors did not find any statistically significant differences between the number of usability issues identified in both cases [1]. We used Loop11, an online platform that allows for user studies, in which participants interact with the tested environment and respond to the instrument's questions; the platform provides metrics, such as participant's response time for each task.

To facilitate this study, we created a user guide [12] for the system and a video [10] that were provided to the participants when they were invited to the study. These materials allowed participants to familiarize themselves with the use of Diaforá as well as the proposed visualizations that would be integrated in the CMV environment.

We evaluated the following metrics:

- **Effectiveness**: Percentage of correct answers. Given a set of different tasks that require the user to identify differences between alternative versions of a taxonomy, we want to measure the participants effectiveness rates between both the original Diaforá and the proposed CMV Diaforá.
- **Efficiency**: Mean completion time and the mean times per task for each visualization environment.

- **Confidence Rate**: A self-reported confidence rate will be asked to the users on a 3-point Likert scale (high, medium, low).
- **Overall Engagement Rate**: A self-reported overall engagement with each environment. This will measure the cognitive connection between the user and the given visualization. It will be evaluated by using a 5-point Likert scale (Strongly disagree, Disagree, Neither agree nor disagree, Agree, Strongly agree). We propose to use the reaction cards proposed in [5] to evaluate this metric and then present the percentages across all tasks per visualization.

5   Stage I: Selecting the *other* view

The design criteria for the *other* view were based on the idea of reducing cluttering and minimize scrolling; thus we focused on designs that would show only the nodes affected by changes, assuming that the nodes that did not change were not so relevant to visualize in the *other* view. We designed two visualization prototypes that focused on showing only the nodes that had some change between 2012 and 2019 versions of the taxonomy. One prototype was based on the indented method of hierarchy visualization of the original Diaforá version. The other prototype was designed considering a radial representation of the hierarchy. These prototypes were evaluated through a user study in order to determine which one presents the information in an easier-to-understand way, and then incorporate the design were participants had the best performance into a coordinated multiple views environment with the original Diaforá's visualization. The following subsections describe the two prototype designs, the participants on this part of the study, the tasks that participants performed, the procedure, and the results of the prototypes evaluation.

5.1   The Designs for the Alternative View

A. Indented design (visualization of only affected nodes)
One of the designs presented the hierarchy using indentation, which followed a similar design pattern of the Diaforá tool, but that shows only the differences. Figure 2 illustrates the visualization displaying the differences between two alternative versions of the *Lycopodiopsida* taxonomy taken from the 2012 *Catalogue of Life* checklist and the 2019 *Catalogue of Life* checklist. This design includes a summary overview of the changes. As shown by the number 1 in the figure, there is a bar indicator that gives the user a quick overview of the changes at each taxonomic level; additionally, the number 2 illustrates a pop-up window that appears when hovering the mouse cursor over a taxon, showing a detailed distribution on the selected taxonomic level.

B. Radial design (visualization of only affected nodes)
The second prototype presents the hierarchy in a radial representation. It is focused to improve screen space usage by providing a visualization that does not require scrolling to see the complete difference tree. Similar to the *Indented Tree* explained above this visualization provides insight into the distribution of

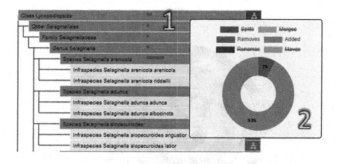

Fig. 2. Indented Tree Taxonomic Differences Visualization

changes for every level of the taxonomy. Notice that the outer ring can be a full ring when the user selects the root of the hierarchy or an arc if the user is hovering over an internal level of the taxonomy, we call this *satellite bar*. The satellite bar shows the distribution of changes on any particular level of the biological taxonomy. For example, Fig. 3 shows the distribution of changes for the *Isoetaceae* family, indicating that 84.7% of changes are additions and the rest 15.3% of changes are exclusions. Also, the intensity of the color background for every node in a gray color scale from white to black indicates the total amount of changes for every node at any particular level. That means the darker nodes are the ones with more changes at their taxonomic level.

5.2 Participants for the Prototypes Evaluation

We recruited six advanced students from the Forestry Engineering career of the Costa Rica Institute of Technology, who had experience working with biological taxonomies. Table 1 summarizes the participants' profiles. We assigned a unique id to each participant.

Table 1. First stage - participants' profiles

Participant ID	Major	Experience (in years)
TS1	Forestry engineering	3
TS2	Forestry engineering	4
TS3	Forestry engineering	4
TS4	Forestry engineering	5
TS5	Forestry engineering	3
TS6	Forestry engineering	5

Fig. 3. Radial visualization showing the satellite bar indicator for the Family *Isoetaceae*

5.3 Tasks for the Prototypes Evaluation

Participants were asked to perform a series of tasks guided by questions. The questions are intentionally not exactly the same for each prototype evaluation since we wanted to avoid the learning effect bias; therefore, we designed equivalent questions. Table 2 lists the questions used to evaluate the radial prototype and Table 3 lists the questions used to evaluate the radial prototype.

5.4 Prototypes Evaluation Procedure

Participants first tested the indented visualization and then proceeded to evaluate the radial visualization. The platform allowed participants to be guided through the study in a way that they could make use of each prototype and answer the questions in the same session. After each question, participants were presented to a 3-level Likert scale evaluation about the confidence rate of giving the correct response. After finishing the 9 questions, there was a 5-level Likert scale evaluation about the user's satisfaction level with the visualization, which means an engagement rate with the visualization according to Attfield [2]. The Loop11 platform automatically calculates the metrics.

5.5 Results of Stage I

This section presents the results obtained of Stage I of the user study, which goal was to select the best of the proposed designs for the *other* visualization that would be included fo conform the coordinated multiple view (CMV) environment.

Completion Time: As Fig. 4 shows, the tasks were completed with the Indented design in a mean time of 27 min whereas with the radial prototype tasks were completed in a mean time of 13.88 min (52% faster than the indented tree representation). The mean difference between the two prototypes was -12.774 min.

Table 2. Taxonomic differences indented tree prototype evaluation questions.

Index	Question
1	What is the most common change in this taxonomy? (The one with most occurrences)
2	Please identify the three families that have the largest amount of split changes
3	What is the most common change in the *Tubificidae* family
4	Select Merges and look up for species *Megasyllis inflata*. Which species did merged for species *Megasyllis inflata*?
5	Look at renames and please enter the new name for the renamed species *Quistadrilus multisetosus*
6	Please enter the family with the most new species added
7	What is the percentage of added species in family *Naididae*?
8	How many changes are in the genus *Paranais*?
9	Please enter the complete taxonomy tree (Phylum to Species) of new species *Syllis boggemanni*
10	Please rate this visualization prototype. Did you like the data visualization? (5-level Likert scale)
*	After every question, there is a 3-level Likert scale rating question about a confidence level that the given answer was correct

Table 3. Taxonomic differences radial tree prototype evaluation questions.

Index	Question
1	What is the most common change in this taxonomy? (The one with most occurrences)
2	What is the percentage of splits in the *Alcyonidae* family?
3	Which is the most common change in the *Nephtheidae* family?
4	Select Merges and look up for species *Lytreia plana*. Which species did merged for species *Lytreia plana*?
5	Look at renames and please enter the new name for the renamed species *Orbicella franksi*
6	Please enter the family with the most new species added
7	What is the percentage of added species in family *Actiniidae*?
8	How many changes are in the genus *Actinia*?
9	Select Splits and look for species *Aegina citrea*. Please enter the complete taxonomy tree (Phylum to Species) of species *Aegina citrea*
10	Please rate this visualization prototype. Did you like the data visualization? (5-level Likert scale)
*	After every question, there is a 3-level Likert scale rating question about a confidence level that the given answer was correct

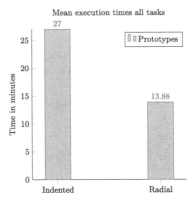

Fig. 4. Indented tree tasks were completed in a mean time of 27 min, and the Radial tree tasks were completed in a mean time of 13.88 min.

Effectiveness Rate: Figure 5 presents the effectiveness rate of participants when using each design (68% effectiveness with the indented visualization and 76% with the radial visualization). The error rate was 8% lower while using the radial tree. By running a Cochran Q test on this results we obtained the values **P= 0196706, Q = 1.66667 with** $\alpha = 0.05$, which means there is no statistically significant difference between both prototypes and can be considered equivalent on the error rate metric.

Confidence: Figure 6 shows the results on participants self-reported confidence for each visualization. Participants rated their confidence on the answers they provided with each design as 1: low, 2:medium, 3: high. Notice that confidence is relatively high for both prototypes but slightly 3% higher on the Radial tree visualization. A Friedman test on these results gave **Q = 0.00000, p = 1.00000**, which indicates that both prototypes provide similar self-perceived confidence.

Engagement Rate: Figure 7 shows that both prototypes obtained a very positive evaluation on engagement rate. Participants reported a slightly better evaluation (33% for strongly agree) to the radial design; however, adding both *Agree and Strongly agree* answers, each design obtained the same 83% positive evaluation. By performing a Friedman test on these results, we obtained **Q = 0.00000, p = 1.00000**, which means there is no statistically significant difference between the engagement rate values of both prototypes.

As can be seen, the participants performed better with the *Radial design*, so this was the one selected at this stage.

6   Stage II: Evaluation of the CMV

Figure 8 presents the Diaforá system displaying both visualization mechanisms, that is, the *edge drawing* and the selected *radial tree design*. These visualizations

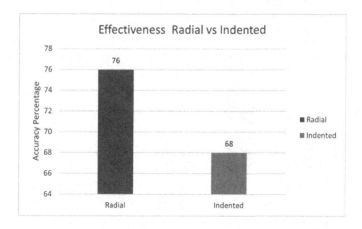

Fig. 5. Effectiveness rate, indented tree vs radial tree designs

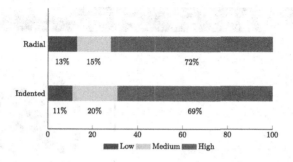

Fig. 6. Self reported confidence for each prototype

are coordinated, that means that the user actions like navigation across taxa affects both visualizations. That means that the users always have the detail and context about the node of the taxonomic group of interest.

6.1 Participants for the CMV Evaluation

Participants for the evaluation of the original edge-drawing Diaforá version against the CMV version, were advanced students of the Forestry Engineering career at the Costa Rica Institute of Technology, a plant taxonomist, and a biologist. Table 4 summarizes the participant's profiles for this stage.

6.2 Procedure for the CMV Evaluation

Since the purpose of the study is to determine whether a coordinated multiple view environment allows a better user performance than a single view user performance, participants were asked to perform a series of tasks with the original version of the Diaforá system, and then perform other equivalent tasks with the

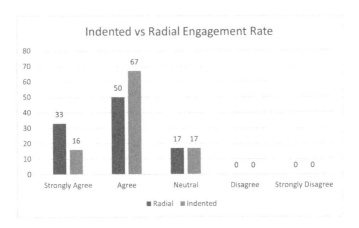

Fig. 7. Engagement rate, indented tree vs radial tree design

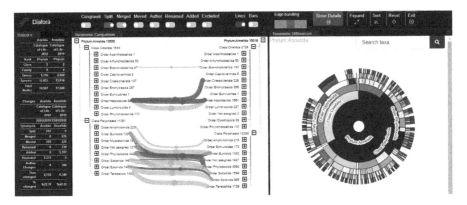

Fig. 8. Diaforá Coordinated Multiple Views environment displaying Annelida's phylum taxonomy comparison

extended CMV version of Diaforá. For this part of the study we also used the Loop11 platform that was prepared in advance to guide participants throughout the study. Participants had to perform tasks for the identification of differences as well as provide feedback on their confidence and engagement levels with each environment. Participants sessions were recorded on videos [11].

6.3   Tasks for the CMV Evaluation

For Stage II evaluation, the approach was similar to the one used for the evaluation of the prototypes, we designed an instrument with 10 questions, which

Table 4. Participants' profiles

Participant ID	Major	Experience (in years)
TS1	Forest engineering student	5
TS2	Forest engineering student	4
TS3	Forest engineering student	6
TS4	Forest engineering student	6
TS5	Forest engineering student	6
TS6	Forest engineering student	5
TS7	Plant taxonomist	24
TS8	Forest engineering student	5
TS9	Biologist	2
TS10	Forestry Engineer	21

was revised by the expert taxonomist. Table 5 presents the questions used to evaluate the original version of Diaforá system (the one that uses *edge drawing* exclusively), and Table 6 enumerates the questions used for the extended CMV Diaforá environment version.

Table 5. Original Diaforá evaluation questions.

Index	Question
1	What is the number of excluded taxa in this taxonomic comparison?
2	What is the percentage of renames in the *Nephtheidae* family?.
3	Which is the most common change in the *Anthozoa* class?
4	Select Moves and look up for genus *Cubaia*. What is the original family and the new family to which this genus belongs?
5	Look at Added and look up for the Order *Leptothecata*, How many added taxa are included in this order?
6	Please enter the family with the most new species added.
7	What is the percentage of excluded species in family *Acroporidae*?
8	How many changes are in the genus *Acropora* ?
9	Select Renames and look for species *Dipsastraea rotumana*. Please enter the complete taxonomy tree (Phylum to Species) of species *Dipsastraea rotumana*.
10	Please rate this visualization prototype. Did you like the data visualization? (5-level Likert scale).
*	After every question, there is a 3-level Likert scale rating question about a confidence level that the given answer was correct

Table 6. Extended Diaforá evaluation questions.

Index	Question
1	What is the number of merged taxa in this taxonomic comparison?
2	What is the percentage of merges in the *Clausophyidae* family?.
3	Which is the most common change in the *Halcampidae* family?
4	Select Moves and look up for genus *Euphellia*. What is the original family and the new family to which this genus belongs?
5	Look at Added and look up for the Order *Zoantharia*, How many added taxa are included in this order?
6	Please enter the family with most species excluded?
7	What is the percentage of excluded species in family *Ellisellidae*?
8	How many changes are in the genus *Viminella* ?
9	Select Renames and look for species *Filigorgia schoutedeni*. Please enter the complete taxonomy tree (Phylum to Species) of species *Filigorgia schoutedeni*.
10	Please rate this visualization prototype. Did you like the data visualization? (5-level likert scale).
*	After every question, there is a 3-level likert scale rating question about a confidence level that the given answer was correct

7   Results

For the analysis of effectiveness, we used Cochran's Q-test method, which is commonly used when you have a group of people performing a series of tasks where the outcome is dichotomic, which means it can be a failure or success. The levels of satisfaction and certainty of the participants are analyzed using the Friedman test method that is suitable for ordinal data like the Likert scale questions [22].

Completion Time: Figure 9 shows the mean time for the single view Diaforá version and for the CMV version. Notice that participants completed the tasks 48.72% faster with the CMV version than with the single view version of Diaforá.

Effectiveness: Figure 10 shows the participants effectiveness rate obtained when comparing Diaforá single view version with and the results for Diaforá CMV version. The error rate was 35% lower while using the CMV environment. By performing a Cochran's Q-test on the obtained results we obtained the values of $\mathbf{P = 0.000002}$, $\mathbf{Q = 22.26087}$, with $\alpha = 0.05$ that means there is a statically significant difference between the two Diaforá system versions.

Confidence: Figure 11 shows the participants self-reported confidence for each Diaforá version. Confidence is higher while using the CMV version (9% higher).

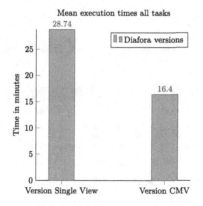

Fig. 9. Completion time results, single view vs CMV.

By calculating a Friedman test on this result we obtained $\mathbf{Q = 2.5, \ p = 0.11385}$, **with** $\alpha = 0.05$ meaning that we did not detect a statistically significant difference on this metric. In practice, we found that users tended to give a high confidence value when selecting the "don't know" option in the exercises.

Engagement: As shown in 12 the CMV version was notably better rated for engagement. A Friedman Test was conducted on the obtained results for the engagement rate. Results showed that the system version used led to statistically significant differences in the engagement rate values $(\mathbf{Q = 10, \ p = 0.00157,}$ **with** $\alpha = 0.05)$.

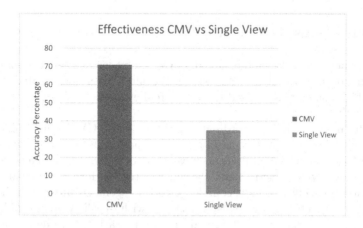

Fig. 10. Effectiveness rate, Diaforá single view version vs Diaforá CMV version

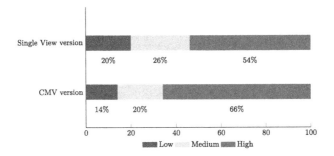

Fig. 11. Self reported confidence for each version

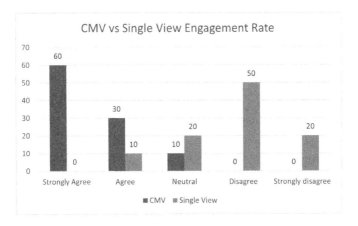

Fig. 12. Engagement rate, Diaforá Single view version vs Diaforá CMV version

8   Conclusions and Future Work

The first stage of the user study was important in order to select a visualization that could be integrated into an extended Diaforá with coordinated multiple views, which could then be used to do the final evaluation (single view vs coordinated multiple views).

Results indicate that a coordinated multiple view (CMV) environment can provide a significant improvement in the overall performance of users when identifying differences between alternative versions of a biological taxonomy than having a single view. Participants showed a better performance in all metrics (completion time, effectiveness, confidence and engagement rates) with the CMV version. Participants performed the tasks faster with the CMV version of Diaforá, Statistical test on effectiveness strongly points to participants having a better performance with the CMV version. The engagement metric result is an indication that participants were satisfied with the CMV. The confidence metric did not show any statistically significant difference between both version, which is an indication that participants rely similarly in both versions.

The proposed CMV visualization design provides a quick way to interact with large taxonomies and extract statistical information like the distribution of changes and its impact on alternative versions of biological taxonomies. The visual identification of taxonomic differences with the CMV Diaforá environment could be useful for database cleanup; for instance, when different databases spell differently the name of authors. For this research we used data from the Catalogue of Life database, in spite that this is a very complete database, there are other sources and providers that use different data formats; future work could focus on a solution that allows data from several sources. Future work could also be directed to correct the taxonomy online and generate a newer version with all fixes that can be uploaded or shared with other taxonomic systems.

Finally, we believe it could be of value to use the CMV Diaforá environment for further user studies that involve other types of hierarchical information.

Acknowledgment. The authors would like to thank all of our test subjects, and the Forestry Engineering School of the Costa Rica Institute of Technology for their support with this research. We also like to thank botanist Armando Estrada-Chavarría, our expert advisor in biological taxonomies, for his collaboration throughout the design of the study.

References

1. Alghamdi, A., Al-Badi, A., Alroobaea, R., Mayhew, P.: A comparative study of synchronous and asynchronous remote usability testing methods. Int. Rev. Basic Appl. Sci. (IRBAS) **1**, 61–97 (2013)
2. Attfield, S., Kazai, G., Lalmas, M., Piwowarski, B.: Towards a science of user engagement (Position Paper). In: WSDM Workshop on User Modelling for Web Applications (2011)
3. Avise, J.C., Liu, J.X.: On the temporal inconsistencies of Linnean taxonomic ranks. Biol. J. Linnean Soc. **102**(4), 707–714 (2011)
4. Ball-Damerow, J., et al.: Research applications of primary biodiversity databases in the digital age. PLOS ONE **14**, e0215794 (2019)
5. Barnum, C.M., Palmer, L.A.: More than a feeling: Understanding the desirability factor in user experience. In: CHI 2010 Extended Abstracts on Human Factors in Computing Systems, CHI EA 2010, pp. 4703–4716. Association for Computing Machinery, New York (2010). https://doi.org/10.1145/1753846.1754217
6. Bjork, S., Redstrom, J.: Redefining the focus and context of focus+context visualization. In: IEEE Symposium on Information Visualization 2000, INFOVIS 2000, Proceedings, pp. 85–89 (2000). https://doi.org/10.1109/INFVIS.2000.885094
7. Boukhelifa, N., Roberts, J., Rodgers, P.: A coordination model for exploratory multiview visualization. In: Proceedings International Conference on Coordinated and Multiple Views in Exploratory Visualization - CMV 2003, pp. 76–85 (2003)
8. Card, S., Mackinlay, J., Shneiderman, B. (eds.): Readings in Information Visualization: Using Vision to Think. Morgan Kaufmann, Burlington (1999)
9. Dong, Y., Fauth, A., Huang, M., Chen, Y., Liang, J.: Pansytree: merging multiple hierarchies. In: 2020 IEEE Pacific Visualization Symposium (PacificVis), pp. 131–135 (2020)
10. Figueroa, M.: Demostración de uso sistema Diaforá. https://bit.ly/3Gj6PFB

11. Figueroa, M.: Diaforá system test subjects videos. Shorturl.at/cvEIS
12. Figueroa, M.: Guía para las pruebas con usuarios para el sistema Diaforá. https://bit.ly/3lyLIH6
13. Forsell, C., Johansson, J.: An heuristic set for evaluation in information visualization. In: Proceedings of the Workshop on Advanced Visual Interfaces AVI, pp. 199–206 (01 2010)
14. Gleicher, M.: Considerations for visualizing comparison. IEEE Trans. Visualization Comput. Graph. **24**(1), 413–423 (2018)
15. Graham, M., Kennedy, J.: Visual exploration of alternative taxonomies through concepts. Ecol. Inf. **2**, 248–261 (2007)
16. Graham, M., Kennedy, J.: A survey of multiple tree visualisation. Inf. Vis. **9**(4), 235–252 (2009)
17. Holten, D., Wijk, J.J.V.: Visual comparison of hierarchically organized data. Comput. Graph. Forum **27**(3), 759–766 (2008)
18. Lee, B., Robertson, G., Czerwinski, M., Parr, C.: CandidTree: visualizing structural uncertainty in similar hierarchies. Inf. Vis. **6** (2007)
19. Li, G.: BarcodeTree: scalable comparison of multiple hierarchies. IEEE Trans. Vis. Comput. Graph. **26**(1), 1022–1032 (2020)
20. Lin, C., Wang, J.: Taxonomic tree tool (2020). http://ttt.biodinfo.org/
21. Linnaeus, C.: Systema Naturae per Regna Tria Naturae, Secundum Classes, Ordines, Genera, Species, cum Characteribus, Differentiis, Synonymis, Locis. Editio Decima 1, Holmiae **1**, 634–635 (1758)
22. MacKenzie, I.S.: Human-Computer Interaction: An Empirical Research Perspective. Elsevier, Amsterdam (2013)
23. Moraes, G.J.D.: Importance of taxonomy in biological control. Int. J. Trop. Insect Sci. **8**(4-5-6), 841–844 (1987)
24. Munzner, T., Guimbretiere, F., Tasiran, S., Zhang, L., Zhou, Y.: TreeJuxtaposer. ACM Trans. Graph. **22**, 453 (2003)
25. Nielsen, J.: How many test users in a usability study? (2012). https://www.nngroup.com/articles/how-many-test-users/
26. Ondov, B., Jardine, N., Elmqvist, N., Franconeri, S.: Face to face: evaluating visual comparison. IEEE Trans. Vis. Comput. Graph. (2018)
27. Regan, H., Colyvan, M., Burgman, M.: A taxonomy and treatment of uncertainty for ecology and conservation biology. Ecol. Appl. **12**, 618–628 (2002)
28. Roberts, J.: State of the art: coordinated 'I&' multiple views in exploratory visualization. In: Coordinated and Multiple Views in Exploratory Visualisation, Zurich, Switzerland, 2 July 2007, pp. 61–71 (2007)
29. Roskov, Y., et al.: Species 2000 'I&' ITIS catalogue of life. In: 2019 Annual Checklist (2019). http://www.catalogueoflife.org/annual-checklist/2019. Accessed 23 Aug 2020
30. Sancho-Chavarría, L., Gómez-Soza, C., Beck, F., Mata-Montero, E.: Diaforá: a visualization tool for the comparison of biological taxonomies, pp. 423–437 (2020)
31. Sancho-Chavarría, L., Beck, F., Mata-Montero, E.: An expert study on hierarchy comparison methods applied to biological taxonomies curation. PeerJ Comput. Sci. **6**, e277 (2020)
32. Tominski, C., Gladisch, S., Kister, U., Dachselt, R., Schumann, H.: interactive lenses for visualization: an extended survey. Comput. Graph. Forum **36** (2016)

33. Väätäjä, H., et al.: Information visualization heuristics in practical expert evaluation. In: Proceedings of the Sixth Workshop on Beyond Time and Errors on Novel Evaluation Methods for Visualization, BELIV 2016, pp. 36–43. Association for Computing Machinery, New York (2016). https://doi.org/10.1145/2993901.2993918

34. Wang Baldonado, M.Q., Woodruff, A., Kuchinsky, A.: Guidelines for using multiple views in information visualization, pp. 110–119 (2000). https://doi-org.ezproxy.itcr.ac.cr/10.1145/345513.345271

35. Ware, C.: Information Visualization: Perception for Design, 3rd edn. Morgan Kaufmann Publishers Inc., San Francisco (2012)

36. Zuk, T., Schlesier, L., Neumann, P., Hancock, M., Carpendale, S.: Heuristics for information visualization evaluation. In: Proceedings of the 2006 AVI Workshop on BEyond Time and Errors: Novel Evaluation Methods for Information Visualization, pp. 1–6 (2006)

Analysis and Application of a Batch Arrival Queueing Model with the Second Optional Service and Randomized Vacation Policy

Kai-Bin Huang[✉]

Department of Business Administration, Fu Jen Catholic University, New Taipei, Taiwan
152400@mail.fju.edu.tw

Abstract. This paper aims to investigate $M^{[X]}/(G_1, G_2)/1/VAC(J)$ queuing system with a random(p) vacation policy and optional second service, where X is the batch arrival number of customers. When no customers are in the system, the server immediately goes on vacation. And when the server returns from a vacation and finds that at least one customer is waiting in the system, the server will immediately provide the First Essential Service (FES). After customers complete the first essential service, some will continue to receive the Second Optional service (SOS). After the customer completes the FES, some customers will continue to choose to accept the second additional equipment adjustment or maintenance service (the probability is θ). In addition, when the server returns from vacation and finds that no customers are waiting for service in the system, the server will be idle in the system with a probability of p waiting for customers to enter the system for service, but there will be a probability of (1-p) to continue vacation. This pattern will continue until the number of server vacations reaches J times. Suppose the server returns to the system after the Jth vacation and finds that no customers are waiting for service in the system; the server will always be idle in the system waiting for customers to enter the system for service. This paper consider the servers are unreliable and can be repaired immediately, and establish the supplementary variables of the system as well as use the supplementary variables to construct the Kolmogorov forward equation that governs the system, and then use the supplementary variable techniques to derive the expected number of customers, the expected waiting time and other important system characteristics in the proposed queueing system. The relevant results can be used as the service performance evaluation and decision-making tools that require secondary optional services and regular maintenance in practical applications of queueing models.

Keywords: Queueing theory · Randomized vacation policy · Batch arrival queueing model · Second optional service · Random breakdown

1 Introduction

Over the years, many queuing system analysis methods with vacation policies have been proposed and widely used in different queuing systems. In practical production or operating environments, managers must consider related production costs or the status of

© Springer Nature Switzerland AG 2023
F. Fui-Hoon Nah and K. Siau (Eds.): HCII 2023, LNCS 14039, pp. 320–333, 2023.
https://doi.org/10.1007/978-3-031-36049-7_24

the service system itself. In practice, the equipment must be regularly maintained or calibrated to confirm the equipment's accuracy, which makes the service provider temporarily leave the main service system and perform other work (vacations) under normal conditions. According to different vacation purposes, service providers will temporarily leave the system according to different principles, so many different vacation strategy queuing systems have been proposed, and such queuing systems have been analyzed. Relevant research results have been successfully applied in various fields, such as product inventory, communication, and computer network systems (Doshi [1]). Yechali [2] and Takagi [3] are the earlier extensive and excellent research results on the vacation queuing model. Generally speaking, service providers only partially stop their services on vacation, but perform other supplementary tasks, such as maintenance or periodic calibration of instruments and equipment. Servi and Finn [4] first studied the M/M/1 queuing system with working vacations. Working vacations mean that service providers will provide services at different service rates during the vacations, and at the end of the working vacations, the service items will be retrieved. After arriving at the queuing system, if it is found that there are no customers in the system, the working vacation will continue. Wu and Takagi [5] extended the M/M/1/WV queuing model of Servi and Finn [4] to the M/G/1/WV model. The M/G/1/WV model of Wu and Takagi [5] Assume that the time of the server in the general service state, working vacation state, and vacation state is subject to the general allocation, and assume that if the server returns to the system after working vacation and finds that there are still customers who need to be served in the system, the server will Switch to normal service state to continue serving customers.

Regarding of research on vacation policy queuing in which the server is reliable (assuming that the system is performing services, no unexpected failures will occur), Baba [6] first proposed that the arrival of customers is a batch arrival. The server will perform multiple The $M^{[X]}$/G/1 queuing model of the vacation strategy, Kella [7] first proposed a queuing model with a control strategy and a vacation strategy. In addition, in practice, the same queuing system may have a different vacation policy. Takagi [3] first proposed a queuing model with variable vacation policy (for the promotion of single and multiple vacation policy), Zhang and Tian [8] People have carried out the Ge/G/1 vacation policy queuing model related to discrete time. Ke [9–14] and others used the supplementary variable technique to study the characteristics of the queuing system in the random vacation strategy, and the number of vacations of the server has a fixed upper limit (at most J vacations), and respectively proposed that the server is reliable and unreliable. And the queuing system performance evaluation of the randomized vacation strategy may delay repairs and further proposes the reliability performance index of the queuing system, which can be used as a reference for managers for equipment work scheduling and maintenance time, thereby improving the operational resilience of the system. The so-called random vacation strategy means that when the server returns from vacation and finds that no customers are waiting for service in the system, the server will have a p probability of being idle in the system and waiting for customers to enter the system for service. Still, there will be (1-p) probability of continuing to the next vacation. This pattern will continue until the server takes the specified number of vacations. Ke and Huang [15] analyzed the transient and static (steady-state) behavior of the $M^{[X]}$/G/1 model with a random vacation strategy using the supplementary variable technique and derived related reliability indicators. Recently, Ke [16] et al. proposed M/M/c queuing model with balking, retry and vacation policy. They analyzed the queuing model of single and multiple vacation policy. In addition to using the matrix method, the system characteristics and cost function are deduced, and the operating conditions of the optimal parameters of different vacation policy are compared by numerical analysis.

The so-called queuing model paper with optional secondary services is that all customers who arrive at the system must accept the first necessary service. When the customers have finished receiving the first necessary service, some customers will likely continue to accept the second kind of service. In practice, many systems have the characteristics of the queuing model of the second optional service. For example, the equipment to be tested (the piece to be tested) is sent to the laboratory to be tested by the testing equipment (standard parts), and the piece to be tested is being tested (necessary service). During the process, problems such as excessive errors may be detected. At this time, customers can choose whether to perform additional instrument adjustment services (selection services). The above-mentioned is an example of a two-phase optional queuing system structure. In recent years, many studies on optional secondary services have been proposed. For instance, Madan [17] analyzed that the service ti me of the first necessary service is assumed to be a general distribution, and the service time of the second optional service is assumed to be an exponential distribution. M/G/1 queuing model with secondary selection service and discusses related practical application scenarios. Medhi [18] extended the model of Madan [17] and assumed the service time of the second optional service as a general distribution.

Krishna Kumar et al. [19] first considered the M/G/1 retry queuing system with optional secondary service and assumed that in the first stage of service, the server might suspend the current customer's service and start serving another customer with high priority. The interrupted customer will enter a retry queue and try to re-enter the first stage of service at any time. Artalejo and Choudhury [20] extensively analyze M/G/1 retry queuing systems with quadratically selectable services and propose some interesting applications. Atencia and Moreno [21] proposed a discrete-time version of the analysis based on Artalejo and Choudhury [20], assuming the system is unreliable. Wang [23] analyzed the queuing system under the condition that the server is unreliable and the second optional service is subject to index distribution. The research results can also be directly carried out under the assumption that the second optional service is subject to the general distribution. More general analysis. Choudhury and Tadj [24] based on the results of Wang [23] and added the assumption that repairs may be delayed after a service failure. Choudhury et al. [25] further considered the analysis of a retry queuing system with optional secondary services, service interruptions, and customers arriving for batches.

The batch arrival queuing model refers to the number of customers arriving each time is a batch, and the size of each batch may be different. The batch arrival queuing model is widely used in practical fields such as manufacturing, computer networks, and communication systems. [26] The Poisson arrival queuing model with a fixed batch size was considered, and [27] also conducted related research. They both consider the batch arrival queuing model with a single working vacation and analyze the distribution of waiting numbers in the queuing system using matrix analysis and probability generating function methods. [28] and [29] analyzed batch arrival queuing models with different types of working vacations. [30–33] deduce the probability generation function and performance indicators related to the queuing model when the server is in different operating states.

The batch service queuing model is also widely used in practice, such as eliminating defective items in the manufacturing system and testing electronic equipment samples in chemical societies. Many queuing systems for batch services have been proposed. [34] studied the batch arrival queuing model with different types of servers by using shift operators and recursive techniques. [35] considered a queuing model with a single server and additional batches of services, and the servers would take vacations. More recently, [36] considered a batch queuing system with quadratically selectable services and analyzed the transient and static queuing behavior.

The rest of the paper is organized as follows. In Sect. 2, we give a description of our proposed queueing system and introduce the mathematical assumptions and notations used to derive the system characteristics. Section 3 establish the supplementary variables of the system and use the supplementary variables to construct the Kolmogorov forward equation that governs the system, and then utilize the supplementary variable techniques to derive the probability distribution of the system size distribution at stationary point of time. Section 4 give a brief conclusion and the contribution of the proposed model.

2 System Description and Notations

As mentioned above, this paper is motivated by a real-world operation management problem, and this section offers an example of the application of management for testing and calibration laboratories.

The operation process of the general testing and calibration laboratory can be briefly addressed below: the customer delivers the device under testing (DUT) to the laboratory after the administrative staff receives the DUT, the laboratory supervisor dispatches the work, and then sends the DUT to the laboratory for testing or calibrating. The customer's DUT may be delivered in batches (batch arrival), and in the process of testing (necessary service), the DUT may be detected errors. At this time, customers can choose whether to perform additional DUT adjustment services (optional services), which is the so-called structure of the two-phase optional queuing system; that is, all DUT (customers) receive an essential first phase of regular service (FES). As soon as the FES of a customer is completed, then the customer may leave the system with probability $1-\theta$ or may be provided with the second phase of optional service (for example, the adjustment of equipment) with probability θ.

Furthermore, when there is no DUT to be tested or calibrated (there is no customer in the system), in order to maintain and confirm the accuracy of laboratory standard instruments (servers), the laboratory usually arranges servers for maintenance or measurement accuracy confirmation work, the work at this time is not the general measurement of the DUT sent by the customer, so it can be regarded as the service provider's vacation, and when the servers leave is completed and returned to the formal working environment of the laboratory, if it is found that there are DUTs waiting for testing or calibrating services, then the testing or calibrating work will be carried out immediately, and if there is no DUT to be tested at this time, there will be a probability p that the servers will be idle in the laboratory and wait for the DUT to enter for testing or calibrating services, but there will be (1-p) probability for another servers maintenance or measurement accuracy confirmation work (for the next vacation), this pattern will continue until the number of servers vacations reaches J.

Suppose the standard part returns to the system after the Jth vacation and finds that there is no DUT waiting for testing service in the working environment. In that case, the standard part will always be idle in the system waiting for customers to enter the system for service. During the testing process of standard parts, due to improper operation of personnel or equipment failure, etc., the standard parts cannot continue to provide normal testing services. At this time, the standard parts must be sent for repair. The equipment is still pending confirmation and other reasons, and there may be delays in repairing (cannot be sent for repair immediately).

Suppose the above-mentioned practical situation is constructed into a theoretical model. In that case, this research project is theoretically to explore $M^{[X]}/(G_1, G_2)/1/VAC(J)$ with a random (p) vacation strategy and optional secondary services Queuing system, where X is the number of customers arriving in a batch. When there are no customers in the system, the server will immediately go on vacation. And when the server comes back from vacation and finds that there is at least one customer waiting for service in the system, the server will immediately serve the customer. The first essential service (FES, First Essential Service) provided, after customers accept the first essential service, some customers will choose to continue to receive the second additional service (SOS, Second Optional Service). When customers have finished receiving the first necessary service, some customers will continue to choose to receive the second additional equipment adjustment or maintenance service (the probability is θ).

In addition, when the server comes back from vacation and finds that no customers are waiting for service in the system, the server will be idle in the system with a probability of p waiting for customers to enter the system for service. Still, there will be a probability of (1-p) continuing vacation. This pattern will continue until the number of server vacations reaches J. Suppose the server returns to the system after the Jth vacation and finds that no customers are waiting for service in the system. In that case, the server will always be idle in the system waiting for customers to enter the system for service. Please refer to Fig. 1 for the relevant queuing system mechanism diagram. This paper will consider the following queuing system: servers that fail and can be repaired immediately.

The following notations are used in this paper, where sub-index $i = 1$ (respectively $i = 2$) denotes FES (respectively SOS).

1. $N(t) \equiv$ the number of customers in the system.
2. $\lambda \equiv$ Poisson arrival rate.
3. $X_k \equiv$ The number of customers belonging to the kth arrival batch.
4. $\theta \equiv$ The probability that the customer may be provided with a second optional service
5. $\alpha_i \equiv$ Poisson breakdown rate.
6. $S_i, S_i(t), S_i^*(\varphi) \equiv$ Service time random variable, service time distribution function and LST of $S(t)$, respectively.
7. $R_i, R_i(t), R_i^*(\varphi) \equiv$ Repair time random variable, repair time distribution function and LST of $R(t)$, respectively.

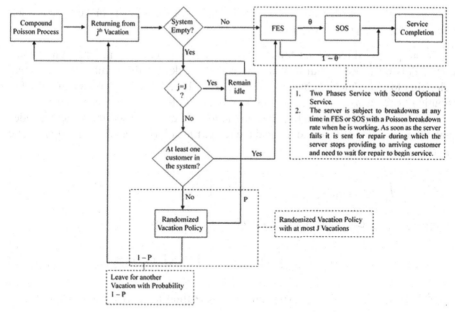

Fig. 1. The concept diagram of $M^{[X]}/(G_1, G_2)/1/VAC(J)$ queuing system with second optional service and randomized vacation policy

8. $V, V(t), V^*(\varphi) \equiv$ Vacation time random variable, vacation time distribution function and LST of $V(t)$, respectively.
9. $\mu_i(x)dx \equiv$ The first order differential (hazard rate) functions of S.
10. $\eta_i(y)dy \equiv$ The first order differential (hazard rate) functions of R.
11. $\omega(x)dx \equiv$ The first order differential (hazard rate) functions of V.
12. $S^-(t) \equiv$ The elapsed service time. (supplementary variable)
13. $R^-(t) \equiv$ The elapsed repair time. (supplementary variable)
14. $V^-(t) \equiv$ The elapsed time of the jth vacation. (supplementary variable)
15. $P_0(t) \equiv$ The probability that the server is idle but available at time t.
16. $P_{i,n}(x, t)dx \equiv$ The probability that there are n customers in the system when the server is busy at time t.
17. $Q_{i,n}(x, t)dx \equiv$ The probability that there are n customers in the system when the server is under repair at time t given that the elapsed service time is x.
18. $\Omega_{j,n}(x, t)dx \equiv$ The probability that there are n customers in the system when the server is on the jth vacation at time t.
19. $P_0 \equiv$ The steady state probability that the server is idle but available.
20. $P_{i,n}(x) \equiv$ The steady state probability that there are n customers in the system when the server is busy.
21. $Q_{i,n}(x) \equiv$ The steady state probability that there are n customers in the system when the server is under repair given that the elapsed service time is x.

22. $\Omega_{j,n}(x) \equiv$ The steady state probability that there are n customers in the system when the server is on the jth vacation.

In our proposed model, customers arrive in batches occurring according to a compound Poisson process with arrival rate λ. Let X_k denote the number of customers belonging to the kth arrival batch, where $X_k, k = 1, 2, 3, \ldots$, are with a common distribution $Pr(X_k = n) = X_n, n = 1, 2, 3, \ldots$ We also define G as the generalized service time random variable representing the completion of a customer service, which includes both the service time of a customer and the repair time of a server. The LST of G can be expressed as follows:

$$G_i^*(\varphi) = \int_0^\infty \sum_{n=0}^\infty \frac{e^{-\alpha_i t}(\alpha_i t)^n}{n!} e^{-\varphi t} \left[R_i^*(\varphi) \right]^n dS_i(t) = S_i^* \left(\varphi + \alpha_i \left(1 - R_i^*(\varphi) \right) \right) \quad (1)$$

From (1), we can obtain the first moment of G given by

$$E[G_i] = -\frac{d}{d\varphi} \left[G_i^*(\varphi) \right]|_{\varphi=0} = E[S_i](1 + \alpha_i E[R_i]) \quad (2)$$

where $E[S_i] = -\frac{d}{d\varphi} \left[S_i^*(\varphi) \right]|_{\varphi=0}$ is the mean service time, and $E[R_i] = -\frac{d}{d\varphi} \left[R_i^*(\varphi) \right]|_{\varphi=0}$ is the mean repair time.

3 Analysis of Stationary Distribution

This section aims to develop the equations and solutions, we first introduce the following random variable for further development of the variant second optional service and randomized vacation queueing model.

$$Y(t) = \begin{cases} 0, \text{ if the server is idle in the system at time } t, \\ 1, \text{ if the server is busy with FES at time } t, \\ 2, \text{ if the server is busy with SOS at time } t, \\ 3, \text{ if the server is unde rrepair during FES at time } t, \\ 4, \text{ if the server is under repai rduring SOS at time } t, \\ 5, \text{ if the server is on the } 1^{th} \text{ vacation at time } t, , \\ \vdots \\ j + 4, \text{ if the server is on the } j^{th} \text{ vacation at time } t, \\ \vdots \\ J + 4, \text{ if the server is on the } J^{th} \text{ vacation at time } t. \end{cases}$$

The supplementary variable $S^-(t)$, $R^-(t)$ are introduced in order to obtain a tri-variate Markov process $\{N(t), Y(t), \delta(t)\}$ where $\delta(t) = 0$ if $Y(t) = 0$, $\delta(t) = S_1^-$ if $Y(t) = 1$, $\delta(t) = S_2^-$ if $Y(t) = 2$, $\delta(t) = R_1^-$ if $Y(t) = 3$, $\delta(t) = R_2^-$ if $Y(t) = 4$. These variables serve as crucial components in our modeling framework, enabling us to more accurately capture the complex dynamics of the system under investigation.

Next, the following limiting probabilities are defined:

$$P_{i,n}(x,t)dx = \lim_{t\to\infty} P\{N(t) = n, \delta(t) = S_i^-(t); x < S_i^-(t) \le x + dx\}, x > 0$$

$$Q_{i,n}(x,y,t)dx = \lim_{t\to\infty} P\{N(t) = n, \delta(t) = R_i^-(t); y < R_i^-(t) \le y + dy | S_i^-(t) = x\}, x > 0, y > 0$$

$$\Omega_{j,n}(x,t)dx = \lim_{t\to\infty} P\{N(t) = n, \delta(t) = V_j^-(t); x < V_j^-(t) \le x + dx\}, x > 0, 1 \le j \le J$$

and

$$\mu_i(x)dx = \frac{dS_i(x)}{1 - S_i(x)},$$

$$\eta_i(y)dy = \frac{dR_i(y)}{1 - R_i(y)},$$

$$\omega(x)dx = \frac{dV_j(x)}{1 - V_j(x)},$$

$$\text{for } i = 1, 2; 1 \le j \le J,$$

where $\mu_i(x)dx$ can be interpreted as the conditional probability function of time for completing the FES or SOS service, given that the elapsed time is x. The $\eta_i(y)dy$ and $\omega(x)dx$ can be referred to as corresponding repair and vacation density.

According to Cox [37], we can write the Kolmogorov forward equations that govern the system under steady-state conditions for sub-index $i = 1$ (respectively $i = 2$) denotes FES (respectively SOS) as follows:

$$\lambda P_0 = \int_0^\infty \Omega_{J,0}(x)\omega(x)dx + p \sum_{j=1}^{J-1} \int_0^\infty \Omega_{j,0}(x)\omega(x)dx, \tag{3}$$

$$\frac{d}{dx}P_{i,n}(x) + [\lambda + \alpha_i + \mu_i(x)]P_{i,n}(x) = \lambda \sum_{k=1}^n X_k P_{i,n-k}(x) + \int_0^\infty Q_{i,n}(x,y)\eta_i(y)dy, x > 0, y > 0, n \ge 1, \tag{4}$$

$$\frac{d}{dx}Q_{i,n}(x,y) + [\lambda + \eta(y)]Q_n(x,y) = \lambda \sum_{k=1}^n X_k Q_{n-k}(x,y), x > 0, y > 0, n \ge 1, \tag{5}$$

$$\frac{d}{dx}\Omega_{j,0}(x) + [\lambda + \omega(x)]\Omega_{j,0}(x) = 0, x > 0, 1 \le j \le J, \tag{6}$$

$$\frac{d}{dx}\Omega_{j,n}(x) + [\lambda + \omega(x)]\Omega_{j,n}(x) = \lambda \sum_{k=1}^n X_k \Omega_{j,n-k}(x), x > 0, n \ge 1, 1 \le j \le J. \tag{7}$$

We solve the above equations by utilizing the following boundary conditions at $x = 0$

$$P_{1,n}(0) = \sum_{j=1}^J \int_0^\infty \Omega_{j,n}(x)\omega(x)dx + (1-\theta)\int_0^\infty P_{1,n+1}(x)\mu_1(x)dx + \theta\int_0^\infty P_{2,n+1}(x)\mu_2(x)dx + \lambda X_n P_0, n \ge 1, \tag{8}$$

$$P_{2,n}(0) = \theta \int_0^\infty P_{1,n}(x)\mu_1(x)dx, n \ge 1, \tag{9}$$

$$\Omega_{1,n}(0) = \begin{cases} (1-\theta)\int_0^\infty P_{1,n}(x)\mu_1(x)dx + \theta\int_0^\infty P_{2,n}(x)\mu_2(x)dx, & n = 0, \\ 0, & n \geq 1. \end{cases} \quad (10)$$

$$\Omega_{j,n}(0) = \begin{cases} \overline{p}\int_0^\infty \Omega_{j-1,n}(x)\omega(x)dx, & n = 0, j = 2, 3, \ldots, J \\ 0, & n \geq 1, j = 2, 3, \ldots, J. \end{cases} \quad (11)$$

and at $y = 0$ and fixed values of x,

$$Q_{i,n}(x, 0) = \alpha_i P_{i,n}(x), \quad x > 0, n \geq 1 \quad (12)$$

with the normalization condition

$$P_0 + \sum_{i=1}^2\left[\sum_{n=1}^\infty\int_0^\infty P_{i,n}(x)dx\right] + \sum_{i=1}^2\left[\sum_{n=1}^\infty\int_0^\infty\int_0^\infty Q_{i,n}(x, y)dxdy\right] + \sum_{j=1}^J\left[\sum_{n=0}^\infty\int_0^\infty \Omega_{j,n}(x)dx\right] = 1. \quad (13)$$

Let
us define the probability generating functions (PGF) for $\{X_n\}$, $\{P_{i,n}(\cdot)\}$, $\{Q_{i,n}(\cdot)\}$ and
$\{\Omega_{j,n}(\cdot)\}$ as follows for $i = 1, 2i = 1, 2$
$X(z) = \sum_{n=1}^\infty z^n X_n, |z| \leq 1,$
$P_i(x; z) = \sum_{n=1}^\infty z^n P_{i,n}(x), |z| \leq 1,$
$Q_i(x, y; z) = \sum_{n=1}^\infty z^n Q_{i,n}(x, y), |z| \leq 1,$
$\Omega_j(x; z) = \sum_{n=0}^\infty z^n \Omega_{j,n}(x), |z| \leq 1, 1 \leq j \leq J.$
Now multiplying (4) by z^n ($n = 1, 2, 3, \ldots$) and then adding the equations up term
by term, we obtain (for $i = 1, 2$).

$$\frac{\partial P_i(x; z)}{\partial x} + [a(z) + \mu_i(x) + \alpha_i] + P_i(x; z) = \int_0^\infty Q_{i,n}(x, y)\eta(y)dy, \quad (14)$$

where $a(z) = \lambda(1 - X(z))$. Similarly proceeding in the usual manner with (5)-(8), we
can finally get the PGF from below procedure (for $i = 1, 2$).

$$\frac{\partial Q_i(x, y; z)}{\partial x} + [a(z) + \eta_i(x)] + Q_i(x, y; z) = 0, \quad (15)$$

$$\frac{\partial Q_i(x; z)}{\partial x} + [a(z) + w(x)] + Q_i(x; z) = 0, \quad (16)$$

and

$$P_1(0; z) = \sum_{j=1}^J\int_0^\infty \Omega_j(x; z)w(x)dx + \frac{(1-\theta)}{z}\int_0^\infty P_1(x; z)\mu_1(x)dx + \frac{\theta}{z}\int_0^\infty P_2(x; z)\mu_2(x)dx + \lambda X(z)P_0$$
$$- \sum_{j=1}^J \Omega_j(0; z) - \lambda P_0. \quad (17)$$

Similarly, from (9), it yields

$$P_2(0; z) = \theta\int_0^\infty P_1(x; z)\mu_1(x)dx \quad (18)$$

which leads to

$$P_2(0; z) = \theta P_1(0; z)S_1^*(A_1(z)) \quad (19)$$

Solving the differential Eqs. (14)–(16), we have

$$P_i(x; z) = P_i(0; z)[1 - S_i(x)]e^{-A_i(z)x}, \tag{20}$$

$$Q_i(x, y; z) = Q_i(x, 0; z)\big[1 - R_i(y)\big]e^{-a(z)y}, \tag{21}$$

and

$$\Omega_j(x; z) = \Omega_j(0; z)[1 - V(x)]e^{-a(z)x}, \tag{22}$$

where $A_i(z) + a(z) + \alpha_i(1 - R_i^* a(z))$.

Solving the differential Eq. (6), it yields

$$\Omega_{j,0}(x) = \Omega_{j,0}(0)[1 - V(x)]e^{-\lambda x}, j = 1, 2, \ldots J \tag{23}$$

Multiplied (23) by $\omega(x)$ on both sides for $j = J$ and integrating with x from 0 to ∞, we get

$$\int_0^\infty \Omega_{j,0}(x)\omega(x)dx = \Omega_{J,0}(0)\gamma_0, \tag{24}$$

where $\gamma_0 = V^*(\lambda)$.

Inserting (24) in (11), we can recursively obtain

$$\Omega_{j,0}(0) = \frac{\Omega_{J,0}(0)}{(\overline{p}\gamma_0)^{J-j}}, j = 1, 2, \ldots J - 1. \tag{25}$$

From (25) and (3) we obtain

$$\Omega_{J,0}(0) = \frac{\lambda P_0}{\gamma_0\left[1 + \frac{p(1-(\overline{p}\gamma_0)^{J-1})}{(\overline{p}\gamma_0)^{J-j}(1-\overline{p}\gamma_0)}\right]}. \tag{26}$$

Finally, we obtain the following from (25) and (26)

$$\Omega_j(0; z) = \Omega_{j,0}(0) = \frac{\lambda P_0}{(\overline{p}\gamma_0)^{J-j}\gamma_0\left[1 + \frac{p(1-(\overline{p}\gamma_0)^{J-1})}{(\overline{p}\gamma_0)^{J-1}(1-\overline{p}\gamma_0)}\right]}, j = 1, 2, \ldots J. \tag{27}$$

It should be noted that from (22) and (27), we can determine $\Omega_j(x; z)$
Using (20), (22) and (27) in (17) we have

$$P_1(0; z) = \frac{\lambda P_0(1-(\overline{p}\gamma_0)^J)V^*(a(z))}{\gamma_0\left[(\overline{p}\gamma_0)^{J-1}(1-\overline{p}\gamma_0)+p(1-(\overline{p}\gamma_0)^{J-1})\right]}$$
$$+ \frac{(1-\theta)P_1(0;z)S_1^*(A_1(z))}{z} + \frac{\theta^2 P_1(0;z)S_1^*(A_1(z))}{z} + \lambda X(z)P_0 - \sum_{j=1}^J \Omega_{j,0}(0) - \lambda P_0. \tag{28}$$

Solving $P_1(0; z)$ from (28) and using (27), we obtain

$$P_1(0; z) = \frac{\lambda z P_0\left(\frac{(1-(\overline{p}\gamma_0)^J)(V^*(a(z))-1)}{\gamma_0[(\overline{p}\gamma_0)^{J-1}(1-\overline{p}\gamma_0)+p(1-(\overline{p}\gamma_0)^{J-1})]} - 1 + X(z)\right)}{z - S_1^*(A_1(z))(1 - \theta(1 - \theta))}. \tag{29}$$

From (20) and (29), it yields

$$P_1(x; z) = \frac{\lambda z P_0 \left(\frac{(1-(\overline{p}\gamma_0)^J)(V^*(a(z))-1)}{\gamma_0 \left[(\overline{p}\gamma_0)^{J-1}(1-\overline{p}\gamma_0)+p(1-(\overline{p}\gamma_0)^{J-1}) \right]} - 1 + X(z) \right)}{z - S_1^*(A_1(z))(1 - \theta(1-\theta))} \times [1 - S_1(x)]e^{-A_1(z)x},$$

(30)

which leads to

$$P_1(z) = \int_0^\infty P(x; z)dx = \frac{\lambda P_0 z \left(\frac{(1-(\overline{p}\gamma_0)^J)(V^*(a(z))-1)}{\gamma_0 \left[(\overline{p}\gamma_0)^{J-1}(1-\overline{p}\gamma_0)+p(1-(\overline{p}\gamma_0)^{J-1}) \right]} - 1 + X(z) \right)}{z - S_1^*(A_1(z))(1 - \theta(1-\theta))} \times \frac{1 - S_1^*(A_1(z))}{A_1(z)}.$$

(31)

Similarly, from (19) we can find that

$$P_2(x; z) = \theta S_1^*(A_1(z)) \times \frac{\lambda z P_0 \left(\frac{(1-(\overline{p}\gamma_0)^J)(V^*(a(z))-1)}{\gamma_0 \left[(\overline{p}\gamma_0)^{J-1}(1-\overline{p}\gamma_0)+p(1-(\overline{p}\gamma_0)^{J-1}) \right]} - 1 + X(z) \right)}{z - S_1^*(A_1(z))(1 - \theta(1-\theta))} \times [1 - S_2(x)]e^{-A_2(z)x}$$

(32)

which leads to

$$P_2(z) = \theta S_1^*(A_1(z)) \times \frac{\lambda z P_0 \left(\frac{(1-(\overline{p}\gamma_0)^J)(V^*(a(z))-1)}{\gamma_0 \left[(\overline{p}\gamma_0)^{J-1}(1-\overline{p}\gamma_0)+p(1-(\overline{p}\gamma_0)^{J-1}) \right]} - 1 + X(z) \right)}{z - S_1^*(A_1(z))(1 - \theta(1-\theta))} \times \frac{1 - S_2^*(A_2(z))}{A_2(z)}.$$

(33)

Furthermore, since $Q_i(x, 0; z)(i = 1, 2)$ can be expressed as.

$$Q_i(x, 0; z) = \alpha_i P_i(x; z).$$

(34)

Utilizing (20) and (34) in (21), we get

$$Q_i(x, y; z) = \alpha_i P_i(0; z)[1 - S_i(x)]e^{-A_i(z)x}[1 - R_i(y)]e^{-a(z)y}.$$

(35)

Inserting (29) into (35), we obtain.

$$Q_1(x, y; z) = \frac{\alpha_1 \lambda P_0 z \left(\frac{(1-(\overline{p}\gamma_0)^J)(V^*(a(z))-1)}{\gamma_0 \left[(\overline{p}\gamma_0)^{J-1}(1-\overline{p}\gamma_0)+p(1-(\overline{p}\gamma_0)^{J-1}) \right]} - 1 + X(z) \right)}{z - S_1^*(A_1(z))(1 - \theta(1-\theta))} \times [1 - S_1(x)]e^{-A_1(z)x}[1 - R_1(y)]e^{-a(z)y}.$$

(36)

Proceeding with a similar procedure, we have

$$Q_2(x, y; z) = \alpha_2 \theta S_1^*(A_1(z)) \times \frac{\lambda z P_0 \left(\frac{(1-(\overline{p}\gamma_0)^J)(V^*(a(z))-1)}{\gamma_0 \left[(\overline{p}\gamma_0)^{J-1}(1-\overline{p}\gamma_0)+p(1-(\overline{p}\gamma_0)^{J-1}) \right]} - 1 + X(z) \right)}{z - S_1^*(A_1(z))(1 - \theta(1-\theta))} \times [1 - S_2(x)]e^{-A_2(z)x}[1 - R_2(y)]e^{-a(z)y}$$

(37)

Evaluating the double integral $\int_0^\infty Q_i(x, y; z)dxdy$, we finally obtain

$$Q_1(z) = \frac{\lambda P_0 z \left(\frac{\left(1-(\bar{p}\gamma_0)^J\right)(V^*(a(z))-1)}{\gamma_0\left[(\bar{p}\gamma_0)^{J-1}(1-\bar{p}\gamma_0)+p\left(1-(\bar{p}\gamma_0)^{J-1}\right)\right]} - 1 + X(z) \right)}{z - S_1^*(A_1(z))(1 - \theta(1-\theta))} \times \frac{1 - S_1^*(A_1(z))}{A_1(z)} \times \frac{\alpha_1\left[1 - R_1^*(a(z))\right]}{a(z)} \quad (38)$$

$$= P_1(z) \times \frac{\alpha_1\left[1 - R_1^*(a(z))\right]}{a(z)}$$

Similarly,

$$Q_2(z) = \theta S_1^*(A_1(z)) \frac{\lambda P_0 z \left(\frac{\left(1-(\bar{p}\gamma_0)^J\right)(V^*(a(z))-1)}{\gamma_0\left[(\bar{p}\gamma_0)^{J-1}(1-\bar{p}\gamma_0)+p\left(1-(\bar{p}\gamma_0)^{J-1}\right)\right]} - 1 + X(z) \right)}{z - S_1^*(A_1(z))(1 - \theta(1-\theta))} \times \frac{1 - S_2^*(A_2(z))}{A_2(z)} \times$$

$$\frac{\alpha_2\left[1 - R_2^*(a(z))\right]}{a(z)} = \theta S_1^*(A_1(z)) P_1(0;z) \times \frac{1 - S_2^*(A_2(z))}{A_2(z)} \times \frac{\alpha_2\left[1 - R_1^*(a(z))\right]}{a(z)}.$$

$$(39)$$

Using (22) and (27) we have

$$\Omega_j(z) = \frac{P_0(V^*(a(z)) - 1)}{[X(z) - 1](\bar{p}\gamma_0)^{J-j}\gamma_0\left[1 + \frac{p(1-(\bar{p}\gamma_0)^{J-1})}{(\bar{p}\gamma_0)^{J-1}(1-\bar{p}\gamma_0)}\right]}, j = 1, 2, 3, \ldots J \quad (40)$$

The unknown constant P_0 can be determined by using the normalization condition (13), which is equivalent to $P_0 + \sum_{i=1}^2 P_i(1) + \sum_{i=1}^2 Q_i(1) + \sum_{j=1}^J \Omega_j(1) = 1$

Finally, let $\Phi(z) = P_0 + \sum_{i=1}^2 P_i(z) + \sum_{i=1}^2 Q_i(z) + \sum_{j=1}^J \Omega_j(z)$ be the probability generating function of the system size distribution at stationary point of time, we can then quickly obtain the expected number of customers in the system, the expected waiting time and other important system characteristics by utilizing the probability generating function developed in this paper.

4   Conclusions

The Kolmogorov forward equation and the probability distribution of the system size distribution for the $M^{[X]}/(G_1, G_2)/1/VAC(J)$ queuing system with a randomized vacation policy and optional second service has been developed by using the supplementary variable techniques. The relevant results can be used as the basis for further deriving the expected number of customers, the expected waiting time and other important system characteristics in our proposed queuing system, and can also be used as the service performance evaluation and decision-making tools that require secondary optional services and regular maintenance in practical applications.

References

1. Doshi, B.T.: Queueing system with vacation-a survey, Que. Syst. 1, 29–66 (1986)
2. Levy, Y., Yechiali, U.: Utilization of idle time in an M/G/1 queueing system. Manag. Sci. 22, 202–211 (1975)

3. Takagi, H.: Queueing Analysis: A Foundation of Performance Evaluation. Vol. I, Vacation And Priority Systems, Part I, North-Holland, Amsterdam (1991)
4. Servi, L.D., Finn, S.G.: M/M/1 queues with working vacations (m/m/1/wv). Perform. Eval. **50**(1), 41–52 (2002)
5. Wu, D.A., Takagi, H.: M/G/1 queue with multiple working vacations. Perform. Eval. **63**(7), 654–681 (2006)
6. Baba, Y.: On the $M^{[X]}$/G/1 queue with vacation time. Operat. Res. Lett. **5**, 93–98 (1986)
7. Kella, O.: The threshold policy in the M/G/1 queue with server vacations. Naval Res. Logist. **36**, 111–123 (1989)
8. Zhang, Z.G., Tian, N.: Discrete time Geo/G/1 queue with multiple adaptive vacations. Que. Syst. **38**, 419–429 (2001)
9. Ke, J.-C., Huang, K.-B., Pearn, W.L.: The randomized vacation policy for a batch arrival queue. Appl. Math. Model. **34**(6), 1524–1538 (2009)
10. Ke, J.-C., Huang, K.B., Pearn, W.L.: Randomized policy of a Poisson input queue with J vacations. J. Syst. Sci. Syst. Eng. **19**(1), 50–71 (2010)
11. Ke, J-C., Huang, K.-B., Pearn, W.L.: A batch arrival queue under randomized multi-vacation policy with unreliable server and repair. Int. J. Syst. Sci. **43**(3), 552–565 (2012)
12. Ke, J.-C., Huang, K.-B.: Analysis of an unreliable server M[X]/G/1 system with a randomized vacation policy and delayed repair. Stoch. Model. **26**(2), 212–241 (2010)
13. Ke, J.-C. Huang, K.-B., Pearn, W.L.: The performance measures and randomized optimization for an unreliable server $M^{[X]}$/G/1 vacation system. Appl, Math. Comput. **217**(21), 8277–8290 (2011)
14. Ke, J.-C., Huang, K.-B.: Analysis of batch arrival queue with randomized vacation policy and an un-reliable server. J. Syst. Sci. Complexity **25**, 759–777 (2012)
15. Ke, J.-C., Huang, K.-B., Kuo, C.-C.: Reliability-based measures for a batch-arrival queue with an unreliable server and delayed repair under randomised vacations. Int. J. Indust. Syst. Eng. **27**(4), 500–525 (2017)
16. Ke, J.C., Chang, F.-M., Liu, T.-H.: M/M/c balking retrial queue with vacation. Qual. Technol. Quan. Manage. **26**(1), 54–66 (2019)
17. Madan, K.C.: An M/G/1 queue with second optional service. Que. Syst. **34**, 37–46 (2000)
18. Medhi, J.: A single server Poisson input queue with a second optional channel. Que. Syst. **42**, 239–242 (2002)
19. Krishna Kumar, B., Vijayakumar, A., Arivudainambi, A.: An M/G/1 retrial queueing system with two phase service and preemptive resume. Ann. Oper. Res. **113** 61–79 (2002)
20. Artalejo, J.R., Choudhury, G.: Steady state analysis of an M/G/1 queue with repeated attempts and two phase service. Qual. Technol. Quant. Manag. **1**, 189–199 (2004)
21. Moreno, A.P.: Geo/G/1 retrial queue with second optional service. Int. J. Oper. Res. **1**, 340–362 (2006)
22. Wang, J., Zhao, Q.: A discrete time Geo/G/1 retrial queue with starting failures and second optional service. Comput. Math. Appl. **53**, 115–127 (2007)
23. Wang, J.: An M/G/1 queue with second optional service and server breakdowns. Comput. Math. Appl. **47**, 1713–1723 (2004)
24. Choudhury, G., Tadj, L.: An M/G/1 queue with two phases of service subject to the server breakdown and delayed repair. Appl. Math. Model. **33**, 2699–2709 (2009)
25. Choudhury, G., Tadj, L., Deka, K.: A batch arrival retrial queueing system with two phases of service and service interruption. Comput. Math. Appl. **59**(1), 437–450 (2010)
26. Oduol, V.K., Ardil, C.: Transient analysis of a single server queue with fixed size batch arrivals. Int. J. Electr. Comput. Eng. **6**(2), 253–258 (2012)
27. Xu, X., Tian, N., Zhang, Z.: Analysis for the *M/M/*1 working vacation queue. Int. J. Inf. Manage. Sci. **20**, 379–394 (2009)

28. Laxmi, P.V., Rajesh, P.: Analysis of variant working vacations on batch arrival queues. Opsearch **53**(2), 303–316 (2015). https://doi.org/10.1007/s12597-015-0236-3
29. Vijaya Laxmi P., Rajesh, P., Kassahun, T.W.: Performance measures of variant working vacations on batch queue with server breakdowns. Int. J. Manag. Sci. Eng. Manag. **14**(1), 53–63 (2018)
30. Chandrika, U.K., Kalaiselvi, C.: Batch arrival feedback queue with additional multi optional service and multiple vacation. Int. J. Sci. Res. Publ. **3**(3): 1–8 (2013)
31. Kirupa, K., Chandrika, K.U.: Batch arrival retrial queue with negative customers, multi optional service and feedback. Commun. Appl. Electr. **2**(4), 14–18 (2015)
32. Suganya, S.: A batch arrival feedback queue with M-optional service and multiple vacations subject to random breakdown. Int. J. Sci. Res. **3**(11), 1877–1881 (2014)
33. Vinnarasi, S., Maria Remona, J., Julia Rose Mary K.: Unreliable batch arrival queueing system with SWV. Int. J, Innov. Res. Sci. Eng. Technol. **5**(3), 2884–2889 (2016)
34. Sree Parimala, R., Palaniammal, S.: An analysis of bulk service queueing model with servers' various vacations. Int. J. Adv. Res. Technol. **4**(2), 22–33 (2015)
35. Kalyanaraman, R., Marugan, S.P.B.: A single server queue with additional optional service in batches and server vacation. Appl. Math. Sci. **12**(56), 2765–2776 (2008)
36. Vijaya Laxmi, P., George, A.A., Girija Bhavani, E.: Performance of a single server batch queueing model with second optional service under transient and steady state. RT&A **4**(65) (2021)
37. Cox, D.R.: The analysis of non-Markovian stochastic processes by the inclusion of supplementary variables, Proc. Camb. Philos. Soc. **51**, 433–441 (1955)

Communicating Sustainability Online: A Soft Systems Methodology and Corpus Linguistics Approach in the Example of Norwegian Seafood Companies

Nataliya Berbyuk Lindström[1]([✉]) and Cheryl Marie Cordeiro[2]

[1] University of Gothenburg, 41296 Gothenburg, Sweden
nataliya.berbyuk.lindstrom@ait.gu.se
[2] RISE Research Institutes of Sweden, 50115 Borås, Sweden
cheryl.marie.cordeiro@ri.se

Abstract. This article presents a qualitative case study of the Norwegian seafood industry's sustainability communication on corporate websites. The research questions focused on how sustainability is communicated, and the communication channels employed by Norwegian seafood companies. The study found that sustainability is communicated through ecological certification, awareness towards the environment and social compliance, and engagement in knowledge exchange. The study highlights the need to create standardized language and a coherent discourse for competitive advantages in ecologically value-added products and digital services. Companies can use underutilized digital resources such as podcasts and direct online sales to consumers to improve stakeholder engagement. The study suggests the direct and active involvement of consumers in designing products that consumers desire, which can increase market share for the Norwegian seafood industry. The limitations of the study are the lack of feedback from small to medium-sized companies, limiting the generalizability of the findings. The study recommends that the Norwegian seafood industry supports developing digital service resources for small and medium-sized companies to remain competitive in the long term.

Keywords: sustainability · corporate online communication · seafood · soft systems methodology · corpus linguistics

1 Introduction

As more consumers take an active interest in responsible seafood consumption, there is a growing need to understand how seafood companies communicate their sustainable seafood practices to consumers through their corporate websites. This study uses as a case example, the seafood industry in the Nordic countries, investigating how Nordic seafood companies communicate the concept of sustainability [1, 2].

© Springer Nature Switzerland AG 2023
F. Fui-Hoon Nah and K. Siau (Eds.): HCII 2023, LNCS 14039, pp. 334–351, 2023.
https://doi.org/10.1007/978-3-031-36049-7_25

The Nordic countries have a long history of fisheries and aquaculture, with Norway being a major producer of Atlantic salmon, rainbow trout, and Arctic charr. Aquaculture is the fastest-growing food-producing sector and plays a central role in meeting the increasing demand for fish and fish products. The sustainable seafood movement, together with ecolabeling of seafood products, has made consumers increasingly aware of responsible seafood production and consumption. Sustainable seafood is defined as seafood that is caught or farmed in ways that consider the long-term vitality of the harvested species, the well-being of the oceans, and the livelihoods of the fisheries-dependent communities [3]. The Norwegian seafood industry is expected to develop and play an important role in the global food production network and ecosystem services. The importance of producing sustainable seafood was identified as a full pillar of the new reformed European Union Common Fisheries Policy, and several intergovernmental and non-governmental frameworks of support exist to promote sustainable fishing practices.

Sustainability communication and corporate responsibility in the seafood industry are essential to achieve socio-economic and ecological synergies [4]. Digital technologies and a corporate web presence are crucial for companies to communicate their sustainability practices and engage with consumers in a multi-actor ecosystem framework. As sustainable seafood production is currently high on the agenda of the Norwegian government, the study focuses on investigating how sustainability is communicated online through corporate websites of companies in the Norwegian seafood industry. The main research questions (RQs) investigated in this study are:

RQ1: How is the concept of sustainable aquaculture communicated by Norwegian seafood companies?
RQ2: Which are the current communication channels of choice, commonly employed by Norwegian seafood companies?

This paper is organized as follows. The Introduction section introduces the subject of study, its background context, and defines the research challenges with corresponding research questions. The Literature Review sets the context of the study of the need to build a common industry or corporate language around the subject of 'sustainability'. This is followed by an outline of the theoretical framework used, which is systems thinking (Soft Systems Methodology, SSM), and a corpus linguistics framework of analysis. We then describe in the Data and Method section, how data for this study was collected, organized, and analyzed. Finally, we present a corpus-driven set of results in the Findings and Discussion section, with some concluding thoughts on the practical and social implications of the findings.

2 Literature Review

2.1 Communicating Sustainability

This section of literature review discusses the challenges in communicating sustainability within the seafood industry. As the world becomes more aware of sustainability practices and ecosystem services, corporate communication on sustainability practices is increasingly necessary to keep stakeholders informed on the sustainability status of the supply chain of products and services offered. Efforts to communicate sustainability have been seen in various industries, including tourism, higher education institutions, the food and beverage industry, and the field of digital communication and media for rural communities [5].

Within the seafood industry, there has also been a growing awareness of sustainability and sustainable seafood amongst industry stakeholders. Sustainability communication empowers stakeholders by providing information and knowledge about where their seafood originates and how it is treated before reaching the dining tables. It also informs interested stakeholders about how the seafood industry is complying with global sustainability and ecological conservation efforts and puts at ease the ecological anxiety of consumers so they can trust that they are consuming responsibly [6].

However, the seafood industry faces challenges in addressing sustainability and sustainable seafood consumption due to the diverse and often low general knowledge about the origin of the product, food authenticity, and food safety [7, 8]. The different rates of maturity of various markets, technologies, and social values have also led to a collection of complementary but fragmented discourse around the concept of sustainability and sustainable seafood [9].

Additionally, due to the nature of the product, whether wild captured from oceans or farm-raised, the discourse of sustainability in the seafood sector remains challenging to address, particularly across regional and global markets [10]. It takes a concerted communication strategy to create a consistent and coherent discourse around sustainability towards multiple stakeholders with different agendas. Bridging industry, corporate, and cultural values require constant meaning negotiation towards a common understanding between stakeholders.

As such, creating a common language towards a collective understanding of sustainability and ecosystem services in the seafood industry requires constant and targeted communication. The challenges in addressing sustainability and sustainable seafood consumption require a collaborative effort from various stakeholders to develop a sustainable seafood industry.

2.2 Seafood from Norway

This section of the literature review discusses the importance of sustainability in the Norwegian seafood industry, which is the country's second-largest export industry. To achieve socio-economic and ecological synergies at different levels of business operations, a systems approach to sustainability is necessary, which involves leveraging synergies between multiple stakeholders and industry sectors. However, this is inherently complex and difficult to put into business practice [11].

Digital technologies and establishing a corporate web presence allow businesses to shift their focus from delivering physical products to value-added product services that fulfill the demands of customers and clients [12]. In the Norwegian seafood industry, this transformative change in business models may not be explicitly strategic or conscious but rather one that evolves in relation to an evolving business environment.

While the industry has developed multi-tiered business models that encompass a variety of stakeholders, a challenge remains in communicating sustainable business practices, as this has been viewed primarily from the enterprise perspective and tends to exclude consideration of the potentially influencing role of social communities and other sectors of society [13]. The evolving socio-technical landscape of the food industry has placed corporate responsibility pressure on Norwegian seafood companies to communicate their sustainability practices in fisheries and aquaculture, and to provide transparency in their business processes [14]. Seafood companies are no longer solely providing raw produce but are also incorporating social responsibility by communicating about sustainability and sustainable business practices [15].

3 Theoretical Framework

3.1 Systems Thinking

The increasing influence of technology on socio-cultural and ecological systems has led to the need for a systems thinking approach applied to solving challenges, which in this study refers to the context of communicating sustainability in the seafood industry. A systems thinking approach considers all the facets and variables that link the social to the technological [16, 17]. The systems approach could be said to have its roots in the fields of operations research and engineering in the early 20th century, where technology was viewed as an assemblage of components that influenced each other. It gained more ecological and organic focus in the 1980s, which applied systems thinking to the study of emerging ecological paradigms in the field of biology. Systems thinking involves investigating parts, wholes, sub-systems, system boundaries, environments, structure, process, and emergent properties, as a hierarchy of systems with positive and negative feedback systems of the observer/actors [18, 19]. In highly urbanized sociocultural environments, individuals belong to various circles of activities and acquaintances in their lives, and the complexity of social hierarchies leads to a more diverse range of institutions and groups which individuals can belong and identify themselves with. Within the field of international business studies, systems thinking has been applied to organizational, management science, management and cybernetics research, and studies on organizing for digital servitization and service ecosystems [20, 21].

3.2 Soft Systems Methodology and Corpus Linguistics

Within the systems thinking approach, soft systems methodology (SSM) is a conceptual framework that can help contextualize the business environment in which sustainability is communicated and illustrate the complex, interrelated processes of sustainable development. Grounded at the intersections of business process modelling and the field of systems engineering, SSM sees computer-based information systems as systems that serve purposeful human action. Developed as a methodology within systems engineering primarily by Peter Checkland [22], SSM takes a critical perspective on traditional 'hard' systems thinking that is usually quantitatively defined, but rather places emphasis on the social environments in which events occur. SSM aims to capture in a qualitatively and holistic manner, the subtle boundaries of social reality and how businesses narrate their corporate values. The framework considers human actors as observers of the phenomenon and thus influencers and co-creators of the language community, the resulting experiences, and events [23, 24]. In this sense, SSM is closely aligned to the systemic functional perspective of language in use.

Language is a systemic functional tool used in daily human transactions, conveying realities as well as creating perspectives of realities through narratives and storytelling [25, 26]. Yet, how language is used in the context of international business has often been under researched in terms of its influence on enterprise performance, competitiveness, and social impact [27]. Corporate narratives, for example, have been found to be important in positioning the company within the business sector through identification of its values as well as communicating its corporate brand and identity. Communication has also been found to be a critical component in facilitating sustainable development goals.

Several research studies on sustainable development indicate that successful sustainable development requires a multi-stakeholder, multi-perspective, or systems approach with three main functional factors in the implementation process of: (i) participants – individual stakeholders, business owners, entrepreneurs, activists, leaders, etc.; (ii) institutions at different levels of societal function – neighbourhood groups, associations, non-governmental organizations, governmental organizations, national, regional, and global institutes; and (iii) communication channels, broadly encompassed in technology – mainstream media, internet, and social media platforms, face-to-face, etc. [28–31].

When investigating the tri-pronged functions towards the successful communication of sustainability, SSM and corpus linguistics are frameworks and methods that allow for uncovering the market communication strategies of Norwegian seafood companies, as they strive to create a communicative balance between participants, institutions, and relevant communication channels [32, 33].

4 Data and Method

4.1 Companies and Corpus Data Creation

Information on Norwegian seafood enterprises was retrieved from the Norwegian Seafood Federation (NSF), that is affiliated with the Confederation of Norwegian Enterprise (NHO). The NHO is the main representative body for Norwegian employers with a current membership of over 28 000 enterprises that range from small to medium family-owned enterprises (SMEs) to multinational enterprises (MNEs). The NSF represents the interests of about 800 member companies in the fisheries and aquaculture sectors in Norway. These enterprises often have global presence and markets, and their business transactions occur along the entire seafood value chain from fjord to dining table.

There are currently more than 400 seafood export registered companies in Norway. Based on annual turnover and number of employees, 17 Norwegian enterprises were identified including 11 MNEs. The group of MNEs include the 4 largest enterprises in salmon, the 4 largest enterprises in salmon, white fish, and pelagic fish, and the 3 largest enterprises for white fish and pelagic fish. In addition to the 11 MNEs, 6 SMEs were identified for this study that include the 3 smallest enterprises for salmon and the 3 smallest enterprises for white fish and pelagic fish.

To obtain a comprehensive understanding of how sustainability is communicated in the Norwegian seafood industry, the researchers used multiple data sources for analysis. These sources included publicly accessible information from corporate websites, annual reports, and brochures, as well as conducting 12 qualitative interviews with individuals in top leadership positions in these enterprises. The interviewees included individuals with extensive experience in corporate leadership, such as Founders, Owners, CEOs, Managing Directors, Directors of Global Marketing and Sales, and Directors of Communications. Each interview lasted approximately 45 to 60 min, with a total interview time of around 750 min. The interviews were transcribed in accordance with the Gothenburg Transcription Standard version 6.4 and cross-checked with audio recordings. The transcription conventions used are indicated in Table 1 and reflected in the text examples used in the study. This combination of data sources was chosen to provide a more complete understanding of the complex phenomenon of sustainability communication from the perspectives of various companies and stakeholders.

The methodology used in this study involved several steps for managing and analyzing the data. First, the text, images, and audio recordings that were collected were checked for accuracy. Next, the collected texts were translated in format and stored in a machine-readable format (i.e.,.txt files). Then, a small corpus was created from a corpus, which allowed for analysis with the concordance software AntConc version 3.5.8. The combined corpus for the large companies had 614,925 words, while the combined corpus for the small companies had 3,234 words. This small corpus data set falls within the definition of a specialized small corpus. Creating this corpus will be useful for text analysis on specific topics of interest. The names of the respondents were anonymized in the text examples provided.

The use of AntConc ensures corpus-driven findings, with as little subjective inference from the researcher's perspective as possible [34–37]. Other functions of AntConc include word frequency counts to uncover salient content and themes in the data. Keyword-in-context (kwic) searches can also help identify in a systematic and systemic manner, how the saliently identified themes and topics were used in context. It is in the context of word use that we see how SSM and corpus linguistics frameworks of analysis converge and complement each other. SSM as a method focuses on the sociocultural context in which events and happenings influence each other toward a certain outlook and outcome. In complement, kwic analysis in AntConc helps uncover nuances in meaning, in how specific keywords such as "sustainability" or "sustainable" are used by the enterprises and conveyed in corporate values. Concordance plots reflect the saliency of the use of keywords so that it helps identifying which enterprises are most sustainable consumption focused. For corporate websites, each corporate website was investigated for their various communication channels. The words "communication channel" are defined as the various means, manner, platforms, or modes by which sustainability can be communicated such as conferences, community programs, open dialogue sessions, financial reports etc.

Table 1. The Gothenburg Transcription Standard 6.4 transcription conventions applied in the text examples in Sect. 4

Symbol	Explanation
$P	participant, initial/s of participant
/, //, ///	a short, intermediate and a long pause respectively

4.2 Ethical Considerations

The research conducted was part of the project "Market Access: Communicating Sustainability," which is part of the Nofima strategic research program called Market Access, supported by the Norwegian Ministry of Trade, Industry and Fisheries, and has project number 12361. The study was approved by the NSD (the Norwegian Center for Research Data, organization number NO 985 321 884 MVA) and conducted in accordance with ethical guidelines set by the NSD to ensure confidentiality, consent, information, and autonomy in research. The consent form emphasized anonymity, and participants were given the right to withdraw from the study at any time. To preserve individual integrity, the names and other identifying features of participants were removed or anonymized, as were the names of companies.

5 Findings and Discussion

5.1 Soft Systems Methodology Findings

Focusing on the corporate websites of the Norwegian seafood companies and how they are designed and constructed, corporate websites give users an immediate impression upon landing at the opening webpage. Images 1 and 2 are example screenshots of the landing webpage of a Norwegian seafood MNE (Image 1), and a Norwegian seafood SME (Image 2). Both screenshots offer a glimpse into the products and services of the seafood enterprise, with the MNE distinctly focusing on offering more information and a broader range of products and services such as their explicit corporate position about sustainability, as well as financial investment reports. The SME example distinctly goes into direct sales contacts, with notably less information on sustainability issues, possibly due to that in this example (Image 2), 'sustainability' is implicit in their production process because their product availability is regulated by government fisheries licenses and quotas that ensure wild fish stock sustainability during harvest. A notable similarity between the MNE and SME is the availability of information in English and in Norwegian, where Norwegian seafood enterprises, regardless of size, have global market outreach. As indication of the global market reach of Norwegian seafood, other seafood enterprises in the sample set have multiple languages in which their websites can be read, depending on their market operations that include Chinese (Mandarin texts), French, German, Japanese, Portuguese, and Spanish. While English as lingua franca is used by most Norwegian seafood companies in their corporate language practices for their commercial websites in an understanding that the choice of language/s have considerable influence over the enterprise's global outreach, language choice seems to remain functional and market oriented. The multi-language use orientation in Norwegian seafood enterprises reflects similar findings how global brands transcend borders to connect with citizens around the world as an interconnected global economy (Figs. 1 and 2).

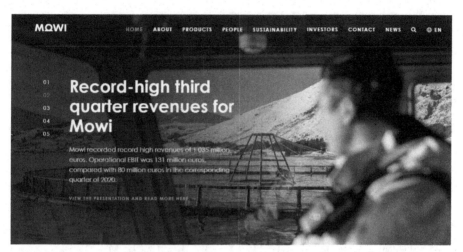

Fig. 1. Example screenshot of a Norwegian seafood multinational enterprise that showcases their products, services, and communication channels.

Fig. 2. Example screenshot of a Norwegian seafood small to medium sized enterprise that showcases their products, services, and communication channels.

The website design, including the use of colors, fonts, and images, gives visitors an initial impression of the corporate branding and values of both MNEs and SMEs. Table 2 presents the list of Norwegian seafood companies studied in this research, along with the communication channels they use to convey the concept of sustainable seafood and their business practices, addressing the first research question. The names of the companies have been anonymized.

Table 2. Communication channels and market communication strategies of Norwegian seafood companies

No.	Company	Corporate Communication of Sustainability												
		Conferences/Tradefairs	Dialogues	Initiatives	News	Partnerships	Programs	Report	Social Media	Video	Webcast	Webpage	Webpage: multilingual	Webshop / online sales
4 largest companies in Salmon														
1	Company Alpha	1	1	1	1	1	1	1	1	1	1	1	1	
2	Company Beta	1	1	1	1	1	1	1	1	1	1	1	1	
3	Company Gamma	1	1	1	1	1	1	1	1	1		1	1	
4	Company Delta	1	1	1	1	1	1	1	1	1	1	1	1	
4 largest companies in Salmon, Pelagic Fish and Whitefish														
5	Company Epsilon	1	1	1	1	1	1	1	1	1	1	1	1	
6	Company Zeta	1			1	1			1	1		1	1	
7	Company Eta	1				1	1	1	1	1		1	1	
8	Company Theta				1	1						1		
3 largest companies in Pelagic Fish and Whitefish														
9	Company Iota	1	1	1	1	1		1	1			1	1	
10	Company Kappa		1	1	1			1	1	1		1	1	
11	Company Lambda											1		
3 smallest companies in Salmon														
12	Company Mu											1	1	
13	Company Nu													
14	Company Xi													
3 smallest companies in Pelagic and Whitefish														
15	Company Pi		1			1		1				1		1
16	Company Rho											1	1	
17	Company Sigma					1			1			1	1	
	Communication strategies strength total	8	8	7	9	11	6	9	10	8	4	15	12	1

There are thirteen potential communication channels or channels that corporations can employ that target different business and social networks. The communication channels are often reflected on corporate webpages, and they include conferences / trade fairs, dialogues (engagement with local communities or political leaders), initiatives (sports sponsorship or student involvement), news, partnerships (with local community associations or universities for research and development), programs (related to environmental awareness or supporting economic activities of local fishing communities), report (financial reports, corporate reports), social media, video (presence on YouTube or Vimeo), webcast, webpage, webpage: multilingual and webshop / online sales. Taken in totality and in overview, these communication channels target different functional factors in the systems approach SSM framework of reaching out to different participants and institutions using social media on the Internet and face-to-face meetings.

The web-based presence of the Norwegian seafood companies is a digital service provided for the purposes of stakeholder engagement. In address to RQ2, the last line of Table 2 indicates the total number of communication channels used by the companies as reflected on their corporate webpages. Higher total scores indicate a greater preference or use of a particular communication channel.

For both MNEs and SMEs, having a corporate webpage (15 of the 17 companies studied) is an important channel for outreach. This communication channel form however is less interactive than active engagement in? dialogues such as presence at conferences / trade fairs (8 of 17 companies) and social dialogues with local communities or schools (8 of 17 companies). As an indication of the international nature of the Norwegian seafood industry, 12 of 17 webpages had multilingual options, which indicates the breadth and scope of market in which they operate. What seems highly under-utilized as communication channels are online web shops as market access, only 1 of 17 companies studied offered direct online sales of their products in a distributor-to-consumer (D2C) context.

Other audio and digital communication channels not currently employed by the industry are podcasts and the sharing of recipes (i.e., how should people use their product in home cooking? How can households eat "nose-to-tail" and use wholefood products with as little waste as possible, towards sustainable and responsible food consumption?). Podcasts can be useful to keep the general population up to date on industry issues or general food trends. There could be avenues for cooperation with chefs and restaurants in podcasting recipe sharing too so that sustainable consumption could be discussed alongside favourite or innovative ways to cook fish and seafood.

Prominent types of information found in larger Norwegian multinational enterprises for fisheries and aquaculture webpages are:

i. financial information to stakeholders, for example, "Operation EBIT of the EUR 99 million in the second quarter for Company Alpha"
ii. large enterprise industry influence on industry standards, for example, "The 2020 Company Alpha Industry Handbook is out" and
iii. large enterprise industry standing by size and strength, for example, "We are the world's largest supplier of farm-raised salmon"

Corporate values and branding can reflect sustainability and sustainable seafood practices [37, 38]. One way this is expressed is through an emphasis on responsible seafood production, highlighting the location of production, for example, in Northern Norway, and consistent quality of products that are "delicious, nutritious and healthy". This is in line with research on product location branding, which suggests that consumers associate certain regional qualities with the place of origin of a product. To achieve sustainable seafood, research has linked the ethical treatment of fish during harvest, including better feed and less bruising of the fish, along with a quick despatch in the process, to a higher quality product that results in less food waste. Studies have also supported the concept that sustainable seafood practices contribute to the production of high-quality products.

Prominent information found in small-to-medium sized enterprise websites is the visual display of the location of their operations and how customers can use their products even if there are no recipes shared. In Company Sigma for example, which is a relatively young company founded in 2011 by a sixth-generation fish buyer, sustainability is communicated to stakeholders in visual images (Image 3) of their finished food products on the website in the form of plated meals, ready-to-eat meals, and pre-packaged single portion foods. While many Norwegian seafood enterprises have global markets, short food supply chains are also part of their operative and communicative strategies where the presenting Nordic style dishes on their websites encourages home-based (domestic and regional) end-market consumers to purchase from local sources that makes for sustainable food consumption.

SMEs also focus on sustainability and sustainable seafood by highlighting the location of their operations in remote and well-maintained environments. The Nordic countries, particularly Norway, are distinctive due to their geographical location, with fish growing and living in Arctic waters that are clean and unpolluted (Fig. 3).

Stockfish	Wet salted fish	Fresh fish
We sell both traditional stock fish - unsalted fish that is naturally dried by the sun and the wind on racks - and industrially dried fish, fish heads, bones and APAMA	During the winter, we produce significant quantities of wet salted cod. We export this to Spain and Portugal.	During the seasonal cod fishery in February / March, we pack fresh cod, cod roe and cod tongues for the European market.

Fig. 3. A corporate website screen shot of Company Sigma, a small to medium sized (SME) orwegian seafood company, on the descriptions of products and accompanying images in the form of plated meals. The text on the website can be found in both English and Norwegian.

Relative to MNEs, there is greater personal communication from SMEs directly to consumers by giving ideas as to how their products can be turned into food at the table (Image 3). Smaller companies also have a more personalized blog-format narrative, in which they converse with their website visitors, sharing stories, inspirations and aspirations. In this sense, smaller Norwegian companies seem to capitalize on their small firm status to be more personal with their users. Personalized narratives draw in customers and stakeholders to the company's sustainability and responsible seafood production and consumption narrative. This active engagement of customers as co-creators of their products shares the responsibility of bringing sustainably produced seafood to market.

Distinctively absent however for SMEs are some forms of socio-economic outreach, and digital resources, relative to their larger MNE counterparts. Exclusive to MNEs seem to be comprehensive societal or research programs targeted at environmental conservation or socio-economic causes (Companies Alpha, Beta, Gamma, Delta, Epsilon and Theta have societal and environmental related outreach programs). Few SMEs also seem to be consistently active in multichannel social media channels such as presence on YouTube, Instagram, Facebook, Tiktok etc. An interesting find is that one SME, Company Pi, has a webshop, where the larger MNEs do not provide such a digital online sales service selling director from producer to consumers on the market.

5.2 Corpus Linguistics Findings: Interviews

Generally, according to the respondents, sustainability is communicated to stakeholders in three main ways that include:

i. ecological certification
ii. raising ecological awareness in stakeholders and consumers and
iii. active engagement with local, regional, and international governance bodies towards better industry practices

In concrete measures that benefit stakeholders, particularly consumers who buy Norwegian salmon in the grocery stores, most companies convey sustainability primarily in the form of (i) ecological certification on product labels for their products. Environmentally friendly certified aquaculture, usually by regional or international standards such as the salmon Global Gap certificate or the ASC (Aquaculture Stewardship Council) certificate for responsible aquaculture are explicit means by which Norwegian seafood companies convey their commitment towards marine resources and conservation. In the Norwegian seafood industry, a sustainably produced seafood product is mandatory. As respondent SM from an MNE related:

> $SM: even if we are a middle sized company / we are also hooked on sustainability / and for us we were quite early having our salmon global gap < good agricultural practice > certified / and also nowadays we are having not all / but a very big part of our production is also asc < aquaculture stewardship council > certified / so the sustainability issue is of course important for everyone / both for ourselves and our customers / so i think that is something we have to live up to

Apart from concrete certifications that appear on product labels for consumers, a broadening of sustainability as a concept to include (ii) ecological awareness towards the environment and social compliance is also developing in the Norwegian seafood sector. As the fisheries sector is highly dependent on weather conditions and fish migration patterns for harvest, employment within the industry is seasonal with most employed on temporary working contracts. With a rising awareness on humanitarianism and proper working conditions for those in the offshore fisheries sector, humane practices for both fish harvesting, and human labour are beginning to be increasingly important. Global events, and happenings in the network value chains have influences and impact on the Norwegian seafood industry. As respondent M relates on the company's values to providing fish in accordance with market demand, and being a responsible employer:

$M: so there's been a limit to how much fish we could take / but i would try to make the fishing season longer / and more market oriented / market oriented fishing and then of course sustainability / we need to work with sustainability also as a social compliance / of sustainability at the work place / and that is also becoming more and more important for the retail sector / there was a very bad case in thailand where people were kidnapped in myanmar and were in forced labour / and the worst situation was they were actually tied up by chains to the vessel / so if the vessel would / yeah / it's like a horror movie /

There are many stakeholders who participate at various levels of society for which sustainable seafood and sustainability are important. The multi-levelled, multistake-holder interests for sustainable Norwegian seafood has been acknowledged by Norwegian seafood companies. The third way in which sustainability is communicated is by (iii) active engagement and looking into a 'higher agenda' for knowledge exchanges in different social and political spheres. Due to a long history and expert knowledge in aquaculture practices, Norway is often seen as advisor to other emerging international seafood industries and companies, with possibility of influencing national and international laws for the industry. As respondent K relates:

$K: i met a person from vietnam at the last aquaculture conference who was very interested in the governmental part / how the government can establish a framework regarding sustainability regulations / so i think that we're in the same boat because we are in salmon export for norway / and i think that the salmon industry is important for society and the population / so both from the political side / the governmental side and the industry side for better regulations / which is a pre-requisite for being able to grow / you can't grow / you can't produce more salmon if it is going against the sustainability regulations / and these regulations are getting tighter and tighter / so i think as an industry as a whole / i'm not familiar with aquaculture regimes in asia or australia or tasmania or other countries / but to me it seems we have the strictest regulations in norway when it comes to this / and that drives the industry technology wise and practices wise / to improve the sustainability aspects regarding the industry from fish welfare to environment to food safety and of course the safety of our personnel also which is a part of our sustainability area / and of course < name of company > as a whole / we have a higher agenda / because all of us want to be a part of a company that takes sustainability seriously and we also think about the future when it comes to the future to be leading in this / both on national level but also global level /

6 Conclusion

6.1 Sustainability Communication

This study of communicating sustainability by investigation of the corporate websites of Norwegian seafood companies has uncovered various corporate communication strategies used by seafood businesses to market seafood. In addition, seafood companies also advertise corporate social priorities such as community involvement and marine resource conservation. These findings align with previous studies in the past decade that emphasize the need for investing in both online presence and community dialogue regarding sustainable consumption [39–42].

It was challenging to get in touch with SMEs in the Norwegian seafood industry, and that might prove to be a study limitation due to the lack of adequate feedback or voice from Norwegian seafood SMEs. Most replies given were that they did not see themselves as having any voice or impact on 'sustainability' practices, even though all their products were certified sustainable. As such, the data collected was necessarily from a niche set of MNEs and SMEs, the purpose of which was to do a general cross-company comparison of how sustainability is communicated on corporate websites.

Still, the interviewees acknowledged the significant role of Norway's seafood industry in the country's economic development, as well as the importance of maintaining a healthy ecosystem to support global seafood provision. Our findings suggest that all Norwegian seafood enterprises view sustainability as a crucial component of their business model, even though the concept of sustainable seafood is relatively new in the industry. This finding is also aligned with the fact that Norway has ambition to be a global producer of sustainable seafood products [14, 43, 44]. Generally, sustainability is communicated to various stakeholders in three ways:

i. ecological certification on product labels for their products. Environmentally friendly certified aquaculture, usually by regional or international standards such as the salmon Global Gap certificate or the ASC (Aquaculture Stewardship Council) certificate for responsible aquaculture are explicit means by which Norwegian seafood companies convey their commitment towards marine resources and conservation.
ii. ecological awareness towards the environment and social compliance is also developing in the Norwegian seafood sector.
iii. active engagement and looking into a 'higher agenda' for knowledge exchanges in different social and political spheres. Due to a long history and expert knowledge in aquaculture practices, Norway is often seen as advisor to other emerging international seafood industries and companies, with possibility of influencing national and international laws for the industry.

In terms of sustainability communication, variations can be observed among different companies based on their size. These differences can be attributed to the various communication channels utilized, such as social media platforms, internet resources (such as webpages), news outlets, reports, and invitations to on-site arena meetings. Such channels allow for interactive communication between the company and its audience. The global outreach of the Norwegian seafood industry, even under global pandemic conditions, is evident from the multilingual options available on many of their corporate webpages, with larger multinational enterprises offering a greater choice of languages in which their webpage content can be viewed by stakeholders [45, 46].

The SSM and corpus linguistics framework of analysis has highlighted that although most large companies have established an online presence, there remains a gap in industry knowledge and practice concerning "value-added" and "co-created" goods. This discrepancy may have practical implications for the industry, prompting the need for a standardized language and coherent discourse to secure a competitive advantage in the market for seafood as an ecologically value-added products and digital services.

Value-added goods refer to products that have been further processed from their initial state, such as "ready-to-eat" or "ready-to-cook" items. "Co-created" goods, on the other hand, refer to products that have been co-developed with clients, customers, or consumers, where ideas are exchanged collaboratively rather than being kept within the company. These goods are consumer-centric, with companies soliciting ideas directly from customers.

6.2   Practical Industry Implications

This corpus driven study demonstrates the importance of exploring and utilizing diverse communication channels for sustainability communication. Business-to-business (B2B), business-to-consumer (B2C), and distributor-to-consumer (D2C) networks have different types of industry stakeholders where various communicative strategies can be implemented. The Norwegian seafood industry needs to develop digital service resources for its SMEs to maintain competitive advantage in the long term. A current challenge is that SMEs often lack time and resources for website development and social media engagement.

The findings suggest that prioritizing communication of high-quality products and value-added product-services in relation to sustainable consumption could benefit all levels of networks and enhance industry competitiveness. Online sales, web shops, and building B2C and D2C networks are identified as specific communication channels for stakeholder engagement and market outreach. To exploit digital resources, companies can explore underutilized audio and digital resources such as podcasts, direct online sales, and instructional videos. Additionally, the study identifies the potential for informal networking and community building by sharing podcasts, instructional videos, and recipe sharing for domestic and professional cooks. To brand the Norwegian seafood industry regionally and globally, combining podcasts with cooking can also provide opportunities for stakeholder engagement.

6.3   Social Implications and Further Research

The identification of a gap in the Norwegian seafood industry goods and services offerings suggests that consumers could play a greater role in designing products that meet their needs, potentially increasing domestic and international market share for the Norwegian seafood industry. As the concepts of sustainable seafood and sustainability are relatively new, future research could explore the evolving narrative and discourse on this topic and provide better practice guidelines for the industry. Further research could also explore the future of digital web presence for Norwegian seafood companies and the portfolio of digital services offered.

References

1. FAO: World fisheries and aquaculture: the state of sustainability in action (2020). https://doi.org/10.4060/ca9229en

2. Garcia, S.M., Rosenberg, A.A.: Food security and marine capture fisheries: characteristics, trends, drivers and future perspectives. Philos. Trans. R. Soc. B: Biol. Sci. **365**(1554), 2869–2880 (2010). https://doi.org/10.1098/rstb.2010.0171

3. Okafor-Yarwood, I.: Illegal, unreported and unregulated fishing, and the complexities of the sustainable development goals (SDGs) for countries in the Gulf of Guinea. Marine Policy **99**, 414–422 (2019). https://doi.org/10.1016/j.marpol.2017.09.016

4. Kuntsman, A., Rattle, I.: Towards a paradigmatic shift in sustainability studies: a systematic review of peer reviewed literature and future agenda setting to consider environmental (un)sustainability of digital communication. Environ Commun **13**(5), 567–581 (2019). https://doi.org/10.1080/17524032.2019.1596144/SUPPL_FILE/RENC_A_1596144_SM8315.PDF

5. Shahzalal, M., Hassan, A.: Communicating sustainability: using community media to influence rural people's intention to adopt sustainable behavior. Sustainability **11**(3), 812 (2019). https://doi.org/10.3390/SU11030812

6. Margariti, K.: 'White' space and organic claims on food packaging: communicating sustainability values and affecting young adults' attitudes and purchase intentions. Sustainability **13**(19), 11101 (2021). https://doi.org/10.3390/SU131911101

7. Bailey, M., Packer, H., Schiller, L., Tlusty, M., Swartz, W.: The role of corporate social responsibility in creating a Seussian world of seafood sustainability. Fish Fish. **19**(5), 782–790 (2018). https://doi.org/10.1111/FAF.12289

8. Cochrane, K.L.: Reconciling sustainability, economic efficiency and equity in marine fisheries: has there been progress in the last 20 years? Fish Fish. **22**(2), 298–323 (2021). https://doi.org/10.1111/FAF.12521

9. Winson, A., Choi, J.Y., Hunter, D., Ramsundar, C.: Ecolabeled seafood and sustainable consumption in the Canadian context: issues and insights from a survey of seafood consumers. Marit. Stud. **21**, 1–15 (2021). https://doi.org/10.1007/s40152-021-00245-y

10. Richter, I., Thøgersen, J., Klöckner, C.A.: Sustainable seafood consumption in action: relevant behaviors and their predictors. Sustainability **9**(12), 2313 (2017). https://doi.org/10.3390/SU9122313

11. Kittinger, J.N., et al.: Committing to socially responsible seafood: ocean science must evolve to meet social challenges in the seafood sector. Science **356**(6341), 912–913 (2017). https://doi.org/10.1126/SCIENCE.AAM9969/SUPPL_FILE/AAM9969.KITTINGER-SM.PDF

12. Creutzig, F., et al.: Digitalization and the Anthropocene. **47**, 479–509 (2022). https://doi-org.ezproxy.ub.gu.se/10.1146/annurev-environ-120920-100056. https://doi.org/10.1146/ANNUREV-ENVIRON-120920-100056

13. Pounds, A., et al.: More than fish—framing aquatic animals within sustainable food systems. Foods **11**(10), 1413 (2022). https://doi.org/10.3390/FOODS11101413

14. Abualtaher, M., Rustad, T., Bar, E.S.: Systemic Insights on the integration of UN sustainable development goals within the Norwegian salmon value chain. Appl. Sci. **11**(24), 12042 (2021). https://doi.org/10.3390/APP112412042

15. Seafood From Norway: Nurturing growth: the Norwegian way (2017)

16. Checkland, P.: Systems thinking, systems practice. Wiley, Hoboken (1999). https://www.wiley.com/en-al/Systems+Thinking,+Systems+Practice:+Includes+a+30+Year+Retrospective-p-9780471986065. Accessed 25 Oct 2021

17. Cabrera, D., Cabrera, L., Powers, E.: A unifying theory of systems thinking with psychosocial applications. Syst. Res. Behav. Sci. **32**(5), 534–545 (2015). https://doi.org/10.1002/sres.2351

18. Capra, F.: Criteria of systems thinking. Futures **17**(5), 475–478 (1985). https://doi.org/10.1016/0016-3287(85)90059-x

19. Hossain, N.U.I., Dayarathna, V.L., Nagahi, M., Jaradat, R.: Systems thinking: a review and bibliometric analysis. Systems **8**(3), 23 (2020). https://doi.org/10.3390/SYSTEMS8030023

20. Kosalge, P.U., Crampton, S., Kumar, A.: Evaluating web-based brainstorming for design thinking in businesses. Int. J. Bus. Inf. Syst. **39**(4), 445–471 (2022)

21. Elia, G., Margherita, A., Secundo, G.: Project management canvas: a systems thinking framework to address project complexity. Int. J. Manag. Proj. Bus. **14**(4), 809–835 (2020). https://doi.org/10.1108/IJMPB-04-2020-0128/FULL/PDF

22. Checkland, P.: Soft systems methodology: a thirty-year retrospective. Syst. Res. Behav. Sci. **13**(4), S11–S58 (2000)

23. Augustsson, H., Churruca, K., Braithwaite, J.: Re-energising the way we manage change in healthcare: the case for soft systems methodology and its application to evidence-based practice. BMC Health Serv. Res. **19**(1), 1–11 (2019). https://doi.org/10.1186/S12913-019-4508-0/TABLES/3

24. Riza, F., Wijaya, C.: Application of soft system methodology for modelling institutional strengthening of salt farmers. Technium Soc. Sci. J. 36 (2022). https://heinonline.org/HOL/Page?handle=hein.journals/techssj36&id=75&div=7&collection=journals. Accessed 18 Feb 2023

25. Halliday, M.A.K., Matthiessen, C.M.I.M.: Construing experience through meaning **1** (1999). https://doi.org/10.1017/CBO9781107415324.004

26. Halliday, M.A.K., Matthiessen, C.M.I.M.: Halliday's Introduction to Functional Grammar, 4th edn. Routledge, London (2014)

27. Brannen, M.Y., Piekkari, R., Tietze, S.: The multifaceted role of language in international business: unpacking the forms, functions and features of a critical challenge to MNC theory and performance. J. Int. Bus. Stud. **45**(5), 495–507 (2014). https://doi.org/10.1057/jibs.201 4.24

28. Ericsson, S., Wojahn, D., Sandström, I., Hedvall, P.O.: Language that supports sustainable development: how to write about people in universal design policy. Sustainability (Switzerland) **12**(22), 1–20 (2020). https://doi.org/10.3390/SU12229561

29. Hilborn, R., Fulton, E.A., Green, B.S., Hartmann, K., Tracey, S.R., Watson, R.A.: When is a fishery sustainable? Can. J. Fish. Aquat. Sci. **72**(9), 1433–1441 (2015). https://doi.org/10.1139/CJFAS-2015-0062

30. Iue, M., Makino, M., Asari, M.: The development of 'Blue Seafood Guide', a sustainable seafood rating program, and its implication in Japan. Mar. Policy **137**, 104945 (2022). https://doi.org/10.1016/J.MARPOL.2021.104945

31. Jovanova, K.: Sustainable governance and knowledge-based economy-prerequisites for sustainable development of the developing and transitional economies. Athens J. Bus. Econ. **7**(1), 67–84 (2021). https://doi.org/10.30958/ajbe.7-1-3

32. Stefanowitsch, Corpus linguistics A guide to the methodology. 2020

33. Hinton, M.: Corpus linguistics methods in the study of (Meta) argumentation. Argumentation **35**(3), 435–455 (2021). https://doi.org/10.1007/S10503-020-09533-Z/TABLES/4

34. Barron, A., Schneider, K., Anthony, L.: Developing AntConc for a new generation of corpus linguists. In: Proceedings of the Corpus Linguistics Conference (CL 2013), July 22–26, 2013. Lancaster University, UK, pp. 14–16 (2013). http://www.antlab.sci.waseda.ac.jp/. Accessed 12 May 2022

35. Anthony, L.: AntConc (Version 3.5.8) Computer Software. Software (2019). Waseda University, Tokyo, Japan. https://www.laurenceanthony.net/software/antconc/. Accessed 15 Nov 2019

36. Anthony, L.: AntConc: design and development of a freeware corpus analysis toolkit for the technical writing classroom. In: 2005 IEEE International Professional Communication Conference Proceedings (2005)

37. Melewar, T.C., Gotsi, M., Andriopoulos, C.: Shaping the research agenda for corporate branding: avenues for future research. Eur. J. Mark **46**(5), 600–608 (2012). https://doi.org/10.1108/03090561211235138

38. Bobrie, F.: Visual representations of goods and services through their brandings: the semiotic foundations of a language of brands. Rech. Appl. Mark. **33**(3), 122–144 (2018). https://doi.org/10.1177/2051570718791784

39. Wei, L.: Examining corporate communications of environmental responsibility on corporate websites: main themes, linguistic features, and text reuse. J. Promot. Manag. **26**(7), 1013–1037 (2020). https://doi.org/10.1080/10496491.2020.1746467

40. Reilly, H., Hynan, K.A.: Corporate communication, sustainability, and social media: it's not easy (really) being green. Bus. Horiz. **57**(6), 747–758 (2014). https://doi.org/10.1016/J.BUSHOR.2014.07.008

41. Bögel, P.M.: Processing of CSR communication: insights from the ELM. Corp. Commun. **20**(2), 128–143 (2015). https://doi.org/10.1108/CCIJ-11-2013-0095

42. Høvring, M.: Corporate social responsibility as shared value creation: toward a communicative approach. Corpor. Commun. Int. J. **22**(2), 239–256 (2017). https://doi.org/10.1108/CCIJ-11-2016-0078

43. Press Release: Lerøy Seafood Group ASA: Acquisition of shares in Havfisk and Norway Seafoods approved by Norwegian Ministry of Trade, Industry and Fisheries. Dow Jones Institutional News, August 2016

44. McDonagh, V.: Norway seafood exports set to hit 2021 record – Fish Farmer Magazine. Fish Farmer Magazine (2021). https://www.fishfarmermagazine.com/news/norway-seafood-exports-set-to-hit-2021-record/. Accessed 26 Jan 2022

45. Straume, H.M., Anderson, J.L., Asche, F., Gaasland, I.: Delivering the goods: the determinants of Norwegian seafood exports. Mar. Resour. Econ. **35**(1), 83–96 (2020). https://doi.org/10.1086/707067/ASSET/IMAGES/LARGE/FG1.JPEG

46. Nyu, V., Nilssen, F., Nenadic, O.: International journal of export marketing managing global supply chain disruptions under the COVID-19 pandemic: the case of the Norwegian seafood industry managing global supply chain disruptions under the COVID-19 pandemic: the case of the Norwegian seafood industry. Int. J. Export Mark. **5**(1), 73–102 (2022). https://doi.org/10.1504/IJEXPORTM.2022.10051069

Comparative Evaluation of User Interfaces for Preventing Wasteful Spending in Cashless Payment

Daisuke Mashiro[1], Yoko Nishihara[2]([✉]), and Junjie Shan[3]

[1] Graduate School of Information Science and Engineering, Ritsumeikan University,
Shiga, Japan
is0430fe@ed.ritsumai.ac.jp
[2] College of Information Science and Engineering, Ritsumeikan University,
Shiga, Japan
nisihara@fc.ritsumei.ac.jp
[3] Ritsumeikan Global Innovation Research Organization, Ritsumeikan University,
Shiga, Japan
shan@fc.ritsumei.ac.jp

Abstract. More and more people pay using cashless payment methods rather than paying cash. In addition to online stores, more and more stores in the real world accept cashless payments, such as Apple Pay, Google Pay, etc. However, people notice that using cashless payments makes them spend more than necessary unconsciously. Some applications would notify the amount of payment on the cashless payment, but they are just after-notices and rarely draw people's feeling to the money spent. A better method is required to prevent wasting money before payment. In this paper, we implement eight user interfaces to avoid waste on mobile payment applications. These interfaces would inhibit users' payment actions for several seconds before they make the final confirmation. We compared the eight user interfaces through evaluation experiments with 25 participants. We found features of the user interfaces that were suitable for preventing waste. In evaluation experiments, we gave a shopping scenario with wasting, and we asked the participants to use the implemented user interface. The participants answered the degree of effect of wasting prevention about the implemented user interface. They also answered the degree of frustration in using the assigned user interface. Experimental results showed that the participants would escape from wasting more if the user interface gave a pleasant task in specific periods.

Keywords: Cashless Payment · Prevention of Wasteful Spending · Distraction for a Few Seconds · Pleasant Task

1 Introduction

More and more people own their smartphones. Cashless and mobile payment applications are becoming commonplace. People use Paypal, Apple Pay, and Google Pay as their cashless payment methods. As it is possible to pay without

© Springer Nature Switzerland AG 2023
F. Fui-Hoon Nah and K. Siau (Eds.): HCII 2023, LNCS 14039, pp. 352–363, 2023.
https://doi.org/10.1007/978-3-031-36049-7_26

cash and reduce the time of payment by cashless payment methods, these benefits make cashless payments welcome to users.

Meanwhile, just because cashless payment methods do not involve direct cash payments, people have less sense of having money spent. This can cause people to spend more money than they thought [1]. Cashless payments are so convenient that it is hard to refuse to use them. As a result, the requirement of how to reduce unnecessary and wasteful spending while using mobile payments is becoming a growing concern.

To prevent wasting money on cashless payment methods, users should carefully consider the item's necessity. If the item is unnecessary, they should decide not to buy it now. If they can launch the mobile payment application immediately at any moment they want to buy something, they would not have enough time to think about that seriously. We speculate that taking the extra time to launch the mobile payment application can bring them the opportunity to reconsider the purchase, just like the original cash payment process of taking bills out of pocket. People are encouraged reconsideration if they temporarily draw attention away from the purchase. It is known that 2.8 s of inhibition has the effect of breaking concentration and distracting people [2]. However, people will not want to continuously use such a user interface that will inhibit their actions. Therefore, a user interface is required to prevent waste without causing too much stress from being interfered.

In this paper, we propose a user interface to prevent wasteful payment on the mobile payment application. The proposed interface inhibits wasteful payment and has less frustration for the users.

2 Related Work

We show existing work of wasteful spending prevention. One of the wasteful spendings is food spending. Methods for prevention of food wasteful spending were proposed [3,4]. A recipe recommendation method is proposed that shares recipes within a group of members prevent food waste at home [5]. There is also a study to prevent the overuse of smartphones [6]. This study proposes a method to prevent overspending through cashless payments.

We show existing work of behavioral disruption. Behavioral disruption often occurs when the behavior is a nuisance to others. To prevent users from posting comments containing toxic words on social networking sites, a method for filtering toxic word posts has been proposed [7]. This method can estimate and filter words that correspond to toxic words, words that are inferred to be toxic words based on context, and words that are cryptic or coined. Another example is the Speech Jammer that inhibits human behavior that interferes with others [8]. In the study, a system that interferes with a specific speaker's speech from a remote location controls the manner and rules of conversation and presentation training. The system uses a mechanism of delayed auditory feedback. A directional microphone is used to delay the speech picked up by a directional microphone and output from a directional speaker. This system can inhibit the speaker's speech without physical pain. This study is similar to the present study in that it inhibits a person's behavior. These studies are considered to be methods to inhibit the actions of those who want to act and protect others. In contrast, our study would inhibit the actions of the users, and the users are protected.

3   Proposed Method

We implement eight user interfaces to prevent users from extravagance when shopping. This chapter illustrates a model case at first that shows a targeted user and a shopping situation. Then, we show our approach to preventing extravagance. After that, we show eight interfaces implemented in the study.

3.1   Targeted User and a Shopping Scenario

John is a university student. He likes comic books and owns many of them. He likes not only one kind of comic book but also several kinds of them. Whenever he finds an enjoyable comic, he positively buys it. Therefore, he often needs help with his money for eating.

One day, John decided to buy some comic books because new editions were being released. He decided to buy four comic books, so he set his budget at 2,000 JPY (one comic cost about 500 JPY). He came to a bookstore and found the four comic books. Unfortunately, he found a new enjoyable comic that he has not seen before. He really wanted the comic book. So, he took five comic books in total to the cashier. He paid for them with cashless payment and got the five comics. When leaving the bookstore, he realized he had squandered, so he could not have a lunch meal.

3.2   Our Approach to Preventing Wasteful Spending

In the above case described in Sect. 3.1, an extravagance was made by buying a comic book that had not yet to be decided to buy before coming to the bookstore. This extra expense made him lose his lunch, which he should have expected before paying. He should temporarily stop his purchase for further consideration. However, it is difficult to inhibit his action when he is focusing on the comic book and the payment application. Therefore, external inhibition is required to prevent extravagance [8]. From past research, we noticed that a person would be released from concentration when inhibited from his actions for about 2.8 s [2]. So, we take the approach that external inhibition is given to temporarily stop his/her desire to buy it. The external inhibition diverts his/her attention elsewhere so that he/she might consider buying it. Moreover, the external inhibition must be less frustrating as well as prevent extravagance.

3.3   Implemented User Interfaces

We implement several user interfaces according to the approach in Sect. 3.2. The user interface is designed to be less frustrating and prevents extravagance. In this section, we implement eight user interfaces. The user interfaces would be

Fig. 1. UI(1): Multiple taps for launching. It requires tapping multiple times to launch a payment application.

Fig. 2. UI(2): Finding a correct button from dummies for launching. It requires finding out the correct button among many fake ones.

launched before the mobile payment application launches. Users need to take the action required by the interface before the mobile payment application can start.

UI(1): Multiple Taps for Launching. Figure 1 shows U1(1), which requires multiple taps to launch a payment application. The number of taps is determined at random when launching a payment application. The more the number of taps, the longer the period for launching.

UI(2): Finding a Correct Button from Dummies for Launching. Figure 2 shows UI(2) that requires finding a correct button from dummies to launch a payment application. Nine buttons are displayed, while one is

Fig. 3. UI(3): Waiting for a fixed period to tap a button for launching. It requires waiting a fixed period before tapping the button to launch the payment application.

Fig. 4. UI(4): Watching a movie before launching. It requires watching a movie to launch a payment application.

correct for launching. If the correct button is tapped, the payment application is launched. However, incorrect buttons do not allow launching.

UI(3): Waiting for a Period to Tap a Button for Launching. Figure 3 shows UI(3) that requires waiting for a period to tap a button to launch a payment application. After the fixed period, the button is activated to be tapped. Before activation, the button does not work.

UI(4): Watching a Movie Before Launching. Figure 4 shows UI(4) that requires watching a movie to launch a payment application. The duration of the video is 10 s. After watching the movie, the payment application will be launched.

Fig. 5. UI(5): Authentication with an e-mail address and password to launch a payment application.

Fig. 6. UI(6): Entering the same string as that given at random. It requires inputting the correct string to launch a payment application.

UI(5): Authentication with E-Mail Address and Password. Figure 5 shows UI(5) that requires authentication, which means a user inputs an e-mail address and password to launch a payment application. Authentication is required for every launch.

UI(6): Entering the Same String as that Given at Random. Figure 6 shows UI(6) that requires entering the exact string as that given at random. The string consists of eight English characters and numbers.

UI(7): Completing a Puzzle Game. Figure 7 shows UI(7) that requires solving a puzzle game by setting each piece to a correct position. The puzzle size is 3 * 3.

Fig. 7. UI(7): Completing a 3*3 puzzle game. It requires completing the puzzle to launch a payment application.

Fig. 8. UI(8): Solving calculation problems. It requires solving three summation problems to launch a payment application.

UI(8): Solving Calculation Problems. Figure 8 shows UI(8) that requires solving calculation problems. The type of problem is a summation. The number of problems is three.

4   Comparative Evaluation Experiments Among Implemented User Interfaces

This section describes comparative evaluation experiments among the implemented eight user interfaces.

4.1   Purpose of the Experiment

We compared the eight interfaces' effectiveness in preventing wasteful spending and analyzed their frustration levels with users.

Table 1. Scenario type used in the experiments.

No	Location of store	Reason to buy
1	Real store	Because he/she is interested in it
2	Online store	Because he/she is interested in it
3	Real store	Because it is provided at reasonable price
4	Online store	Because it is provided at reasonable price

Table 2. Scenario 1: shopping at a real store and buying something that a person is interested in.

You go to buy a comic book that you want. Your budget is 2,000 JPY. At the store, you find another comic book that you are interested in. After some hesitation, you decide to buy the comic books that you originally planned to buy, and you find. You bring them to the cashier.

4.2 Experimental Procedure

The experimental procedure was as follows.

1. Measure participants' stress levels in the general state.
2. Let participants read a shopping scenario.
3. Ask participants to use an implemented user interface and imagine they are actually paying before the cashier in the scenario they read.
4. Measure participants' stress levels after using the user interface.
5. Ask participants to fill out a questionnaire after the experiment.

We recruited 25 university students to participate in the experiment. For stress measurement in 1 and 4, we used Fitbit Charge 5[1]. Fitbit Charge 5 measures a person's stress level by Electro Dermal Activity. We prepared four types of scenarios by considering the location of the store (actual store/online store) and the reason why a person wants to buy it (because he/she is interested in it/because it is provided at a reasonable price.), that are shown in Table 1. Used scenarios are shown in Table 2, 3, 4, and 5. Five participants tested one user interface with one scenario. Each participant answered one question about prevention from wasteful payment with a 5-point scale (5 denoted the participant decided to stop buying it, 1 represented he/she did not.)

4.3 Evaluation Method

We compared the implemented user interfaces with two points. The first one was an average score on the questionnaire. The score of the questionnaire was given a 1 to 5 score. The scores were averaged and compared among the implemented user interfaces.

---

[1] https://www.fitbit.com/global/us/products/trackers/charge5.

Table 3. Scenario 2: shopping at an online store and buying something that a person is interested in.

You find a comic book you want at a discount on the Web store, so you decide to buy it. Your budget is 2,000 JPY. While searching at the Web store, you find another comic book you have wanted for a long time. You add the comic book you find and that you have wanted for a long time in your basket. Then, you move to the payment Web page.

Table 4. Scenario 3: shopping at a real store and buying something that is provided at a reasonable price.

You go to a time sale at a supermarket. Your budget is 3,000 JPY. At the time sale, you find a variety of foods. Everything is sold at a reasonable price. So, you add too many foods to your basket. You bring them to the cashier.

The second was a difference in stress levels before and after using a user interface. If the difference marks plus the score, it means the user interface gives more stress. In contrast, a negative score means the user interface gives less stress.

4.4   Experimental Results

Table 6 shows results about prevention from wasteful payment. The most effective interface in preventing waste was "UI(7): Completing Puzzles for Launching."

Table 7 shows the user interface's stress level. The interface with the lowest stress value was "UI(7): Completing Puzzles for Launching."

5   Discussion

We discuss three points by referring to the experimental results.

Table 5. Scenario 4: shopping at an online store and buying something that is provided at a reasonable price.

You are looking for something good during the Prime Day sale on Amazon.com. Your budget is 5,000 JPY. Then, you find something you have been interested in at a discount. You add it to your basket even though it is over your budget. Then, you move to the payment page.

Table 6. Experimental Results of Preventing Wasteful Spending (Average of 5-point Rating)

	Real store		Online store		
	Interested	Reasonable	Interested	Reasonable	Average
UI(1): Multiple Taps	1.4	1.6	1.2	2.4	1.55
UI(2): Finding a Correct Button	1.8	3.0	2.4	1.8	2.25
UI(3): Waiting for a Period	1.8	1.2	2.0	1.2	1.55
UI(4): Watching a Movie	1.2	2.6	2.4	3.6	2.35
UI(5): Authentication	1.2	2.8	2.4	2.8	2.3
UI(6): Enter the Same String	2.0	3.2	2.0	2.0	2.3
UI(7): Completing a Puzzle	2.0	2.6	2.6	3.4	**2.85**
UI(8): Solving a Calculation	2.8	1.8	1.6	2.8	2.25
Average	2.01	2.35	2.11	2.55	–

Table 7. Experimental Results of Stress using the Implemented User Interface. The Average was the Difference between Before and After Using.

	Real store		Online store		
	Interested	Reasonable	Interested	Reasonable	Average
UI(1): Multiple Taps	−0.2	−0.4	1.2	1.8	0.6
UI(2): Finding a Correct Button	1	1	2.6	1.6	1.55
UI(3): Waiting for a Period	3	−3.4	0.6	2.4	0.65
UI(4): Watching a Movie	0.8	2.2	0.6	1.8	1.35
UI(5): Authentication	−0.8	3.4	1.8	2.6	1.75
UI(6): Enter the Same String	−1.8	−1.4	−2.2	5.4	0
UI(7): Completing a Puzzle	−2.4	0.6	0.2	−5.2	**−1.7**
UI(8): Solving a Calculation	−5.4	2.6	−1.6	2.6	−0.45
Average	−1.22	0.58	0.08	1.48	–

5.1   Prevention of Wasteful Spending

Table 6 shows that UI(7) was the most effective for the prevention of wasteful spending. UI(7) requires solving a puzzle game to launch a payment application. The participants enjoyed solving the game that distracted them from payment. Therefore, they were prevented from wasteful spending.

In contrast, UI(1) and UI(3) were less effective in prevention. UI(1) requires multiple taps, and UI(3) needs to wait for a period to tap for launching. The button to be tapped was for launching a payment application. The participants were engaged with the button until launching. They were not distracted by interruptions from buying the targeted items. Therefore, they were not prevented from wasteful spending.

From the above discussion, we found that wasteful payment could be prevented if people were distracted from launching the payment application by interruption by the user interface.

5.2   Frustration of Using the User Interface

Table 7 shows that UI(7) was the less frustrating to use. As mentioned, UI(7) requires solving a puzzle game to launch a payment application. Solving a puzzle game is generally exciting and makes players fulfilled when completed. Therefore, the participants felt less stress in using the user interface, so the difference in stress levels was lower.

In contrast, UI(5) was the most frustrating to use. UI(5) requires authentication with an e-mail address and password. The authentication needed to be input from the keyboard. The number of inputting would be more significant. The participants needed to concentrate on input, not make typing errors. Therefore, the level of frustration was the highest.

From the above discussion, we found that the frustration level would be suppressed if the task given by the user interface was exciting and fulfilled.

5.3   Difference of Wasteful Spending Prevention Between Scenarios

We discuss the differences in wasteful spending prevention between scenarios.

Table 6 shows that the effect of wasteful spending prevention at an actual store was more significant than that at an online store. Shopping at an actual store is more likely to make wasteful spending than those at an online store [9]. The experimental results denoted that the interruptions by the user interfaces had more effect on shopping at an actual store.

Table 6 also shows that the effect of wasteful spending prevention for items sold at a reasonable price was more significant than that for items interested in by the participants. The participants commented that buying an exciting item was not wasteful, even if their budget was over. Therefore, we must examine intensively how to distract them from buying interested in items when their budgets are over.

6   Conclusion

This paper proposed an approach to prevent wasteful spending in cashless payment applications. The approach uses the external inhibition user interface to distract people's paying behavior for several seconds to make them reconsider their buying decision. In this paper, we implemented eight user interfaces to distract users' attention from the payment process. In the evaluation experiments, we gave a shopping scenario to a participant and asked the participants to use a given user interface. Then, they answered the questionnaire about the prevention of wasteful spending and measured their frustration levels when using the user interface. Experimental results showed that the participants could be prevented from wasteful spending with less frustration if a user interface gave an exciting and fulfilling task.

References

1. Shan, A., Eisenkraft, N., Bettman, J.R., Chartrand, T.L.: "Paper or Plastic": how we pay influences post-transaction connection. J. Cons. Res. **42**(5), 688–708 (2016)
2. Altmann, E.M., Trafton, J.G., Hambrick, D.Z.: Momentary interruptions can derail the train of thought. J. Exp. Psychol. **143**(1), 215–22 (2014)
3. Farr-Wharton, G., Foth, M., and Choi, J. H.: Colour coding the fridge to reduce food waste. In Proceedings of the 24th Australian Computer-Human Interaction Conference, pp. 119–122 (2012)
4. Altarriba, F., Lanzani, S. E., Torralba, A., Funk, M.: The grumpy bin: reducing food waste through playful social interactions. In: Proceedings of the 2017 ACM Conference Companion Publication on Designing Interactive Systems, pp. 90–94 (2017)
5. Yalvaç, F., Lim, V., Hu, J., Funk, M., Rauterberg, M. : Social recipe recommendation to reduce food waste. In: CHI 2014 Extended Abstracts on Human Factors in Computing Systems (2014)
6. Xu, X., et al: TypeOut: leveraging just-in-time self-affirmation for smartphone overuse reduction. In Proceedings of the 2022 CHI Conference on Human Factors in Computing Systems, pp. 1–17 (2022)
7. Omi, R., Nishihara, Y., Yamanishi, R.: Extraction of paraphrases using time series deep learning method. In: International MultiConference of Engineers and Computer Scientists 2019, pp. 276–278 (2019)
8. Kurihara, K., Tsukada, K.: SpeechJammer: a system utilizing artificial speech disturbance with delayed auditory feedback (2012). http://arxiv.org/abs/1202.6106
9. Petro, G.: Consumers are spending more per visit in-store than online. What does this mean for retailers?. Forbes (2019). https://www.forbes.com/sites/gregpetro/2019/03/29/consumers-are-spending-more-per-visit-in-store-than-online-what-does-this-man-for-retailers/?sh=7b0a17a77543. Accessed 6 Feb 2023

Emotional Communication and Interaction with Target Groups Exemplified by Public Educational Institutions

Christina Miclau[1]([⊠]), Josef Nerb[2], and Bernhard Denne[1]

[1] Hochschule Offenburg – University of Applied Sciences, Badstrasse 24, 77652 Offenburg, Germany
christina.miclau@gmail.com
[2] Pädagogische Hochschule Freiburg, Kunzenweg 21, 79117 Freiburg im Breisgau, Germany

Abstract. Public educational institutions are increasingly confronted with a decline in the number of applicants, which is why competition between colleges and universities is also intensifying. For this reason, it is important to position oneself as an institution in order to be perceived by the various target groups and to differentiate oneself from the competition. In this context, the brand and thus its perception and impact play a decisive role, especially in view of the desired communication of the institution's own values and its self-image, the brand identity. To this end, emotions serve as an approach to creating positive stimulation and brand loyalty.

Keywords: Empirical research · Emotional Communication · Emotional Interaction · Target Group Oriented Communication

1 Relevance and Status Quo

Emotions are relevant in verifying a brand's impact and to what extent it is aligned with the brand identity that is being applied. Surveys and interviews with the various target groups are intended to provide information on the emotional impact of the brand of a public educational institution. In addition, the aim is to answer questions about the perception, emotions or the position of educational institutions in their surroundings and to capture the different feelings of the individual target groups for the best possible communication classification.

Over the past few years, competition between universities has intensified, which is why competitive and also intensive behavior, as well as corresponding external communication, are becoming increasingly important. Universities recognize the necessity, not only due to the performance-linked allocation of financial resources, but also existing consequences with regard to the different target groups. The perception of existing and potential target groups has a profound impact on the acceptance and position of the respective university in society [1]. As a result, universities are currently increasingly addressing the fact that the brand plays a significant role in positive perception and

F. Fui-Hoon Nah and K. Siau (Eds.): HCII 2023, LNCS 14039, pp. 364–376, 2023.
https://doi.org/10.1007/978-3-031-36049-7_27

increased competitiveness. In addition to creating a profile at various levels, strategic positioning is an element of distinguishing oneself among the competition. The design, communication and being of the guiding principles, strategy and sense of brand, to name but a few, is considered a requirement for establishing a successful university brand [2].

A brand, meaning the identification or labeling of a product, a company or an institution, not only reflects a specific quality or functional feature, but is also a sign of self-expression, self-presentation or even selfassurance [3]. The self-image of a brand, called brand identity, which reflects the actions of the internal target group, is mutual to the external image of a brand, the brand image, which embodies the external perception [4]. The significance of a brand influences information processing, judgment and, as a result, the behavior as a consequence of brand perception. Accordingly, perception and impact are a key factor for acceptance and awareness [3]. As a result, self-image and image should always correspond in order to express what the brand represents and how it should perform to the public [5]. In the field of brand management, it is therefore important to create a positive brand image and to recognize the need for constant adjustments in the course of change over time [3].

When it comes to determining the perceived impact, it is often easier for people to use self-relevant functions of a brand, i.e. characteristics of the brand personality, for describing it [3] and therefore to express the perceived emotions, which reflect the impact of the external image to a greater extent.

2 Emotions and Their Application in Communication

2.1 Theories of Emotion

Internal and external stimuli induce alterations of the nervous system, which are also known as emotion [6]. According to this, emotions are ubiquitous and describe strong motivational systems for automatic situation assessment and reaction [7], which influence over 70% of our decisions [8].

Despite their ubiquity, there is no single definition for the concept of emotion, yet Kleinginna and Kleinginna (1981) encountered 92 different definitions in the course of their research [9]. Vogel (2007) justifies this in terms of the broad spectrum of emotion theories and the fact of the subjectivity and complexity of emotional phenomena [10].

The two-factor theory according to Schachter/Singer refers to the physiological activation as a result of an emotional situation, which provides information about the existence of an emotion and its intensity [11]. However, Scherer emphasizes, within the framework of his component process model and his appraisal theory, the aspect of subjectivity of emotions triggered by a stimulus. The components (organismic subsystems) he refers to are crucial in determining behavioral responses to events. [12].

Scherer describes emotions as "an episode of interrelated, synchronized changes in the states of all or most of the five organismic subsystems as a reaction to the appraisal of an external or internal stimulus event that is relevant to the organism's main concerns." The states of subsystems, also known as components, include cognition, physiological regulation, motivation, motor expression, and monitoring/feeling, which evaluate external or internal stimulus events triggered by emotions. These events are "analyzed" by the organism in terms of their relevance which leads to a reaction. The requirement for doing so is the relevance of the event and its consequences for the organism and thus the satisfaction of needs and the fulfillment of objectives [12].

According to the appraisal theory, "the importance of subjective evaluations and appraisals" become evident and indicate the influence of personal values, goals, and desires and the neglect of objective perception on the occurrence of emotions [12, 13]. Based on different sets of criteria, the Stimulus Evaluation Checks (SECs), which refer to the emotion eliciting process, emotions "can be seen as a mediating interface between an environmental input and behavioral output" [12]. Scherer states that the central appraisal purposes justify the following four SECs [12]:

1. Relevance: How relevant is this event?
2. Implications: What are the implications of this event and how do they affect the respective goals?
3. Coping potential: How well can the event be coped with?
4. Normative Significance: How important is the event in terms of social norms and values?

Following the subjective assessment of a situation, the individuality of each SEC can be recognized. These impressions are shaped by internal factors, such as the extent and direction of motivation and individual differences, or by external factors, e.g., culture and norms [12].

Besides considering emotional phenomena as they are considered in Scherer's component process model, the involvement of affective phenomena is also important for interdisciplinary research, which is why a differentiation of affective phenomena from emotions should be drawn on the basis of affect types as follows [14]:

- Preferences: Evaluation in terms of liking or disliking a stimulus.
- Attitudes: Enduring beliefs and dispositions; classified into cognitive, affective, and motivational components.
- Moods: Often evoked without apparent cause; typically of low intensity, but in many cases of long duration.
- Disposition to affect: Tendency to experience a mood or type of emotion frequently.
- Interpersonal attitudes: Triggered by specific events and rarely spontaneous assessments.

2.2 Media-Generated Emotions

Emotions can be triggered and thus avoided both passively and actively, which is possible through the use of media. Here, the two approaches, the Construct of (Tele)Presence and the Three-Factor Theory according to Zillmann (2004) shall provide insight. Telepresence describes a certain state during the experience of media content. It involves taking the recipient to a place of virtual reality by generating emotions comparable to real-life situations [15].

Moreover, the Three-Factor Theory indicates memories as an essential component in emotional reactions, i.e., stimulus-response connections, which is why iconic or symbolic or linguistic representations can be used as triggers. The three factors, "dispositional or behavioral, arousing or energizing, and experience components," have different time courses and, because of their interacting effects, produce intensified emotional responses and thus more intense experience [15]. Schwender (2007) also recognizes memories, evoked by specific images and sounds, as sufficient for creating emotions [16].

Emotional elements of images evoke effects such as so-called climatic or experiential effects. Climate effects occur peripherally and lead to an emotional image perception, whereas dominant or centrally placed image elements create experience effects. The use to create emotional effects by means of communication devices is done by a frictionless application of these elements [17].

Within the perception of audio-visual media, the evaluations of an individual are also responsible for emotional reactions, in view of adopting perspectives, the so-called Theory of Mind (putting oneself into the mental state of one's counterpart), empathy and perception, and reaction of the observer [18].

We are able to understand the mental state of our counterpart or of another person. Due to the pure imagination of the state, since it is not possible that one can have an insight of the other person's mind, one speaks of a theory - the Theory of Mind. Through expressions of emotion in the mimics and gestures of a communication partner or a person to be seen, i.e. through signs, the intentions and thoughts could be interpreted. This way, actions in fictional stories can also be interpreted by the viewer in terms of thinking, knowledge, and intentions [18].

Due to the human predisposition and ability to empathize with other people and empathize with their emotional state, feelings are triggered empirically within, which is evoked by witnessing a certain behavior of a third party, e.g., when another person is crying. Accordingly, empathy is linked to the Theory of Mind, which allows us to gain a glimpse of what is occurring in the minds of others. As a consequence thereof and the empathy caused, emotions arise from experiencing emotions of others, "which we can assume are those of the other person". That recreation of emotions provides comprehension of other's sensations, thoughts, and actions [18].

The mental processing of impressions determines the different interpretations of perceptions. Depending on the emotional state, the perception or processing of information received from the environment, such as sounds, sensations of cold or heat, or lighting conditions, varies. Our brain ascribes meaning to the impressions by transferring experiences and predispositions to the perceptions. Emotions that subsequently arise during perception are part of the interpretation of the world's impressions [18].

2.3 Emotional Communication Policy

Communication policy has a major role in marketing and carries out the task of selecting and determining the appropriate conditions and distribution channels. It is essential to provide information to potential customers on the exact products offered and the conditions, as well as to notify as to where and how they can be found.

However, communication is not just about focusing all actions on a particular product, service and potential customer. Proper communication also includes the consideration of the complex environment "with which a company is in constant communication" and which can have a significant impact on the nature of communication and ultimately the perception of the company and its offerings. Furthermore, the environment consists of any business relationships related to the company, which is why the goal of communication policy and the appropriate orientation of all activities is the transmission of all product and company information to existing and potential customers. The core should be the establishment of ideal conditions to satisfy customer needs, such as market transparency or the creation of a decision-making framework [19].

When planning the appropriate communication policy, it is important to consider the following questions [19]:

1. Communication subject: With whom do we want to communicate?
2. Communication object: What do we want to communicate? For example, is it about individual products or about the company as a unit?
3. Communication process: Which procedure do we choose to communicate with our communication partners? How should the communication relationships be structured?

Thereby, it is helpful to work through the elements of communication policy [19]:

- Communication object (Who, Whom or What?)
- Communication target group (To whom?)
- Communication message (What?)
- Communication activity (How?)
- Communication area (Where?)
- Communication timing (When?)

Definition of Target Groups. A target group consists of all people and organizations that should be reached by the company and whose needs should be fulfilled by means of the portfolio. For a target group-oriented approach, addressing the entirety of the stakeholders is generally helpful to divide them into smaller groups that are similar, for example, in terms of their behavior, needs, interests or characteristics. Instead of focusing solely on sociodemographic criteria such as age, gender, and place of residence, it is advisable to include characteristics and traits and also to define a clear line of demarcation to non-customers. The following characteristics can be distinguished, and also combined: demographic characteristics, geographic characteristics, psychological characteristics, group-specific characteristics [19].

Key questions for defining the target group could be [19]:

- Who do we create value for with our portfolio of offerings?
- Who are our most important customers?
- What are their needs?

Determining the target group is about identifying the people who have a need for my offerings and are both willing and able to realise a purchase intention. When placing advertisements, relevant people affected can be divided into the following groups according to the intensity of the advertising effect [19]:

1. Advertising addressees: Actual target group to which the advertising is directed.
2. Advertising-affected persons: Group that has been in contact with the advertising and has thus been reached by the advertising.
3. Advertising impressions: Group that has consciously or unconsciously perceived the advertising.

4. Advertising rememberers: Proportion of those touched by the advertising who can remember the advertising object and its characteristics at a later moment (actively or passively).
5. Advertising agent: Percentage of the advertising audience that actually buys the advertising object.
6. Advertising transmitters: Percentage of advertising recipients who do not buy the advertising object themselves (perhaps because there is no need for it), but who pass on the advertising content.

Nevertheless, one should note that in the field of advertising, the groups mentioned can also overlap and, in addition, a certain amount of loss of intensity cannot be entirely excluded, as it can never be assumed that the entire target group has been reached with the selected advertising. Thanks to the use of online marketing (online advertising, social media marketing, etc.), it is possible, however, to significantly reduce stray losses and, by collecting customer-specific data, to address individual messages, i.e., personalized advertising, to the target groups. This allows real-time advertising based on click behavior, and also results in immediate feedback regarding the success of the advertising and the acceptance of the customers [19].

Precise knowledge of the target group is the basis for the company's strategy, its range of products and services, and the creation and establishment of a brand, which is why the instrument of brand policy plays an important role in the context of communication policy. Understanding what brand identity, i.e. what self-image my company has and how I want to be perceived by the public is essential. Brand identity reflects the internal target group's actions and is interdependent with the brand image, which embodies the external perception. It is important that identity and image correspond in order to create a brand and to strengthen its meaning in the target group's minds.

The significance of a brand has an impact on information processing, judgment and consequently behavior as a result of brand perception, which is why both perception and impact are the basis for the acceptance and awareness of my company and my portfolio. Due to aforementioned, it is the task of brand management to create a positive brand perception and to identify any necessary adjustments as a soon as possible.

3 Measuring Emotional Perception

3.1 Theoretical Models for the Classification of Emotions

There is a wide range of different methods for measuring and classifying emotions. The methods described in this paper can be applied in the context of surveys. Important factors are ease of use and the recognition of concrete emotions.

The Affect Grid by Russel (1989) is a single-item scale, which has the advantage of having a short and repeatable procedure. Other scales, usually multi-item checklists or questionnaires, are comparatively more time-consuming and complex in design. The Affect Grid allows the assessment of two affect dimensions; pleasure and displeasure or arousal and sleepiness. It is suitable for identifying discrepancies and their impact or monitoring mood patterns and effects [20].

The Circumplex Model simplifies individuals' ability to describe the characteristics of, for example, a brand personality by providing predefined terminology. Surveys enable the identification and measurement of emotions by means of self-description, among other methods. People are aware of emotions, which is why they can be defined visually, described verbally or classified according to one of Russell's dimensions (see Fig. 1) [21].

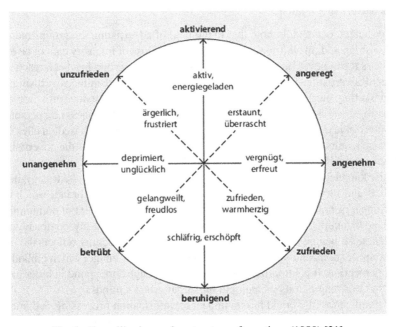

Fig. 1. Russell's circumplex structure of emotions (1980) [21].

Moreover, Peters and Slovic (2007) studied affective asynchrony and among them different types of measures, based on differentiated conceptualizations. This involves considering affective attitudes that are influenced by emotions, such as "holistic reactions to objects" or reactions triggered by spontaneous impressions of the object, "bipolar or unipolar structures, or discrete emotional evaluations, e.g., angry, happy, or a general valence, e.g., good, bad" [22].

Separating positive and negative affects permits better prediction or understanding of attitudes and behaviors. Often, people are not aware of their "own "true" attitude towards an object or situation" and rather construct a, resulting from existing internal and external indicators, value. [22] Past research has been conducted according to the following scheme [22]:

- "Please think about the word 'church' for a moment.
- When you hear the word 'church', what is the first word or image that comes to mind?
- Still thinking about 'church', what is the next word or image that comes to mind?

- What is the third word or image …?"

A further recording of physical changes unifies the three components subjective, physiological and behavioral following Sokolowski [23], one of the most common methods for measuring emotions.

Nevertheless, BIAS effects cannot be excluded, especially in procedures of subjective experience aspects such as questionnaires. There are risks of demand characteristics and the time lag between the emotion's arising and answering or completing a questionnaire, which is why stimuli cannot be directly attributed [24]. Similarly, implicit procedures that take physiological measures into account have the possibility of bias, since, depending on the physical measure, e.g., skin conductance, it is just possible to infer a person's physical reaction and not the emotion that was evoked during perception [10].

3.2 Methodology for the Assessment of Emotional Impact According to Slovic

The affective component of attitudes should be conceptualized and measured. Measurements were made in three studies based on different conceptualizations and were then compared. Affective attitudes can be (1) holistic responses to objects or responses derived from spontaneous images of the objects; (2) bipolar or unipolar in structure; and finally, (3) discrete emotional evaluations (e.g., angry, happy) or more general valenced evaluations (e.g., good, bad). The results also provided support for the hypothesized effect of affective asynchrony. In particular, reduced affective measurement, which required more reflection, was shown to cause the correspondence between the affective component and the intended behaviors/attitudes. The measurements were conducted as follows [22]:

Measure 1. Participants rank their feelings toward the image they have of an object by choosing from a range of numbers −2 (very negative) to + 2 (very positive).

Measure 2. Unipolar picture-taking, picture-based, unipolar, and valenced. Participants describe their feelings about an object on a scale from 0 (not at all negative) to − 2 (very negative) and on a scale from 0 (not at all positive) to + 2 (very positive). This allows one to rank feelings that are both positive and negative about an object.

Measure 3. Holistic, bipolar, and valenced assessment. Participants choose between two bipolar valenced pairs (dislike/like and bad/good) and decide on a scale from −3 to + 3.

Measure 4. Holistic unipolar valenced ratings. Using four unipolar valenced terms (like, good, dislike, and bad), feelings toward the object are to be ranked on a 4-point scale ranging from 1 (disagrees/describes) to 4 (fully describes).

Measure 5. Holistic, bipolar, discrete emotion rating measure. Participants rank their emotions using a 7-point scale ranging from −3 to + 3 for eight bipolar discrete emotion pairs (i.e., love/hate, unhappy/joyful, sad/happy, pleased/annoyed, anxious/calm, relaxed/tense, interested/bored, and angry/satisfied). In addition, the left/right orientation of the positive/negative pairs has been alternated.

This paper includes a consideration of emotional communication and findings from two empirical studies to classify the emotional impact of an exemplary public educational institution. The Peters and Slovic approach was used as a design framework.

Subjects. This initial study was conducted with the university's stakeholders. Internal stakeholders are students of the university as well as employees and professors, whereas external stakeholders are pupils, the society and companies. A total of data has been collected so far from 108 company contacts, 168 pupils, 106 students, 147 employees and professors, and currently 91 people of the society.

Questionnaire. Study 1 served to collect terminology for describing public educational institutions using the example of a given university. In this way, a first impression of the perception on the different target groups, i.e. pupils, students, employees and professors, companies and society, became apparent. The survey, which was conducted with the online tool LimeSurvey, first asked the respondent whether he or she knew the university, so that any subsequent correspondence between knowledge and perception could be investigated. The respondent was invited to think of the institution, describe it in three words, and indicate what emotions would be associated with the institution when thinking of it. The survey resulted in the following collection of terms. They contain of the most recent analysis of data from the surveys of companies with 314 mentioned terms, pupils with 304 mentioned terms, and students with 189 mentioned terms.

Results. We identified the eight most frequently mentioned terms of each target group (see Table 1) and use them as the basis for Study 2, which addresses the emotional classification of these terms. Differing from the research of Peters and Slovic, who had the object and the associated thought evaluated by the same person, an evaluation is provided by the other target group in each case. The reason given is that a questioning of the same person is not possible anyhow due to the anonymous online survey and, in addition, the investigation of a cross-target group effect, and thus not a purely target group-specific classification, is being pursued.

Table 1. Overview of the most frequently mentioned terms describing the respective University.

Term	Number of denominations	Percentage of Frequency by Participants	Percentage of Frequency by Total of Words
Companies			
innovative/innovation	20	18,52%	6,37%
regional	14	12,96%	4,46%
modern	12	11,11%	3,82%
open minded	11	10,19%	3,50%
down-to-earth	11	10,19%	3,50%
professional	11	10,19%	3,50%
organized	12	11,11%	3,82%
reliable	10	9,26%	3,18%

(*continued*)

Table 1. (*continued*)

Term	Number of denominations	Percentage of Frequency by Participants	Percentage of Frequency by Total of Words
Pupils			
modern	43	25,60%	14,14%
organized	28	16,67%	9,21%
informative	16	9,52%	5,26%
interesting	13	7,74%	4,28%
big	13	7,74%	4,28%
friendly	11	6,55%	3,62%
easy to navigate	11	6,55%	3,62%
practice-oriented	10	5,95%	3,29%
Students			
modern	16	15,09%	8,47%
cooperative	8	7,55%	4,23%
diverse	6	5,66%	3,17%
beautiful	6	5,66%	3,17%
technical	6	5,66%	3,17%
friendly	5	4,72%	2,65%
familiar	5	4,72%	2,65%
organized	5	4,72%	2,65%

Study 2 reflects the setup by Peters and Slovic, except for more detailed intros and wording of the questions, see Fig. 2. This is due to the fact that the survey is carried out online and not in person, so any uncertainties and follow-up questions cannot be addressed. In addition, there is an increased risk that the respondents might not understand the intention of the question and therefore be unable to answer it because of their non-specialist background.

While the data from this study is still currently being conducted and analyzed, some challenges and difficulties are already apparent at the time of the surveys and will be considered below.

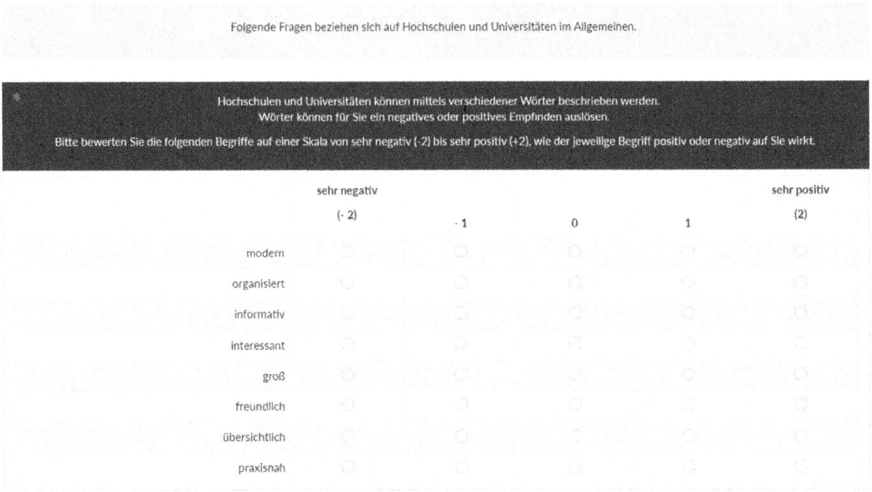

Folgende Fragen beziehen sich auf Hochschulen und Universitäten im Allgemeinen.

Hochschulen und Universitäten können mittels verschiedener Wörter beschrieben werden.
Wörter können für Sie ein negatives oder positives Empfinden auslösen.
Bitte bewerten Sie die folgenden Begriffe auf einer Skala von sehr negativ (-2) bis sehr positiv (+2), wie der jeweilige Begriff positiv oder negativ auf Sie wirkt.

	sehr negativ (-2)	-1	0	1	sehr positiv (2)
modern	○	○	○	○	○
organisiert	○	○	○	○	○
informativ	○	○	○	○	○
interessant	○	○	○	○	○
groß	○	○	○	○	○
freundlich	○	○	○	○	○
übersichtlich	○	○	○	○	○
praxisnah	○	○	○	○	○

Fig. 2. Excerpt of the questionnaire in German.

4   Discussion and Final Evaluation

In the framework of our research on the emotional impact of public educational institutions, two studies were conducted with the different target groups. These are pupils, students, the staff, companies (as an environment and in the context of cooperation) and society. Each stakeholder group perceives the public educational institution and has an influence on its affect. Thus, in order to choose the appropriate communication policy, it is necessary to build an understanding of each, yet so different, target group, their needs and behavior.

To reach the large number of people, we performed a written online survey for both studies, in accordance to Slovic. In the first study, respondents were asked for their image regarding the institution by using adjectives to describe it. Finding terms to describe that image they had of the institution (external image of the brand) proved to be of no difficulty for them. The results also showed an increased similarity in the choice of terms and allow initial assumptions regarding a cross-target group classification of the external image.

The second study, however, was more complex, dealing with the emotional classification of the terms, as described in Sect. 3.2. Because of the written and thus impersonal survey, there were challenges on behalf of the participants in comprehending the questions and therefore a "correct" classification. Talking to some respondents showed that they found it difficult to attribute emotions and emotional orientations to the terms or an object. This would have required more detailed explanations and guidance in responding. However, such risk occurs when conducting an online survey, which is not face-to-face, and must be taken into account when interpreting the results.

Creating a suitable offer, addressing the needs of customers and establishing an external image that matches the self-image, the company's own identity, requires an understanding of the market and, more importantly, of the target group. This paper indicates the relevance of an emotional orientation of the communication policy and the active inclusion of the target group. Beyond that, the use of different methods and models enables target group-oriented communication, a suitable advertising presence and more personal interaction with customers.

However, the research also revealed the complexity of emotional factors and the difficulty in using written surveys. Due to the aggravated conditions of contact restrictions during the realization, future studies should be adapted in their design and procedure.

In the case of public educational institutions, which have a large number of different target groups that are extremely diverse in their characteristics and factors, the focus should be directed at one or two essential target groups that are decisive for entrepreneurial success. By focusing on fewer target groups, a final examination of the effect of implemented communication activities based on the results on this target group is both easier and more effective for the development of optimized marketing activities.

References

1. Erhardt, D.: Hochschulen im strategischen Wettbewerb. Gabler Verlag, Wiesbaden (2011). https://doi.org/10.1007/978-3-8349-7114-2
2. Boos, M., Grubendorfer, C., Mey, D.: Hochschule als Marke: Stand der Diskussion und ein Konzept der Organisationsberatung. Organisationsberatung, Supervision, Coaching **20**(1), 5–15 (2013). https://doi.org/10.1007/s11613-013-0307-3
3. Wänke, M., Florack, A.: Markenmanagement. In: Moser, K. (ed.) Wirtschaftspsychologie, pp. 101–118. Springer, Heidelberg (2015). https://doi.org/10.1007/978-3-662-43576-2_7
4. Burmann, C., Halaszovich, T., Schade, M., Hemmann, F.: Identitätsbasierte Markenführung: Grundlagen - Strategie -Umsetzung - Controlling. Springer Fachmedien Wiesbaden, Wiesbaden (2015). https://doi.org/10.1007/978-3-658-07506-4
5. Kernstock, J., Esch, F.-R., Tomczak, T., Redler, J., Langner, T.: Bedeutung des Corporate Brand Management erkennen und Denkschulen verstehen. In: Esch, F.-R., Tomczak, T., Kernstock, J., Langner, T., Redler, J. (eds.) Corporate Brand Management, pp. 3–26. Springer, Wiesbaden (2014). https://doi.org/10.1007/978-3-8349-3862-6_1
6. Izard, C.E.: Die Emotionen des Menschen. Eine Einführung in die Grundlagen der Emotionspsychologie, Beltz, Weinheim (1981)
7. Pontes, U.: Was ist Emotion? Von Mimik bis Hormon (2011), https://www.dasgehirn.info/denken/emotion/was-sind-emotionen. Accessed 12 Jan 2023
8. Häusel, H.-G.: Limbic success. So beherrschen Sie die unbewussten Regeln des Erfolgs; die besten Strategien für Sieger. 1st edn., Haufe, Freiburg im Breisgau (2002)
9. Kleinginna P. R. Jr., Kleinginna A. M.: A categorized list of emotion definitions, with suggestions for a consensual definition. In: Motivation and Emotion, vol. 5, no. 4, pp. 345–379 (1981)
10. Vogel, I.: Emotionen im Kommunikationskontext. In: Six, U., Gleich, U., Gimmler, R. (eds.): Kommunikationspsychologie Medienpsychologie, pp. 135–157, Weinheim (2007)
11. Müller, A., Gast, O.: Customer-experience-tracking – Online-Kunden conversion-wirksame Erlebnisse bieten durch gezieltes Emotionsmanagement. In: Keuper, F., Schmidt, D., Schomann, M. (eds.): Smart Big Data Management, Berlin, pp. 313–343 (2014)
12. Scherer, K. R., Brosch, T.: Plädoyer für das Komponenten-Prozess-Modell als theoretische Grundlage der experimentellen Emotionsforschung. In: Janke, W., Schmitt-Daffy, M., Debus, G. (eds.): Experimentelle Emotionspsychologie: Methodische Ansätze, Probleme und Ergebnisse, pp. 193–204, Papst, Lengerich (2008)

13. Ellsworth, P. C., Scherer, K. R.: Appraisal Processes in Emotion. In: Davidson, R. J., Scherer, K. R. (eds.): Handbook of Affective Sciences, pp. 572–596, Oxford University Press, Oxford, Oxford (2003)
14. Scherer, K.R.: What are emotions? And how can they be measured? Soc. Sci. Inf. **44**(4), 693–727 (2005)
15. Batinic, B., Appel, M.: Medienpsychologie: Springer Medizin Verlag, Heidelberg (2008)
16. Schwender, C.: Medien und Emotionen. Springer Fachmedien, Wiesbaden (2007)
17. Kroeber-Riel, W.: Bildkommunikation. Imagerystrategien für die Werbung, Vahlen, Munich (1996)
18. Schwender, C.: Medien und Emotionen. In: OSC **10**(1), 5–16 (2003)
19. Thommen, Jean-Paul., Achleitner, Ann-Kristin., Gilbert, Dirk Ulrich, Hachmeister, Dirk, Jarchow, Svenja, Kaiser, Gernot: Allgemeine Betriebswirtschaftslehre: Umfassende Einführung aus managementorientierter Sicht. Springer Fachmedien Wiesbaden, Wiesbaden (2020). https://doi.org/10.1007/978-3-658-27246-3
20. Russell, J.A., Weiss, A., Mendelsohn, G.A.: Affect Grid: a single-item scale of pleasure and arousal. J. Pers. Soc. Psychol. **57**, 493–502 (1989)
21. Mattenklott, A.: Emotionale Werbung. In: Moser, K. (ed.) Wirtschaftspsychologie, pp. 85–106. Springer, Heidelberg (2007). https://doi.org/10.1007/978-3-540-71637-2_6
22. Peters, E.; Slovic, P.: Affective asynchrony and the measurement of the affective attitude component. Cogn. Emotion. **21**(2), pp. 300–329 (2007)
23. Wulf, C.: Emotion. In: Wulf, C., Zirfas, J. (eds.) Handbuch Pädagogische Anthropologie, pp. 113–123. Springer, Wiesbaden (2014). https://doi.org/10.1007/978-3-531-18970-3_9
24. Foscht, T., Swoboda, B., SchrammKlein, H.: Käuferverhalten: Grundlagen - Perspektiven - Anwendungen. Springer Fachmedien Wiesbaden, Wiesbaden (2015). https://doi.org/10.1007/978-3-658-08549-0

A New Perspective on the Prediction of the Innovation Performance: A Data-Driven Methodology to Identify Innovation Indicators Through a Comparative Study of Boston's Neighborhoods

Eleni Oikonomaki[1,2(✉)] [iD] and Dimitris Belivanis[3]

[1] URENIO Research, Aristotle University of Thessaloniki, 541 24 Thessaloniki, Greece
[2] Northeastern Boston University, 360 Huntington Ave, Boston, MA 02115, USA
elenoikonomaki@gmail.com
[3] Dallas, USA
Dbelivanis@stanford.edu

Abstract. In an era of knowledge-based economy, commercialized research and globalized competition for talent, the creation of innovation ecosystems and innovation networks is at the forefront of cities' efforts. The success of both top-down implemented innovation district and community-level innovation ecosystems is complex and has not been well examined. Yet, limited data shed light on the association between indicators and innovation performance on a smaller scale than regional. In this context, public authorities, private organizations, and academics respond to the question of the most promising indicators that can predict innovation with various innovation scoreboards. The present paper initiates the discussion by focusing on the following research query: In what manner can we harness real-time big datasets from unconventional sources to detect novel indicators of innovation and augment the prevailing toolkits for assessing innovation? For this purpose, the city of Boston has been selected as a case study to reveal the importance of its neighborhood's different characteristics in achieving high innovation performance. The study uses a large geographically distributed dataset across Boston's 35 zip code areas, which contains various business, entrepreneurial-specific, socio-economic, and other types of data that can reveal contextual urban dimensions, with the ultimate goal to incorporate the new innovation indicators derived from the non-traditional datasets into an integrated assessment framework. Additionally, to depict the level of innovation in specific zip code areas, novel metrics associated with innovation locations are put forth. The findings of this analysis seek to establish a new innovation index that can generate innovative analysis models applicable in various scenarios. Leveraging the knowledge extracted from non-traditional datasets, the objective is to provide valuable insights for strategic decision-making, resource allocation, and the management of innovation endeavors. This study contributes to the ongoing discourse on innovation by introducing a new theoretical framework that elucidates the connections between urban context, socio-economic characteristics of cities, and innovation performance.

© Springer Nature Switzerland AG 2023
F. Fui-Hoon Nah and K. Siau (Eds.): HCII 2023, LNCS 14039, pp. 377–399, 2023.
https://doi.org/10.1007/978-3-031-36049-7_28

Keywords: Innovation Determinants · Big-Data Analytics · Innovation indicators · Socio-Economic Analysis · Innovation Ecosystems · Spatial Analytics

1 Introduction

Over the last decades, innovation has increased its importance within the pattern of economic growth, moving to the central stage of economists and policymakers concerning the factors that enable the process. According to the "innovation-based growth theory", economic prosperity results from increase in knowledge, scientific and technological improvements, along with the development of an effective private-public partnership. The origin of the innovation-based growth models goes back to Romer [1–3] where growth is driven by specialization and increasing division of labor. Innovation performance is therefore considered a crucial determinant of competitiveness and national progress in the 21st century globalized economy [4]. Cities and innovation are nowadays strongly linked. The emerging trend of innovation districts globally is another indication of cities' efforts to become a favorable globalized environment for innovation to prosper.

Although there are research studies attempting to examine the dynamics that lie behind the creation of an innovation district, less emphasis has been given on the synergies and socio-economic characteristics of the actors coexisting in the district or being involved in its development process. The existing empirical studies are qualitative case studies, such as those by Spigel in Canada [5] and Mack and Mayer in the US [6] and studies measuring innovation ecosystems with quantitative data, such as the study by Ács et al., 2014; and Radosevic and Yoruk, 2013 [7, 8] and several others mostly focusing on national and regional level [9, 10].

This research area still lacks empirical evidence at a smaller than regional scale, and more specifically at the functional region and city level. Several scholars support that there is substantial variation between different localities of the same region [11, 12]. Local innovation indexes are considered to be useful sources of information for policymakers, given the fact that microregions appear to have specific strengths and weaknesses in terms of innovation and therefore they face specific problems and require differentiated policies even within a given region [13, 14].

Autant-Bernard & LeSage have also identified that innovation performances change dramatically between different areas within a region [15]. In the same direction, Porter supports the idea that a country's innovative capacity depends on the more specific innovation environments present in a country's industrial clusters, whether firms invest and compete on the basis of new-to-the-world innovation depends on the microeconomic environment in which they compete, which will vary in different fields [16].

Other studies have also highlighted the significance of location in fostering innovation within an ecosystem. The article "The Geography of Innovation" by Mark P. Feldman, published in 1994, explores the relationship between geography and innovation, focusing on the spatial patterns and processes of innovation [17]. Feldman emphasizes that innovation is not evenly distributed across geographic areas but exhibits spatial concentration. Feldman discusses the role of geography in shaping innovation systems and identifies various factors that contribute to spatial patterns of innovation, but also emphasizes the importance of spatial proximity in facilitating knowledge flows, interactions, and collaboration among actors involved in innovation. Geographic proximity enables the exchange of tacit knowledge, and the development of social networks, which are crucial for innovation.

Feldman's article emphasizes that understanding the geography of innovation requires considering both the internal dynamics of regional innovation systems and the external linkages and interactions with other regions. Globalization and the exchange of ideas, technologies, and resources across regions, institutions and innovation clusters play a significant role in shaping innovation patterns. By studying and leveraging these dynamics, policymakers and researchers can develop strategies and policies to support and enhance innovation within specific regions, leading to economic growth and development.

Audretsch and Feldman support that the geography of innovation and production is influenced by the presence and extent of R&D spillovers [18]. They argue that knowledge spillovers are more likely to occur and have a greater impact when firms or institutions are in proximity to each other. This proximity facilitates face-to-face interactions, knowledge exchange, and collaborative activities. The authors also discuss the role of localized factors, such as the availability of specialized inputs, skilled labor, and supporting infrastructure, in shaping the geography of innovation and production. They argue that these localized factors contribute to the agglomeration of economic activities and create clusters of innovative firms or industries in specific regions.

In the article "The New Economics of Innovation, Spillovers and Agglomeration: A Review of Empirical Studies", Feldman focuses on the concept of knowledge spillovers, which refers to the unintended transfer of knowledge and ideas between firms or individuals [19]. Feldman reviews empirical studies that investigate the extent and impact of knowledge spillovers on innovation and economic performance. These studies provide evidence that proximity between firms or individuals facilitates knowledge flows and enhances innovation outcomes.

In this context, the 'innovation ecosystem' concept has grown in popularity during the last two decades. This term refers to the collective network of researchers, institutions, organizations, and resources within a specific geographical area. The interaction and collaboration among these elements are crucial for innovation to flourish. When researchers can build upon each other's knowledge, inventions, and discoveries, it creates a positive feedback loop where innovation begets more innovation.

For instance, an ecosystem that fosters collaboration and knowledge exchange by providing shared research infrastructure, which allows both university and private sector researchers to work together, can potentially achieve a higher rate of innovative output. This collaborative environment enables researchers to leverage diverse perspectives, expertise, and resources, leading to more breakthroughs and advancements. In contrast, an ecosystem that invests a similar number of resources but lacks effective collaboration and operates with rigid institutional boundaries or silos may impede innovation.

In summary, it gets clear that the success of individuals is influenced not only by their own abilities but also by the collaborative and knowledge-sharing environment created by the ecosystem. The configuration of the ecosystem, including the interactions between its components and the norms that govern them, can significantly impact the overall rate of innovation within that ecosystem.

Boschma and Frenken argue that economic geography should embrace an evolutionary framework to better understand the dynamics and processes of economic development [20]. They highlight that traditional economic geography has mainly focused on static spatial patterns and has not fully captured the dynamic nature of economic systems and their evolutionary trajectories. Furthermore, the authors address the methodological challenges of incorporating an evolutionary perspective into economic geography. They argue for the use of quantitative methods, such as spatial econometrics and agent-based modeling, to capture the complex dynamics and interactions between economic agents and spatial contexts and better understand the processes of economic development and spatial patterns. In the same direction, our study focuses on assessing the dynamic characteristics of the innovation ecosystem, such as the concentration of diverse populations and land uses.

1.1 Research Goals

The current paper firstly discusses literature on innovation assessment in order to define the research question context and the innovation strategies cities put into action to become ecosystems for innovation and entrepreneurship. It builds upon previous efforts to identify new ways and more efficient technological tools to understand the dynamics of neighborhoods with high innovation performance [21]. A previous study suggested that the existing metrics-based toolkits focus mainly on innovation indexes from a top-down perspective and hardly measure at a smaller than the regional scale, almost neglecting the fact that innovation is mainly built from bottom-up processes.

This study focuses on innovation success factors at the urban functional area and city level, given the fact that these are generally seen as the most adequate levels of analysis in terms of proposing policy [22, 23] and nurturing entrepreneurship [24–26]. We aim to develop an integrated assessment framework by incorporating the new innovation indicators derived from the non-traditional datasets into an integrated assessment framework. By combining these new indicators with existing metrics and methodologies, we aim to create a holistic approach to evaluating innovation. This framework should provide a comprehensive view of the innovation landscape and support decision-making processes. Innovation trends evolve rapidly. By utilizing real-time, dynamic big datasets, we could establish a system to monitor the identified innovation indicators on an ongoing basis, then create alerts or dashboards to track changes, detect emerging ecosystems, and identify potential opportunities or risks. The insights gained from the non-traditional datasets can inform strategic decisions, resource allocation, and innovation management efforts, uncover new patterns and trends.

Our analysis section is divided into three parts corresponding to three scales of inquiry. Initially we provide a short literature review on innovation success factors and more specific initiatives implemented by the city of Boston towards this direction. Then Boston's neighborhoods were looked at zip-code level in regard to their business activity. We elaborate on the creation of a dataset consisting of the start-up activity, business permits with the census id of each business, merged with other socio-economic data published by the Census Bureau and other Boston GIS files. We propose a model that captures key factors that have strong influence on innovation performance across the different neighborhoods. After explaining the method, we check the results and generate a few tables and maps emphasizing Boston's neighborhood characteristics and their correlations with high business activity and start-up presence.

382   E. Oikonomaki and D. Belivanis

This study uses a big data analytics strategy as our main method, ultimately resulting in a critical explanation of our own proposals for detecting innovation determinants. In support of this effort, data has been collected from several open-source web mapping and other types of platforms for entrepreneurship, demographics, socio-economic data, and land use.

Finally, in the findings section, we briefly explain the outcome of this paper, to respond to our research question. The immediate locale of each neighborhood was explored to weigh the advantages and disadvantages of their respective environments for entrepreneurs and innovation activity, but also the socio-economic characteristics, in a way to better understand the high concentration of innovation hubs, incubators, accelerators at specific areas of the city. By highlighting the different neighborhoods' distinctive socio-economic features, strengths, and shortcomings, we aim at providing a conceptual discussion of the notion of a new Innovation Index and elaborate why and how entrepreneurship shows systemic characteristics at the local level.

More specifically, the paper sets out the findings of the study as follows: Sect. 1 introduces the topic of the existing innovation indicators and the level of innovation assessment and identifies the gap for empirical evidence at the urban functional area (local) level. (Sect. 1.1); The section also describes the main research goals and proposes a theoretical framework that will provide support to policy makers to improve the current tools they use to better understand the success of innovation districts (Sect. 1.1). Section 2 analyzes the indicators already discussed in literature (Sect. 2.1), which are considered important ingredients for successful innovation districts. It also contains an analysis of Boston's neighborhoods as local clusters of innovation (Sect. 2.2) and discusses previous studies assessing Boston's neighborhoods in terms of their entrepreneurial and socio-economic characteristics. Section 3 attempts at introducing a new method for analyzing innovation determinants by incorporating publicly available datasets with the creation of a customized dataset concerning innovation. For this purpose, information was scrapped from the web and queried from different apis. Those extracted features are used as innovation indicators and their connection with the available socioeconomic data is examined at the zip-code level. It also contains an overview of all open-source datasets and features used for this analysis (Sect. 3.1) and visualizations of all the study's key findings that indicate feature collinearity and neighborhood patterns (Sect. 3.2). Section 4 discusses the main constraints we faced during our research presented in Chapter 3. It concludes with recommendations for improving our method for the prediction of innovation determinants.

2 Previous Studies Assessing Innovation Determinants at the Local Scale: Literature and Theory Framework

After exploring the broader issues related to the necessity of measuring innovation performance at the local level, we will now delve into the specific factors that determine innovation, as mentioned in the literature. A previous study has evaluated various innovation indicators compiled by renowned institutions such as the GEDI Institute (Global Entrepreneurship Index), the European Observatory for Clusters and Industrial Change, the Kauffman Foundation Research Series on City, Metro, and Regional Entrepreneurship, the World Economic Forum, Global Innovation Index (GII). Several institutions provide rankings for countries based on their innovation performance, considering various indicators such as research and development (R&D) investments, patent filings, and knowledge transfer. Even though most of the institutions focus on national, there are few that provide an Innovation Index at the city level, such as the one developed by 2thinknow. This index helps identify cities that foster an environment conducive to innovation and highlights emerging innovation hubs. Additionally, Innovation Cities Program, initiated by the Intelligent Community Forum (ICF), focuses on identifying and promoting cities that exhibit high levels of innovation and sustainable economic growth.

Regarding the successful spatial aspects of innovation, the Brookings Institution offers insights into the emerging urban model known as the "innovation district" [42]. These districts are defined as geographic areas where leading anchor institutions and companies cluster and interact with start-ups, entrepreneurs, business incubators, and accelerators. They are characterized by their compactness, accessibility to transit, technological connectivity, and the availability of mixed-use housing, offices, and retail spaces. Central buildings within the community serve as inclusive hubs for local entrepreneurs, while new housing developments cater to residents of varying income levels. Skill training centers also contribute to the success of these districts by providing residents with technology-focused education that aligns with the employment opportunities in the district.

The strength of innovation ecosystems heavily relies on the attraction and retention of talent, the cultivation of an entrepreneurial and risk-taking culture, the presence of research and innovation infrastructure, and the collaboration of compatible and complementary stakeholders within the system. Key actors, often referred to as orchestrators, play a crucial role in shaping the development of innovation ecosystems, both directly and indirectly. The Quadruple Helix Model of innovation, originally conceived by Elias Carayannis and David Campbell, identifies four major actors in the innovation system: science, policy, industry, and society. On one hand, higher education institutions serve as anchors that foster an entrepreneurial environment. Local governments encourage an entrepreneurial approach that promotes risk-taking and grassroots participation in addressing key city issues. Intermediary actors form a well-established networking structure that supports entrepreneurial collaboration, cross-fertilization, and co-creation at multiple scales. Additionally, local authorities contribute to urban regeneration initiatives that complement economic development efforts and provide civic-led spaces for grassroots collaboration and cooperation within society [27].

Innovation ecosystems are influenced by various policy agendas at the local, regional, national, and international levels, including initiatives like the UN 2030 Agenda for Sustainable Development. Internationalization is another crucial element for developing competitive, sustainable, and successful strategies in local and regional innovation ecosystems [28].

2.1 Innovative Neighborhoods of Boston: Socio-Economic and Entrepreneurial Neighborhood Characteristics

The Greater Boston area is currently one of the most innovative locations in the US local development landscape, thanks to its high agglomeration of educational institutions and industries. The entire urban region, which is recording the highest rate of growth anywhere in the US [29], is increasing its capacity to attract the interest of major investors. Over the last three decades, the cities of Boston and Cambridge, public and private investments have prioritized the boost of sectors such as education, financial services, life sciences, and the high-tech industries, which today represent the main clusters within the entire urban region. Greater Boston is a huge metropolitan area that includes the counties of Suffolk and Middlesex. The municipalities of Boston and Cambridge are located within it. In 2016, the Greater Boston area was the foremost location in the U.S. for fostering entrepreneurial growth and innovation [30]. This ranking is based on how well the top 25 US metropolitan areas "attract talent, increase investments, develop specializations, create density, connect the community and build a culture of innovation" [20].

While the San Francisco Bay Area dominates most categories in various indices, it is noteworthy that "Boston claims the top spot due to its collaborative community and the declining quality of life experienced by many of its residents." The survey also reveals that the Bay Area ranks relatively low in terms of quality of life (22nd), likely influenced by the rising cost of living. Conversely, Boston is the second-ranked city compared to the Bay Area in most traditional startup activity metrics, but local entrepreneurs emphasize stronger connections with universities, institutions, and citizens. The cities of Boston and Cambridge serve as prominent hubs for higher education institutions, startups, tech industries, and research centers, creating an ideal urban ecosystem for innovation and attracting highly skilled creative workers, innovators, and investors. Additionally, the local government plays a crucial role in fostering the growth of the innovation ecosystem by offering services, funding, and resources for businesses and entrepreneurs (e.g., the Cambridge Entrepreneurship Assistance Program, the Cambridge Small Business Enhancement Program). They also implement planning initiatives and establish new spaces to support the local innovation community.

There are several public-private partnerships involving collaboration between Boston city's government and private-sector that aim to enhance innovation. Financing a project through a public-private partnership can allow a project to be completed sooner or make it a possibility in the first place such as the District Hall, Roxbury Innovation Center. On the other hand, Cambridge Innovation Center, Masschallenge are considered successful private –led initiatives, while Kendall Square, Seaport District, Dudley Square have been initiated as public-driven regeneration initiatives.

LifeTech Boston
In 2004, the Boston Planning Development Agency (BPDA), previously known as the Boston Redevelopment Authority (BRA), initiated a policy initiative called 'LifeTech Boston.' This marked a crucial milestone in the development of a new redevelopment model, eventually leading to the establishment of the first Innovation District in Boston. The 'LifeTech Boston' initiative aimed to create a supportive ecosystem for life sciences and technology companies, fostering innovation and economic growth in the city. However, it was the subsequent 'Boston Innovation District' (BID) that emerged as the most significant component of this original strategy. Launched in 2010 under the leadership of the Menino administration, the BID was a comprehensive planning initiative that aimed to transform a specific area of Boston into a vibrant hub for innovation and entrepreneurship [31].

Boston Innovation District

The BID project aims to create a complex neighborhood that mimics the success of 22@Barcelona, with the ultimate goal to attract financers, resources, and talent. The BID project was conceived to redevelop the South Boston Waterfront, an underutilized area of 1,000 acres that previously hosted industrial activities and parking and transform it into a thriving hub of innovation and entrepreneurship together with new residential, commercial, and retail spaces (about 7.7 million sq. ft.) with a mixed-use configuration. The BID sought to leverage the city's existing strengths, such as its renowned higher education institutions, research centers, and entrepreneurial talent, to attract and nurture innovative companies. The Boston Innovation District continues to evolve and progress, with ongoing efforts to shape the area into a dynamic and collaborative environment for businesses and entrepreneurs. The district offers various incentives, resources, and infrastructure to support the growth of innovative industries, including tech startups, creative enterprises, and research-driven organizations.

The establishment of the Boston Innovation District through the 'LifeTech Boston' initiative and subsequent planning efforts has been instrumental in fostering a thriving innovation ecosystem in the city. It has created opportunities for collaboration, knowledge exchange, and economic development, further solidifying Boston's reputation as a prominent center for innovation and entrepreneurship [32].

District Hall

District Hall opened in 2013 and was founded through a unique public-private partnership in response to the City's call. First project in their 23-acre master plan for Seaport Square. The city invited the Cambridge Innovation Center (CIC) to create an active innovation hub concept. CIC then developed the initial concept and design, as well as provided construction management and financial support for the project. CIC has ongoing responsibility and holds the lease for the property. The Venture Cafe' Foundation was involved in the initial planning process and continues to manage operations and programming since 2013. District Hall also plays a role in supporting and promoting Boston's innovation ecosystem. It hosts various programs and initiatives aimed at nurturing entrepreneurship, providing resources and support to startups, and connecting entrepreneurs with mentors and investors. It serves as a platform for showcasing innovative ideas, products, and services, contributing to the growth and vibrancy of Boston's innovation community [33].

Cambridge Innovation Centers

Cambridge Innovation Centre (CIC) and the Venture Cafe' Foundation were both founded by Tim Rowe, the former in 1999 as an incubator, the latter in 2010 as a social experiment. CIC is a private entrepreneurial activity based on renting shared and flexible office spaces with an innovative style. It currently hosts over 700 companies across two buildings, located in Kendall Square and in downtown Boston, about 500 of which are start-ups. One of the key features of CIC is its emphasis on building a vibrant and interconnected community. CIC brings together entrepreneurs, investors, mentors, and industry experts under one roof, facilitating networking opportunities and knowledge sharing. The community aspect of CIC is often viewed as a valuable resource for startups, as it enables them to connect with like-minded individuals, seek advice, and form partnerships. In addition to office space, CIC offers various support services and resources to assist startups and entrepreneurs in their journey. This includes access to educational programs, mentorship initiatives, fundraising assistance, and connections to strategic partners and investors. CIC's goal is to provide an ecosystem that nurtures innovation, accelerates growth, and helps its members thrive [34].

Kendall Square

Kendall Square is a former brownfield located in Cambridge (MA), opposite the Charles River. It started in 1868 as an industrial district and consolidated this function with the opening of the first underground line nearby. The presence of the Massachusetts Institute of Technology dates to 1916. Following the Second World War, the area entered an era of decline, which the Cambridge Redevelopment Authority (CRA), established in 1955, sought to reverse also through the clearance of 29 acres of land for the accommodation of NASA. Kendall Square is home to numerous leading academic and research institutions, including the Massachusetts Institute of Technology (MIT), Harvard University, and several biotechnology and pharmaceutical companies. The area has a rich history of scientific and technological advancements, with many groundbreaking discoveries and innovations originating from institutions based in Kendall Square. The neighborhood has evolved into a hub for startups, entrepreneurship, and venture capital investment. It offers a thriving ecosystem that fosters collaboration, knowledge exchange, and networking among professionals, researchers, and innovators. The close proximity of renowned educational institutions and research centers creates opportunities for partnerships and the commercialization of cutting-edge ideas and technologies. Kendall Square is characterized by its modern architecture, with numerous office buildings, research facilities, and high-tech companies occupying the area. In recent years, Kendall Square has also seen significant urban development, with the addition of residential spaces, retail establishments, restaurants, and green spaces. This mix of amenities creates a lively and dynamic environment, attracting a diverse community of residents, employees, students, and visitors [35].

Nubian Square (Formerly Dudley Square)
The 'Neighbourhood Innovation District' (NID) is an on-going public strategy launched in 2014 by the government of Boston City Council. The main goal of this initiative is to encourage and spread innovation and technology within low-income neighborhoods to improve small business growth and local economic development. Following specific criteria mentioned in the innovation district literature (transit access, affordable office space, arts and cultural amenities, involvement of non-profit organizations) and considering the features of the area (e.g., the presence of higher-education institutions, vacant lots, transportation nodes) the location for the first experiment was chosen to be the 'Dudley Square-Upham's Corner Corridor', a vibrant zone within the Roxbury neighborhood. Today it is known as Nubian Square and it hosts community events, art exhibits, and performances that celebrate the cultural heritage and contributions of the Roxbury community. The area is also home to many local businesses, including restaurants, shops, and community organizations that contribute to the neighborhood's character and vitality [36].

Roxbury Innovation Center
Roxbury Innovation Center (RIC) was created through a public-private partnership with the City of Boston and The Venture Café Foundation. Its mission was to support local economic development, in Roxbury, Dorchester and Mattapan by empowering and guiding innovation and entrepreneurship, as viable career options. Since opening in late 2015, RIC has provided a diverse variety of resources for small business owners of all stages and industries, through instructional workshops and courses, networking events, and office hour mentorship. One of the key objectives of the Roxbury Innovation Center is to promote diversity, equity, and inclusion in the innovation and entrepreneurship ecosystem. The center actively seeks to engage and support underrepresented entrepreneurs, including women, people of color, and individuals from disadvantaged backgrounds. It aims to create an inclusive and accessible environment where all community members can access the resources and opportunities needed to succeed [37].

MassChallenge
Headquartered in the United States with locations in Boston, Israel, Mexico, Rhode Island, Switzerland, and Texas, MassChallenge strengthens the global innovation ecosystem by accelerating high-potential startups across all industries, from anywhere in the world for zero-equity taken. As MassChallenge's flagship location, MassChallenge Boston has brought together corporates, policy makers, and innovation leaders to support and inspire the next generation of innovators. Over the past eight years, it accelerated more than 1,000 startups from across the world. The organization operates multiple programs across different locations, including Boston, Israel, Mexico, Switzerland, and Texas. Each program typically lasts several months and includes workshops, mentoring sessions, pitch competitions, and access to a network of potential investors and partners.

In addition to its accelerator programs, MassChallenge hosts various events and initiatives aimed at fostering collaboration and innovation within the startup ecosystem. This includes conferences, networking events, and hackathons that bring together entrepreneurs, investors, and industry professionals to share ideas and create meaningful connections [38].

3 Innovation Indicators at the Local Urban Scale: Visualizing the Relationships Among Features and Innovation Performance

The MAPS-LED project is a previous study in the same direction, which emphasizes innovation as a critical element for the further development of the cities. More specifically, the project proposes a cluster mapping strategy that tries to localize the economic clusters to understand better innovation at the neighborhood level. The report of this EU funded project published in 2019 marked different areas of Boston's greater area related to the respective cluster hosted. For this current study, we are considering Boston's innovative neighborhoods, and other standard neighborhoods, based on the MAPS-LED project (MAPS-LED project n.d.).

TREnD project, another EU funded project tries to shed light on Boston's resilience strategies, resulting in a better and more inclusive transition process for the adoption of the green innovative technologies. The project "Navigating the green transition during the pandemic equitably: A new perspective on technological resilience among Boston Neighborhoods facing the shock" published in 2023, conceptualizes this transition process as an evolutionary path [25].

3.1 Scope, Methods, and Data Collection

This current study starts with detecting innovation patterns at Boston's functional urban area (FUA), which refers to a geographic area that encompasses a core city and its surrounding commuting zones or hinterland, where people typically reside and commute to the core city for various economic, social, and cultural activities. It aims at contributing to the current research efforts in innovation ecosystems by providing a dynamic model to demonstrate the correlations between the innovation performance and neighborhood characteristics. The project's methodological framework strongly relies on exploring a significant number of observations to diagnose underlying correlations and connections of the socioeconomic data to innovation indicators and illustrate the interdependencies in entrepreneurial activity and social characteristics. The study introduces the use of big data analytics to support the innovation performance measurement at this subregional level. It also explores new ways such analyses may afford more opportunities than were previously examined by innovation assessment toolkits.

Thus, having this in context to answer the research questions stated earlier, we used the Boston Permits dataset published by Boston Area Research Initiative (BARI) to capture appropriate information about neighborhoods' tendency toward investing in business projects. The permit database is an excellent data source to capture the Boston changes during seven years. Benefiting the quantitative and georeferenced observations, it would be facilitating to get the pulse of the city according to the level of projects.

For the purpose of creating an innovation index, available data on the web was utilized. Specifically, the Google Maps api, OpenStreetMap api, and indeed api, were used. For the google api, the keywords shown in Table 2 were used and all the locations were gathered with their corresponding unique id, number of ratings, and average rating measuring user satisfaction and the overall quality of the establishment or location of the building with entrepreneurial activity (ranging from 1 to 5 stars, with 5 stars being the highest possible rating). The data is available for each individual keyword, however the concatenation of all the keywords is considered and presented as the most indicative. It is worth pointing out that each location that corresponds to a unique id was considered only once even though it could appear in multiple keyword searches. The number of locations is the most indicative as it is directly expressing the number of businesses that are relevant to innovation, but the total number of ratings and the mean rating provided additional insights, the average rating of the zip code was weighted by the number of ratings for each location.

In addition, the OpenStreetMap api was used to augment the dataset with locations of innovation. The use of keywords was not possible but rather the category of the building was used specifically the following tags were used to query locations in Boston: (*company = startup, office = coworking, office = research*). Unfortunately, the dataset was not covering the whole range of zip code areas, however as expected the area where those locations existed coincided with the areas which had the highest presence in the Google Map dataset. Even though the dataset is not as complete it is worth investigating as it is an open source that also provides the evolution over time and links the locations of innovations directly and not through a keyword that could be misleading.

Finally, data from the website www.indeed.com was collected, a site where job postings are available. Specifically, all jobs related to technology were collected and they were assigned to a specific zip code where the job was located. Unfortunately, most of the job postings were not including the exact address which were disregarded as there was no available zip code, the solution of geo-decoding was considered but has not been implemented yet. The dataset was not covering all the zip codes of the municipality of Boston, but the location with the most job offering coincided once again with the location where the highest innovation activity was shown from google maps.

Summary of Variables

The building permits include variables that can be grouped into three categories: business characteristics, all of which are original to the data; geographic information, which includes variables introduced by BARI to make the data more easily compatible with other data sources; and types of work, categorized by BARI based on the types of permits granted. For the purpose of this study only permits that were considered commercial or mixed use were considered.

The census data was provided in census tract and it was aggregated to zip code areas, and as for the data collected from the Google and Openstreetmap the exact location was available and therefore a spatial merging was performed with the polygons of Boston zip code areas. An overview of the data used is given in Table 1.

Table 1. Features and source used for supporting the dataset.

Entrepreneurship - Innovation Activity	Social Data	Economic Data
Start-up locations (gmap)	Race (Census Bureau)	Median Income(Census Bureau)
Start-up locations (openstreetmap)	Vacancy (Census Bureau)	Median Household value (Census Bureau)
Start-up locations (crunchbase)		Job openings (indeed)
Business Permits (BARI)		

Table 2. Keywords used for the google search.

innovation hubs	clustering	innovation center	startups
innovation districts	open innovation	tech hub	technology park
incubator	accelerators	regional innovation	co-working space

3.2 Research Findings

The purpose of this study is first to identify and quantify through an index or indexes of the innovation performance of each area. This was achieved through the data gathering for locations with innovation impact, and it was performed as described above. The values collected from Google maps were considered to be the most appropriate as they provide similar results with other proposed values, they are more reliable for areas with low innovation presence and can be reproduced easily for other areas in the USA or in a different country.

Specifically in Fig. 1 the correlation of three proposed innovation metrics with the socio-economic data is presented, those are namely the number of innovation locations, the number of average ratings, and the total number of ratings of all the locations in the zip code area. The first observation that can be made is that there are no high collinearities, this could rise from the heterogeneity of the studied areas as some are purely residential and some are business districts. Once the dataset is augmented with more areas it would be interesting to consider those two types of areas separately.

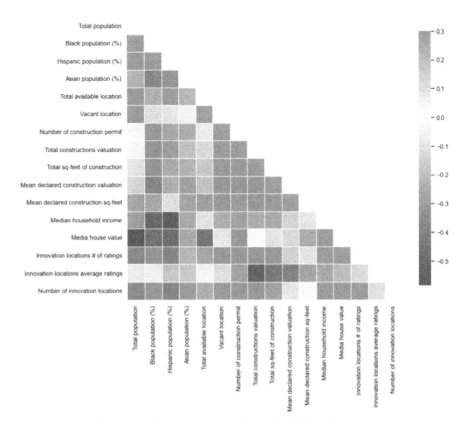

Fig. 1. Matrix. Correlation coefficient is indicated by color.

Even though the correlations are relatively small (-0.5 - 0.3) the correlation matrix provides some interesting insights. There is a negative correlation of Black and Hispanic population with innovation locations, while a positive correlation between White population and innovation locations could be understood better when looking at Fig. 2. The locations where the number of innovation locations are high is in the city center where the business district is and it happens to have a higher percentage of white population as shown in Fig. 4, which can be explained historically and due to the elevated average price of the households in the central area. Another correlation that can be ascribed to the fact that most of the innovation locations are in business districts is its correlation with vacant spaces. Due to the volatility of the real estate market in business areas it

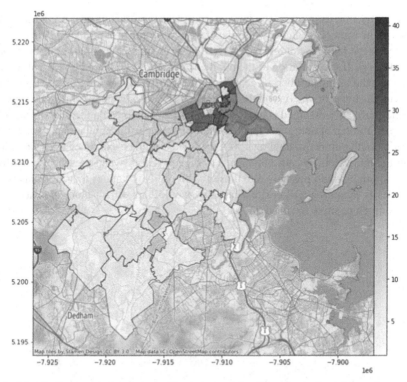

Fig. 2. Map indicating concentration of business activity, intensity indicated by color and the significance level by size (darker color indicates higher number of concentrated businesses).

makes sense that there are more vacant locations, something also seen in Fig. 3 where a map of vacant spaces is presented.

The number of constructions permits is positively correlated with the number of innovation locations, a bidirectional explanation can be provided as the more innovation locations and startup companies are attracted to an area the more construction permits and from the other side the areas with most business activity and therefore construction will attract more companies of which some will be related to innovation. An interesting finding that cannot be easily explained is the negative correlation of the average rating to the construction permit activity of an area, this should be studied further in more areas, as this could be an artifact from cases when the rating is not representative when the number of ratings is small for each location.

It is really interesting how the innovation indexes are correlated among them. The first is that the rating is negatively correlated with the number of ratings, which could be explained as the rating is not representative if there are not enough ratings as mentioned above. The positive correlation of the number of locations and number of total reviews was anticipated as the number of locations increased the bigger the chance of review to be provided. However, the reason the correlation is low 0.3 and not almost collinear as someone could have assumed is because the number of innovation locations is linked to the average number of ratings per location therefore a super-linear relationship of number of locations and number of total reviews exists. This could be explained as collaboration among the innovators, because this leads to bigger visibility and presence in the local community that would possibly rate those locations. This implies that the areas where innovation is blooming will create a hub that would later attract more innovators and this becomes a positive feedback loop.

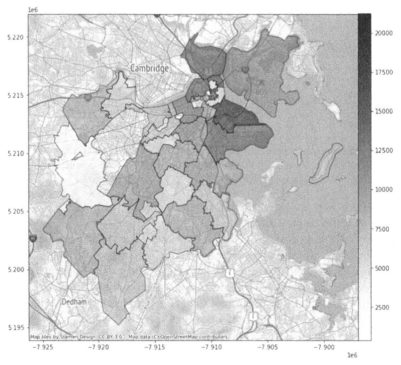

Fig. 3. Map indicating location of vacant units in the municipality of boston; darker color indicates higher number of vacancies.

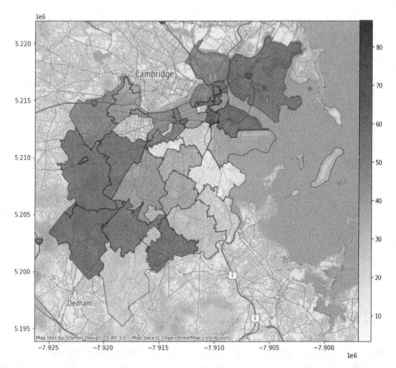

Fig. 4. Map indicating White population as a percentage of total populations.

The fourth map was created in this context to depict concentration of startups' activity across the same neighborhoods but using this time the open-source dataset created for the Start-up cartography project [39]. This map provides a visual representation of the distribution and concentration of start-up activities, such as research and development, and entrepreneurial ventures. While the maps show similar levels of innovation, it suggests that the neighborhoods being compared exhibit comparable levels of innovative output or entrepreneurship. This might indicate that the downtown neighborhoods have similar conditions or factors that foster innovation, such as supportive infrastructure, access to resources, a skilled workforce, or a culture of entrepreneurship. Additionally, similar levels of innovation on the maps may also reflect the presence of effective knowledge sharing and collaboration networks among organizations within these neighborhoods. Overall, the similarity in the levels of innovation observed on the maps implies clear innovative and entrepreneurship capacities and performance among the areas being studied (Figs. 2 and 5).

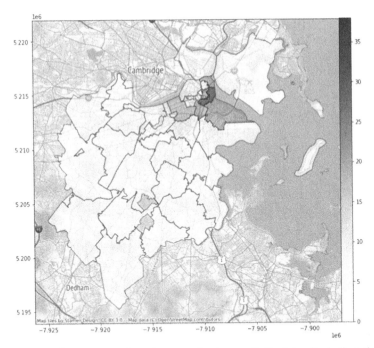

Fig. 5. Map indicating concentration of start-up business activity, intensity indicated by color and the significance level by size (darker color indicates higher number of concentrated entrepreneurship activity), source: Start-up Cartography project.

4 Conclusions and Next Steps

First, we analyzed the data in the spatial context to perform an illustration of innovation activity mapping for policymakers of Boston city. We observed clustering phenomena by downtowns and adjacent neighborhoods' tendency on higher innovation activity among 23 Bostonian neighborhoods. Exercising innovation heat maps, based on data coming from different sources, helped us cross check the communities that outperform and concentrate the highest number and ratings of businesses, highest number of investments in business permits etc. As discussed in Sect. 3.2 we found that Downtown, Wharf District, North End, Back Bay, the top performing zip codes in terms of innovation, are areas where the most civic buildings are located (see Fig. 1).

Second, high innovation/entrepreneurial activity can contribute to more dynamic and appealing communities that can result in greater property values. By equally supporting all neighborhoods to invest in the development of new mixed-use, business projects, the renovation and beautification of existing ones, the city of Boston could mitigate the disparities among districts in the same city. Even though in the past, the city has supported several initiatives to encourage and spread innovation and technology within low-income neighborhoods such as the 'Dudley Square-Upham's Corner Corridor', a vibrant zone within the Roxbury neighborhood, as well as the Roxbury Innovation Center (RIC), these neighborhoods don't appear to have the same dynamic as the downtown located areas.

Third, the maps are created to call for more attention on structurally weak neighborhoods by policymakers and the communities to think again. However, for the next steps of this research, we suggest a comparison analysis approach, which would use the Great Boston metropolitan area and compare it with another area of similar scale. This would allow us to provide more meaningful insights to local authorities. We would also suggest a normalization of the business permit data, using a measure of "x USD per business" spent on business-related projects. In this way, we could capture neighborhoods' size, population, and business density patterns to find a uniform measure in comparing them.

Fourth, we drew districts' socio-economic diagrams capturing vacant units, concentration of White population as a percentage of total populations etc. We believe there is more room for investigating community-level and opportunities at a larger scale, the metropolitan area of Boston due to the diverse populations in terms of ethnicity and race across the different cities. An analysis on the initial features across Boston's neighborhoods revealed significant opportunities for identifying innovation determinants. A question that remains unanswered is what other variables could capture the innovation activity.

Looking only at the white population ratio, vacancy and business concentration mapping is a starting point for our study. In a future study, it would be worth it focusing more on different ethnicities and races, land use and amenity locations. A following study should consider incorporating data extracted by social media activity, such as twitter, foursquare etc., mobility data but also data regarding internet and energy use (Fig. 5).

Author Contributions. Conceptualization, E.O; methodology, E.O, D.B.; investigation, E.O, D.B.; writing - original draft preparation: ,E.O
writing—review and editing E.O; project editing, project administration,
All authors have read and agreed to the published version of the manuscript.
Institutional Review Board Statement: Not applicable.
Informed Consent Statement: Not applicable.

Data Availability Statement.   https://www.google.com/maps
https://www.openstreetmap.org/#map=4/38.01/-95.84
https://www.indeed.com
https://www.census.gov

Conflicts of Interest.   The authors declare no conflict of interest.

References

1. Romer, P.M.: Endogenous technological change. J. Polit. Econ. **98**, S71–S102 (1990)
2. Romer, P.M.: Increasing returns and long-run growth. J. Polit. Econ. **94**, 1002–1037 (1986)
3. Romer, P.M.: Growth based on increasing returns to specialization. Am. Econ. Rev. **77**(2), 56–62 (1987)
4. OECD: Innovation and growth rationale for an innovation strategy. https://www.oecd.org/sti/39374789.pdf
5. Spigel, B.: The relational organization of entrepreneurial ecosystems. Entrep. Theory Pract. **41**(1), 49–72 (2017)

6. Mack, E., Mayer, H.: The evolutionary dynamics of entrepreneurial ecosystems. Urban Stud. **53**(10), 2118–2133 (2016)
7. Ács, Z.J., Autio, E., Szerb, L.: National systems of entrepreneurship: measurement issues and policy implications. Res. Policy **43**(3), 476–494 (2014)
8. Yoruk, E., Radosevic, S.: Entrepreneurial propensity of innovation systems: theory, methodology and evidence. Res. Policy **42**(5), 1015–1038 (2013)
9. European Commission: European Innovation Scoreboard: Innovation performance keeps improving in EU Member States and regions. https://ec.europa.eu/commission/presscorner/detail/en/ip_21_3048
10. Leendertse, J., Schrijvers, M., Stam, E.: Measure twice, cut once: entrepreneurial ecosystem metrics. Res. Policy **51**(9), 104336 (2022)
11. Sternberg, R.: Regional dimensions of entrepreneurship. Found. Trends R Entrepreneurship **5**(4), 211 (2009)
12. Fritsch, M., Wyrwich, M.: The long persistence of regional levels of entrepreneurship: Germany, 1925–2005. Reg. Stud. **48**(6), 955–973 (2014)
13. Anselin, L.V.: Local geographical spillovers between university research and high technology innovations. J. Urban Econ. **42**(3), 422–448 (1997)
14. Bajmocy, Z.: Constructing a local innovation index: methodological challenges versus statistical data availability. Appl. Spat. Anal. Policy **6**, 69–84 (2012). https://doi.org/10.1007/s12061-012-9080-5
15. Autant-Bernard, C., LeSage, J.: Quantifying knowledge spillovers using spatial econometric models. Reg. Sci. **51**(3), 471–496 (2010)
16. Porter, M.E.: The Competitive Advantage of Nations. Free Press, New York (1990)
17. Feldman, M.P.: The Geography of Innovation. Kluwer Academic Publishers, Dordrecht (1994)
18. Audretsch, D.B., Feldman, M.P.: R&D spillovers and the geography of innovation and production. Am. Econ. Rev. **86**(3), 630–640 (1996)
19. Feldman, M.P.: The new economics of innovation, spillovers and agglomeration: a review of empirical studies. Econ. Innov. New Technol. **13**(5), 355–377 (2004)
20. Boschma, R., Frenken, K.: Why is economic geography not an evolutionary science? Towards an evolutionary economic geography. J. Econ. Geogr. **6**(3), 273–302 (2006)
21. Oikonomaki, E.: Enabling real-time and big data-driven analysis to detect innovation city patterns and emerging innovation ecosystems at the local level. In: Streitz, N.A., Konomi, S. (eds.) HCII 2022, vol. 13325, pp. 404–418. Springer, Cham (2022). https://doi.org/10.1007/978-3-031-05463-1_28
22. Bradley, J., Katz, B.: The Metropolitan Revolution: How Cities and Metros Are Fixing Our Broken Politics and Fragile Economy. Brookings Institution Press, Washington, D.C. (2013)
23. Spigel, B.: Entrepreneurial Ecosystems: Theory Practice and Futures. Edward Elgar Publishing (2020)
24. Feld, B.: Startup Communities: Building an Entrepreneurial Ecosystem in Your City. Wiley, Hoboken (2012)
25. Feldman, M.P.: The entrepreneurial event revisited: firm formation in a regional context. Ind. Corporate Change **10**(4), 861–891 (2001)
26. Fazio, C.E., Guzman, J., Liu, Y., Stern, S.: How is COVID changing the geography of entrepreneurship? Evidence from the startup cartography project. Working paper series, vol. w28787 (2021)
27. Carayannis, E.G., Campbell, D.F.: 'Mode 3' and 'quadruple helix': toward a 21st century fractal innovation ecosystem international. J. Technol. Manag. **46**(3/4), 201–234 (2009)
28. Rissola, G., Hardliner, J.: Place-based innovation ecosystems. a case-study comparative analysis. Publications Office of the European Union, Luxembourg, EUR 30231 EN (2020)

29. Kahn, K., Barczak, G., Ledwith, A., Perks, H., Ledwith, N.: An examination of new product development best practice. J. Prod. Innov. Manag. **29**, 180–192 (2012)
30. U. C. o. C. Foundation (2016). "https://www.uschamberfoundation.org/innovation-that-matters
31. LifeTech Boston. https://biotechtuesday.com/content/boston-lifetech
32. https://www.smartcitiesdive.com/ex/sustainablecitiescollective/case-study-boston-waterfront-innovation-district/27649/
33. D. H. Boston. https://districthallboston.org
34. Cambridge Innovation Center. https://cic.com
35. Kendall Square. https://kendallsquare.org
36. Visit Nubian Square. https://www.visitnubiansquare.com
37. Roxbury Innovation Center. https://roxburyinnovationcenter.org
38. MassChallenge. https://masschallenge.org
39. Andrews, R., Fazio, C., Guzman, J., Liu, Y., Stern, S.: The startup cartography project: measuring and mapping entrepreneurial ecosystems. Res. Policy **51**(2), 104437 (2022). ISSN: 0048-7333
40. Sohrabi, P., Oikonomaki, E., Hamdy, N., Kakderi, C., Bevilacqua, C.: Navigating the green transition during the pandemic equitably: a new perspective on technological resilience among Boston neighborhoods facing the shock. In: Bevilacqua, C., Balland, P.-A., Kakderi, C., Provenzano, V. (eds.) New Metropolitan Perspectives: Transition with Resilience for Evolutionary Development, pp. 285–308. Springer, Cham (2023). https://doi.org/10.1007/978-3-031-34211-0_14
41. MAPS-LED project. https://www.cluds.unirc.it/project/project-mapsled/
42. The Brookings, The rise of Innovation Districts (2014). https://www.brookings.edu/articles/rise-of-innovation-districts/

Interruptions in the Workplace: An Exploratory Study Among Digital Business Professionals

Fabian J. Stangl[1]([✉]) and René Riedl[1,2]

[1] Digital Business Institute, School of Business and Management, University of Applied Sciences Upper Austria, Steyr, Austria
{fabian.stangl,rene.riedl}@fh-steyr.at

[2] Institute of Business Informatics – Information Engineering, Johannes Kepler University Linz, Linz, Austria

Abstract. The use and role of digital technologies can bring significant benefits to individuals, organizations, and businesses, such as ubiquitous access to information, rapid communication, and productivity gains. Technological change, however, also poses challenges for Human-Computer Interaction (HCI), including increased and unexpected interruptions caused by digital technologies while performing tasks in the workplace. To mitigate the negative consequences of interruption overload, companies need to develop strategies, so that employees can work effectively and efficiently. Against this background, we conducted an exploratory study to allow digital business professionals to describe their general perception and management of interruptions. The research results reveal that most interruptions can be classified as either an intrusion (35%) or a distraction (28%). Although beneficial outcomes of work interruptions can be observed, they have an overall detrimental impact on both psychological well-being and work performance, with the strongest negative impact on work performance (29%). The analysis also revealed that most digital business professionals predominantly deal with an interruption immediately while the primary activity remains temporarily unfinished (43%), or they finish the primary activity first before starting the interrupting activity (43%). This exploratory study concludes with contributions and implications for research and practice, thus providing a valuable foundation for future research activities on interruption science.

Keywords: Interruption · Work Interruption · Interruption Perception · Interruption Management · Interruption Types · Interruption Consequences

1 Introduction

The use and role of technology has always been at the heart of Human-Computer Interaction (HCI) and has been studied by both researchers and practitioners in a variety of domains [1]. The traditional focus of HCI research is on humans and how humans interact with technology in the best possible way, with the ultimate goal to make technologies intelligent so that perceived ease-of-use reaches a maximum, thereby positively affecting user satisfaction and further downstream variables such as task performance

© Springer Nature Switzerland AG 2023
F. Fui-Hoon Nah and K. Siau (Eds.): HCII 2023, LNCS 14039, pp. 400–422, 2023.
https://doi.org/10.1007/978-3-031-36049-7_29

[2]. Indeed, digital technologies can bring significant benefits for individuals, organizations, and businesses, such as ubiquitous access to information, rapid communication, and productivity gains [3–5]. The technological environment is also one of the areas where the dynamics of digitalization can be noticeably observed, such as the information and communication technologies (ICT) used in the professional environments [6]. For example, during the COVID-19-induced lockdowns, companies had to quickly adapt web-conferencing systems such as Microsoft Teams or Zoom so that employees could continue with their daily tasks and activities [7]. By adopting and leveraging digital technologies, companies can benefit in terms of productivity, effectiveness, and efficiency [8], which has also been demonstrated on a macroeconomic level [9]. To this end, companies need to continuously evaluate and adapt their business infrastructure to take advantage of improvements in digital technologies [10]. This can be particularly challenging for small and medium-sized enterprises, which generally have limited human and financial resources to successfully understand and adopt the uses and role of digital technologies [11].

The rapid pace of technological change and the ubiquity of technology also pose new challenges for HCI research that need to be investigated to ensure beneficial interaction between humans and technology in business, management and organizational contexts [2]. Indeed, due to the development and increasing use of technology and acceleration worldwide [12], the phenomenon of interruptions caused by digital technologies, hereafter referred to as *IT-mediated interruptions*, has received increasing attention in both scientific research and practice. In the following, we present some exemplary research findings to emphasize the importance of knowledge on IT-mediated interruptions [13–18]. For example, research revealed that IT devices interrupt employees four to six times per working hour [19]. Also, case studies in companies have found that information workers switch between tasks on average every 3 [20] to 11 min [21] due to interruptions, with multiple interruptions occurring during the performance of a single task [22]. Aside from the occurrence of interruptions, research indicates that between 25% [18, 21] to 41% [23] of all interruption cases result in an original task not being resumed at all. In this context, Leroy and Glomb [24] tellingly write that "[c]ompleting tasks without interruptions has become a luxury" (p. 380).

A prominent example of IT-mediated interruptions are computer-mediated communication (CMC) interruptions, which can be caused by digital collaboration tools such as Microsoft Teams or Slack [25]. Such interruptions, however, are a common feature of today's workplace, as it enhances communication and facilitates decision making [16, 26, 27]. At the same time, the proliferation of digital technologies in the workplace has also contributed dramatically to work-related interruptions [28], which can lead to higher stress levels and negative effects on performance and productivity (for a review, please see [29]). To mitigate the negative consequences of such overload caused by IT-mediated interruptions, companies need to develop strategies, such as various measures to control the frequency of ICT-induced interruptions, so that employees can work effectively and efficiently [28, 30].

Digital business professionals are a group of human beings who are confronted with the constant influx and rapid spread of digital technologies in recent years. Following Bharadwaj et al. [10] and Woodard et al. [31], we define digital business professionals as individuals in the work area of digital business (e.g., digital business manager) who predominantly use digital resources (e.g., digital collaboration tools) to contribute to value creation. To achieve competitive advantages and strategic differentiation within the digital business strategy, they often focus on design capital, which is typically designs for digital artifacts (e.g., software components and their associated interfaces and data structures), and design moves, which are strategic actions that enlarge, reduce, or modify design capital (e.g., development of a new product or service). Their work is characterized by the continuous and intensive use of ICT, which is why they can be considered ICT competent. This definition also considers two types of users depending on how familiar they are with the digital world, namely digital immigrants (i.e., born before 1980) and digital natives (i.e., born after 1980) [30, 32, 33], originally introduced by Prensky [33]. Digital immigrants are "those who were not born into the digital world – they learnt to use information systems at some stage in their adult lives" [32, p. 711], and digital natives are "those who have grown up in a world where the use of information and communications technology is pervasive and ubiquitous" [32, p. 711]. Note that research has shown mixed results regarding the generational differences of IT use and ICT competency between these two types of users [30]. Given their general expertise in dealing with digital technologies, though, one may assume that digital business professionals have experience in dealing with rapid technological change and the ubiquity of technology. Indeed, they work in an information-rich digital media environment and are constantly exposed to ICT, which seems to be a nurturing ground for HCI research.

The continued growth and use of digital technologies in the workplace has many implications, and also important ones in the organizational context. One phenomenon that has been increasingly observed in recent years in scientific research and practice is the constant-checking of ICT as an individual habit in the search of new information [34] or to stay up to date at work [35, 36]. Constant connectivity through collaboration tools is thereby a factor that can lead to ICT-induced stress, which in turn can reduce job satisfaction and lead to lower organizational and long-term commitment [37]. In this context, higher computer self-efficacy is a crucial difference in the perception and management of interruptions [30, 37], as it may enable individuals to appropriately interact with digital technologies [38]. Against this background, the purpose of this study is to gather a deeper understanding of workplace interruptions among individuals, who are constantly exposed to digital technologies due to their information-rich digital media environment and who are expected to exhibit higher levels of computer self-efficacy compared to other professionals. Hence, we aim to explore the following main research question (RQ) in this paper: **How do digital business professionals perceive and manage interruptions in the workplace?**

The remainder of this paper is structured as follows. Section 2 describes the research background, which consists of a definition of interruption along with the literature background of our study. Section 3 then details the research methodology of our qualitative research design. The corresponding subsections provide information on the sample, the data collection, and the analysis procedure. In Sect. 4, we present the results of our exploratory study. Section 5 follows with a discussion of the contributions and implications along with limitations of our study. Finally, in Sect. 6, we provide a concluding statement.

2 Research Background

To define an interruption, we rely on the definition of Puranik et al. [39], who define an interruption as *"an unexpected suspension of the behavioral performance of, and/or attentional focus from, an ongoing work task"* (p. 817). Notably, the authors derived their definition from an analysis of previous work interruption definitions from 247 publications [39]. However, this definition implies that the primary task of an individual is interrupted by an unexpected event, which requires processing of the interruption at behavioral level (e.g., interruption requires response like cookie acceptance response) and/or cognitive attention (e.g., interruption during text-reading by online advertisements). In this study we explore an individual's overall exposure to interruptions over the course of their work. Our research constitutes exploratory research, and its purpose is to lay a foundation for future research activities on interruption science [40]. Specifically, we aim to address three gaps in the extant literature that are relevant to both scientific research and practice, which we describe below.

Reviewing the previous literature on interruptions reveals that interruptions can be described based on various attributes [40]. Indeed, the origins of research on interruptions go back to the pioneering experimental investigations and publications of Kurt T. Lewin and his experimental research program in the 1920s (e.g., [41–43]). During this long period, different perspectives on the conceptual and descriptive structure of interruptions have emerged, allowing interruptions to be distinguished from each other. For example, interruptions can be distinguished between technology-mediated or non-technology-mediated interruption, high-rated (interruption is critical) or low-rated interruption (interruption is uncritical), during working hours interruption or outside working hours interruption and internally initiated interruption by an endogenous event (e.g., mind wandering) or externally initiated interruption by an exogenous event (e.g., phone ringing) (for an overview of the various attributes of an interruption, please see [40]). To distinguish between these different interruption attributes, there are several classification concepts (for an overview, please see [44]). As an example, Jett and George [45] proposed a typology that categorizes work interruptions in an organizational context into four types. The first type of work interruptions are breaks, which are endogenous events that result in time off from work to accommodate personal needs and the rhythm of the day (e.g., take time to have lunch). The second type are discrepancies, which are endogenous or exogenous events that reveal a discrepancy between task performance expectations and actual performance of primary task activities and draw attention to the source of the inconsistency (e.g., individual receives feedback relevant to task performance). The third type are distractions, which are exogenous events that trigger psychological reactions due to external stimuli or secondary activities that interrupt focused attention on primary task activities (e.g., individual is exposed to noise while performing a task). Finally, the fourth type are intrusions, which are exogenous events that temporarily interrupt the execution of primary task activities due to an event that is irrelevant to task performance (e.g., unscheduled phone call). Another example is Addas and Pinsonneault [13], who proposed a taxonomy of IT-mediated interruptions based on the relevance (i.e., the interruption is relevant or irrelevant to the primary task), and structure of their content (i.e., the interruption is either system-generated or mediated by the system and activates different cognitions and actions). Given the above interruption attributes, we aim to classify the specific types of interruptions that can arise at the workplace of digital business professionals. Hence, we address the following RQ1: **What specific types of interruptions can be classified among digital business professionals?**

Experimental research on interruptions has been conducted for various purposes. As a result, research has revealed detrimental consequences of interruptions related to performance, including forgetting to complete primary task after the interruption, delays in resuming the primary task, and decreased task accuracy [46]. Other studies found that interruption can lead to significant psychological consequences. For example, the pioneering study by Maria A. Rickers-Ovsiankina [42] investigated the resumption of interrupted actions and found that there is a tendency to resume an interrupted action when it has not yet been completed. Such findings can have significant personal (e.g., work-family conflict [47], work-life conflict [48], or workplace exhaustion [49]) and organizational consequences (e.g., increased users' resistance behavior when implementing new information system [50]) if employees are mentally unable to separate themselves from work [49]. For example, it has been observed that employees increasingly use voluntary work-related technology use during non-work time [48, 51, 52] to accomplish unfinished tasks, thus interrupting their recovery processes [53]. Although interruptions are therefore predominantly associated with negative consequences, there is also research that demonstrates that interruptions can also have positive consequences. For example, Zijlstra et al. [54] found that interruptions can break through a growing sense of boredom and can lead to an increase in positive feelings and a decrease in strain. Also, Sonnentag et al. [55] found that perceived responsiveness to CMC interruptions (i.e., online messages) and perceived task accomplishment promoted feelings of having completed work and resulted in positive feelings in the evening. Another study by Murray & Khan [56] found that interruptions at the beginning of a task tend to have no effect on task performance. Hence, we address the following RQ2 to explore the consequences of interruptions on psychological well-being and work performance among digital business professionals: **How do interruptions affect psychological well-being and work performance of digital business professionals?**

Research on interruptions has explored a variety of issues. One of these issues also relates to managing interruptions, to understand how individuals deal with an interruption when their ongoing work needs to be suspended. To this end, research has identified methods of how individuals deal with interruptions. For example, Kirmeyer [57] identified in an observational study on police radio dispatchers three different coordination methods for interruption processing. In sequential processing, the primary activity is processed first, before starting the interrupting activity. In preemption, the interrupting activity is processed immediately, while the primary activity remains temporarily unfinished. Finally, in case of simultaneity, the interrupting activity and the primary activity were processed with rapid task switching to accomplish both activities. Another example of research into coordination methods for dealing interruptions is the research of McFarlane [58, 59], who analyzed existing literature from several different disciplines relevant to the design of user interfaces for HCI. He identified four methods of how individuals deal with interruptions. For example, individuals can deal with the interruption immediately (i.e., the individual deals with the interruption immediately while the primary activity remains temporarily unfinished), negotiated (i.e., the individual can negotiate how to deal with the interruption), mediated (i.e., the individual delegates the interruption to a mediator who decides how to deal with the interruption), or scheduled (i.e., the individual decides when to deal the interruption). Notably, the coordination methods of McFarlane [58, 59] have also been validated in empirical research [58–62]. Since our research is exploratory, we investigate how professionals in digital business deal with interruptions in the workplace. Hence, we address the following RQ3 to explore processing behavior among digital business professionals in more detail: **How do digital business professionals deal with interruptions?**

3   Research Methodology

For our exploratory study of work interruptions among digital business professionals, we chose a qualitative research design with interviews as the data collection method and qualitative content analysis to analyze the interviews. The interview itself consisted of four sets of questions. The first set of questions was related to demographic information about the interview participants, and their use of ICT in the workplace. Here, we also informed participants of an understanding of ICT on the basis of the examples adopted from Ayyagari et al. [63]. In the second set of questions, interviewees were asked to describe their perception of interruptions over the course of their work to classify specific types of interruptions and the resulting consequences on psychological well-being and work performance. The subsequent third set of questions examined how interruptions are managed and how professionals deal with them. Participants were asked how they coordinate interruptions at work and what strategies they generally use to address interruptions. As a foundation for developing interview questions to explore general perceptions of work interruptions and management methods to deal with them, we used a recently developed taxonomy of interruptions, which describes different attributes of an interruption based on research on interruptions [40]. The fourth and final set of questions dealt with an estimation of interruptions and opinions about ICT in the workplace.

3.1   Interview Participants

Prior to the interviews, two additional interviews were conducted with other researchers to validate our interview guide. These researchers are also knowledgeable in the field of digital business. Some minor adjustments were then made based on their suggestions. Appendix 1 provides the final set of interview questions.

Overall, seven interview participants agreed to take part in this study. The interview partners were recruited through the personal network of one of the authors of this paper. Out of these interviewees, 1 was female and the age ranged from 25 to 33 years (mean [$\pm$SD] age: 27,7 $\pm$ 2.56 years; median 27 years). Each participant had a master's degree. The participants were recruited from Austrian companies in different business sectors, including IT (4), industry (2) and consulting (1), and work in a range of roles relevant to digital business, such as Digital Marketing Manager (2), Head of Marketing & PR, Product & Project Manager, Project & Online Marketing Manager, Support for Operations & Sales for E-Business and Technology Consultant (1 in each case). Their work experience ranged from 3 to 15 years (mean [±SD]: 7.86 ± 4.78 years; median 7), with job levels ranging from professional (6) to senior management (1). For ICT use at work, we used a 5-point Likert scale (1 represents "Never", 5 represents "Always"), with the examples adopted from Ayyagari et al. [63]. To assess the total ICT use at work, we conceptualized the variable "Total ICT Use at Work", which contains all the ratings of the different ICTs per participant at daily work. The participants had an overall high level of ICT use in the workplace (mean [±SD] ICT use in number of 4.39 ± 1.15; median 5 in number of ICT use). Table 1 presents the information on demographics and ICT use in the workplace.

Table 1. Demographics and ICT Use in the Workplace of Interview Participants

Information Type	Variable	Distribution	
Demographics	Gender	1 (14%) female; 6 (86%) male	
	Age	Mean [±SD] age: 27,7 ± 2.56 years; median 27 years	
	Education	Master's degree (100%)	
	Business Sector	Consulting	1
		Industry	2
		IT	4
	Work Area	Digital Marketing Manager	2
		Head of Marketing & PR	1
		Product & Project Manager	1
		Project & Online Marketing Manager	1
		Support for Operations & Sales for E-Business	1
		Technology Consultant	1
	Years of Work Experience	Mean [±SD]: 7.86 ± 4.78 years; median 7	
ICT Use in the Workplace	Mobile Technologies	Mean [±SD]: 5 ± 0.0; median 5	
	Network Technologies	Mean [±SD]: 4.86 ± 3.78; median 5	
	Communication Technologies	Mean [±SD]: 4.43 ± 1.13; median 5	
	Enterprise / Database Technologies	Mean [±SD]: 3 ± 1.73; median 3	
	Generic Application Technologies	Mean [±SD]: 4.43 ± 0.78; median 5	
	Collaborative Technologies	Mean [±SD]: 5 ± 0.0; median 5	
	Other Work Specific Technologies	Mean [±SD]: 4 ± 1.41; median 4	
	Total ICT Use at Work	Mean [±SD] ICT use: 4.39 ± 1.15; median 5 ICT use	

3.2 Interview Procedure

Prior to the start of the interviews, all participants were contacted individually by one of the authors of this paper. The main purpose of this briefing contact was to encourage participants to think about the interruptions they experience in their workplace, with a particular focus on how and why they typically perceive and deal with these interruptions, and how often they experience them in an average day. Also, participants were assured complete anonymity throughout the data collection and analysis period and were given the opportunity to clarify any questions they had about the purpose and goal of the study.

To ensure the quality of the data on interruptions, we asked participants to respond in writing to interview questions after a typical workday. This allowed them to freely allocate their time to responding to the questions. This also ensured that the quality of the answers was not affected by lack of time and that the response rate was 100% for all questions in the interview guide. However, participants were also given the opportunity to answer the questions verbally, although each participant preferred to answer the questions in writing. As part of the interview process, additional time was arranged with participants after data collection to allow them to ask questions or make clarifications as needed. This debriefing also served to thank participants once again for their time.

3.3 Interview Data Coding and Analysis

To analyze the collected interview data, a content analysis approach was chosen. The analysis of the interview data was done in a straightforward manner [64] and included the development of a category system as first step [65]. For this purpose, the interview material for the qualitative content analysis was first screened with the aim of extracting a specific structure from the material. During this process, we used the taxonomy of interruptions with the various attributes of an interruption [40] and specific literature on research findings from research on interruptions as a guide for identifying constructs. For example, we used the typology of work interruptions by Jett and George [45] or the different coordination methods for dealing with interruptions by Kirmeyer [57] to develop an initial structure for our category system. This guiding structure was then iteratively applied to the interview material with the goal of developing a preliminary category system. To this end, all textual elements were extracted, resulting in about 100 generated codes. These codes were then structured along the identified context-specific constructs (i.e., interruption types, interruption consequences, and interruption management) with corresponding definitions from research on interruptions and formed into the category system using examples from the material. The procedure for qualitative content analysis of our interview material is thus a theory-oriented and rule-bounded procedure for the systematic investigation [65].

To assess the applicability of the results of our coding process, we asked in the second step an independent expert to assign examples of the interview material to our proposed codes. Among the examples, this independent evaluator disagreed on only one code. The disagreement was then resolved in a brief discussion. Table 2 provides the category system derived from this exploratory study.

In the third step, both authors of this paper independently coded the interviews with the category system and entered the results into a spreadsheet. This research approach is similar to that of Russell et al. [66], who exploratorily investigated how individuals cope with e-mail interruptions by using a spreadsheet and numerically noting all answers that referred to a particular code in their category system. In this way, the number of corresponding answers can be summed with a frequency of reports per code and compared with the other codes. However, as outlined by Russell et al. [66], interview participants can give answers that refer to multiple codes, since codes are not mutually exclusive. Also, each interview participant may have given several answers that can refer to different codes. For this reason, the calculated frequencies only refer to a summary of how many times an answer (i.e., code) was given in the interview data and with which frequencies. Nevertheless, by extracting all textual elements, we were able to consider 100% of the interview material.

The fourth and final step was to compare and verify the results of the coding process, which also allowed an assessment of the intercoder agreement between the two authors of this paper. The intercoder agreement was calculated following established procedures of Cohen [67] and was 92%, above accepted thresholds (< 61%, [68]). However, to reach full consensus on the coding of the interview data, both authors discussed discrepancies in coding and clarified those codes where they disagreed.

Table 2. Overview of Category System on Interruption Constructs

Category	Code	Definition	Reference
Interruption Types	Break	Endogenous event that results in time off from work to accommodate personal needs and the rhythm of the day (e.g., take time to have lunch)	[45]
	Distraction	Exogenous event that triggers psychological reactions due to external stimuli or secondary activities that interrupt focused attention on primary task activities (e.g., individual is exposed to noise while performing a task)	[45]
	Intervention	Exogenous event that reveals a discrepancy between task performance expectations and actual performance of primary task activities (e.g., individual receives feedback relevant to task performance)	[13]
	Intrusion	Exogenous event that temporarily interrupts the execution of primary task activities due to an event that is irrelevant to task performance (e.g., unscheduled phone call)	[13, 45]

(*continued*)

Table 2. (*continued*)

Category	Code	Definition	Reference
Interruption Consequences	Negative Performance Outcomes	The subject's work performance is negatively affected by the interruption (e.g., interruptions lead to decreased task accuracy)	[46]
	Negative Psychological Outcomes	The subject's psychological well-being is negatively affected by the interruption (e.g., interruptions lead to work exhaustion)	[49]
	Neutral Outcomes	The subject's work performance or psychological well-being is not affected by the interruption (e.g., interruptions tend to have no effect on performance at the start of a task)	[56]
	Positive Performance Outcomes	The subject's work performance is positively affected by the interruption (e.g., interruptions lead to a work strategy change)	[54]
	Positive Psychological Outcomes	The subject's psychological well-being is positively affected by the interruption (e.g., interruptions lead to the triggering of positive emotions)	[55]
Interruption Management	Preemption	The interrupting activity is processed immediately, while the primary activity remains temporarily unfinished	[57]
	Sequential Processing	The primary activity is processed first, before the interrupting activity is started	[57]
	Simultaneity	The interrupting activity and the primary activity are processed with rapid task switching to accomplish both activities	[57]

Kindly note that a) we have decided to use the term "intervention" in line with Addas and Pinsonneault [8], who based it on the original definition of "discrepancy" by Jett and George [45]; and b) references to the consequences of interruptions serve to highlight an exemplary finding of research on interruptions.

4   Research Results

In this section, we present the main findings of our content analysis, guided by our three RQs. We also include interviewee statements that offer insights into the general perceptions and management of work interruptions among digital business professionals. Kindly note that the interviews were conducted in German and the exemplary statements were literally translated. Appendix 2 provides additional results, such as interviewees' estimates of frequencies of work interruptions, that were not a primary focus for this exploratory study.

4.1   What Specific Types of Interruptions Can Be Classified Among Digital Business Professionals?

Our qualitative content analysis indicated that most of the interruption types are either intrusions (e.g., unscheduled phone calls during task performance) or distractions (e.g., receipt of online messages perceived as irrelevant during task execution). The result of the content analysis on interruption types is shown in Table 3, together with some exemplary statements.

Table 3. Result of Content Analysis and Exemplary Statements on Interruption Types

Code	Distribution	Exemplary Statement
Break	15%	"Occasional breaks from tasks are also necessary. I like to plan my free time activities during this time."
Distraction	27.5%	"Our company operates a very transparent information policy. As a result, we receive irrelevant online messages around the clock, which regularly interrupts my attention to work."
Intervention	22.5%	"Interruptions can be helpful because they contain, for example, critical feedback for an ongoing project or similarly important information."
Intrusion	35%	"Occasionally, someone comes into my office and says there's cake, or I'm invited for coffee."

4.2   How Do Interruptions Affect Psychological Well-Being and Work Performance of Digital Business Professionals?

Our qualitative content analysis revealed that most of these interruptions were perceived negatively and can affect individual's psychological well-being and work performance. For example, it was mentioned that such interruptions have negative consequences on the workflow. In this context, one interviewee has deactivated communication technologies for several weeks during concentrated work, because otherwise the aspired flow state cannot be achieved. Yet, our data also reveal that interruptions are also perceived positively.

For example, interruptions can improve work performance because they can include relevant information about work tasks. The result of the content analysis on interruption consequences is shown in Table 4, together with some exemplary statements.

Table 4. Result of Content Analysis and Exemplary Statements on Interruption Consequences

Code	Distribution	Exemplary Statement
Negative Performance Outcomes	28.5%	"Interruptions can distract me from the current task and thus disrupt the workflow. This causes the task to take longer."
Negative Psychological Outcomes	25%	"Due to interruptions, I have been deactivating or turning off communication technologies during concentrated work for the past three to four weeks, as I kept getting pulled out of work and not getting into a flow."
Neutral Outcomes	17.9%	"I neither feel disturbed by interruptions, nor do they affect my work performance."
Positive Performance Outcomes	10.7%	"Interruptions are helpful because they usually provide me with additional information about work tasks."
Positive Psychological Outcomes	17.9%	"I'm grateful for interruptions because they give you important information about a project or a current problem."

4.3 How Do Digital Business Professionals Deal with Interruptions?

Our qualitative content analysis indicated that most interviewees predominantly deal with the interruption immediately while the primary activity remains temporarily unfinished (i.e., preemption), or they finish the primary activity first before starting the interrupting activity (i.e., sequential processing). Also, interviewees have a low tendency to

Table 5. Result of Content Analysis and Exemplary Statements on Interruption Management

Code	Distribution	Exemplary Statement
Preemption	43%	"If the interruption is more important than the current activity, I take care of it immediately."
Sequential Processing	43%	"When I finished my work, I take care of the interruption."
Simultaneity	14%	"If my workday allows it or has just started, I take care of the interruption at the same time."

deal with the interrupting activity and the primary activity by switching between the activities (i.e., simultaneity). The result of the content analysis on interruption management is shown in Table 5, together with some exemplary statements.

5   Research Discussion

The research results provide a valuable foundation for future research activities on interruption science [40]. Based on our results, we describe below contributions and implications along with limitations that could provide opportunities for future research.

5.1   Contributions and Implications

We contribute to research by providing a practical perspective on the general perception and management of interruptions by classifying specific types of interruptions, quantifying the resulting consequences for psychological well-being and work performance, and examining how interruptions are managed. In our exploratory study, we focused on digital business professionals because their information-rich digital media environment should give them experience in managing work interruptions that arise primarily from the ubiquitous availability of digital devices and tools and the ambiguous contemporary use of these devices and tools in the workplace. Through qualitative content analysis, it was possible to obtain empirical insights into perceptions and management of interruptions. In this process, we also considered the conceptual and descriptive structure of interruptions derived from prior research [40]. Overall, our research methodology represents a methodologically sound approach to the systematic study of interruptions, parts of which can be used for further research.

The main implication for research is that our developed category system can serve as a foundation for further empirical research. In this context, it can be insightful to focus on similar or different target groups. For example, an empirical study found that medical doctors on average perceive 10.58 interruptions per hour and nurses 11.65 interruptions per hour [69]. Such a high number of interruptions could, for example, lead to a different frequency distribution of the consequences or the management of interruptions, or also to new types in the classification of interruptions. This could make it possible to compare target groups based on the number of interruptions. With the category system we developed (see Table 2), we contribute to the research on interruptions by taking a comprehensive view of interruptions.

From a practical perspective, our exploratory study illustrates the importance of scientific discourse on interruption. Indeed, the results of our qualitative content analysis revealed that in most cases, the execution of the primary tasks of digital business professionals is temporarily interrupted by intrusions or distracted due to external stimuli or secondary activities that interrupt focused attention on primary task activities (see Table 3). It can be said that interruptions can disrupt the routine flow of work as one shifts from automatic execution of routines to conscious processing of information [70]. Csíkszentmihályi [71] indicates that the optimal experience at work is in a psychophysiological flow state, which he describes as "the state in which people are so involved in an activity that nothing else seems to matter; the experience itself is so enjoyable that people will do it even at great cost, for the sheer sake of doing it" (p. 4). In fact, an empirical study has shown that the psychophysiological flow state can also lead individuals to complete a demanding task with moderate parasympathetic activity (decreased HRV activity) [72]. Interruptions, such as incoming online messages via collaboration technologies (i.e., CMC interruptions), disrupt flow, negatively affecting work performance in complex tasks [73] and generating work-related stress [74]. As an implication for practice, research activities and findings from research on interruptions, such as influencing factors (see Table 8) or recommendation (see Table 9) in managing interruptions, are therefore particularly valuable as they can contribute to the experience of flow. As an example, in a recent study by Nadj et al. [75], it was found that even frequent interruptions do not have a negative impact on flow during task performance if they provide task-relevant information. Nonetheless, since this study constitutes exploratory research and recommendations were not the primary focus of this study, the purpose of this study was to provide a foundation for future research activities on interruption science [40] by providing a practical perspective on the general perception and management of interruptions. Thus, further empirical investigation of specific attributes of interruptions can provide society and businesses with important insights into how best to manage interruptions.

5.2 Limitations

Our work is not free of limitations. First, one limitation of our exploratory study is the sample size. However, the approach to developing our category system followed a theory-oriented and rule-bounded procedure to subsequently systematically analyze the data from our collected interviews. We therefore note that a larger sample size could lead to further refinement of the category system in terms of interruption types and interruption management, as well as possible changes in the existing frequency distribution. Second, we asked interview participants to respond in writing to interview questions after a typical workday. Although this procedure allowed them to freely allocate their time to answer the questions, it is possible that certain interruption attributes could not be fully explored. For example, interviewees estimated the average number of IT-mediated interruptions during a typical workday to be quite low, with a maximum of 20 interruptions per day (see Table 7). Empirical research on the interruption frequency revealed that IT devices and programs interrupt an individual about 70 times per day during the completion of actual work tasks [13]. Although examining the frequency of interruptions during a typical workday was not the primary focus of this study, the difference in perception of interruptions may be due to age, gender, or computer self-efficacy in managing IT-mediated interruptions [30]. Future research addressing the above limitations could therefore provide further insights into the perception and management of interruptions.

6 Concluding Statement

The goal of this exploratory study was to allow digital business professionals to describe their general perception and management of interruptions. Specifically, our study presents a theory-oriented research methodology (derived from insights from prior research on interruptions such as the taxonomy of interruptions [40]) along with rules for systematically examining the collected qualitative data by following an established methodology for data coding and analysis [65]. This approach allowed us to classify specific types of interruptions, quantify the resulting consequences for psychological well-being and work performance, and examine how interruptions are managed. Overall, this exploratory study lays a valuable foundation for future research activities on interruption science [40]. Indeed, the systematic investigation of interruptions helps to address "any negative impact on attitudes, thoughts, behaviors, or body physiology that is caused either directly or indirectly by technology" [76 p. 5]. It will be rewarding to see what insights future research will reveal.

Acknowledgments. This research was funded by the Austrian Science Fund (FWF) as part of the project "Technostress in Organizations" (project number: P 30865) and by the Austrian Research Promotion Agency (FFG) as part of the project "Interruption" at the University of Applied Sciences Upper Austria. Also, we would like to thank the digital business professionals who participated in this study.

Appendix 1: Overview of Interview Questions

Table 6. Overview of Interview Questions

Question Set	Question
Demographic Information and ICT Use in the Workplace	Can you please briefly introduce yourself and describe your daily work routine in your company (e.g., job title, work area)?
	With which gender do you identify?
	Please indicate your age
	Do you live in Austria? If no, in which country do you live?
	What is your highest completed education?
	In which business sector do you work?
	How many years of professional experience do you have?

(continued)

Table 6. (*continued*)

Question Set	Question
	Please rate how often do you use the following information and communication technologies (ICTs) in your daily work (5-point Likert scale; 1 represents "Never", 5 represents "Always")? • Mobile technologies (e.g., Cell phone, Laptop) • Network technologies (e.g., Internet, Intranet, VPN) • Communication technologies (e.g., e-mail, voicemail) • Enterprise / Database technologies (e.g., SAP or Oracle applications) • Generic application technologies (e.g., Word, Excel, PowerPoint) • Collaborative technologies (e.g., Instant Messaging, Video Conferencing) • Other work specific technologies
	How many years of experience do you have using ICTs in the workplace?
Interruption Perception	Are you regularly interrupted in your work by ICTs?
	What kind of interruptions do you mainly experience during your working hours? Please give some examples
	What do these interruptions mean to you in a professional life? For example, do these interruptions help you perform your work because they are relevant? Or are these interruptions rather unimportant for you in a professional context? Again, please give a few examples
	How do you feel when you are interrupted in your work? For example, can interruptions affect your work performance, or do you feel disturbed by them? Please give examples
Interruption Management	How do you behave when you are interrupted? For example, do you try to take care of the work and the interruption at the same time, or do you take care of the interruption first before resuming the original task? Please give examples
	Are there also interruptions that you need to take care of immediately? If so, what factors influence this decision? Please give examples

(*continued*)

Table 6. (*continued*)

Question Set	Question
	What do you do if you cannot complete your work in the scheduled working time due to the interruptions? Do you complete these tasks the next day during working hours or on the same day outside regular working hours? Are there any specific factors you consider when making your decision? Please give examples
	Do you have any other tips or tricks on how to best handle interruptions?
Interruption Estimation and Opinion on ICT in the Workplace	On average, how many times per day are you interrupted by ICTs (e.g., an instant message via Slack or Microsoft Team) while completing an ongoing task?
	On average, how much time do you lose from your workday due to interruptions?
	On average, how much time per week do you need to complete your work tasks outside of regular working hours due to interruptions?
	What do you think about the use of ICTs in the workplace in general?

Appendix 2: Additional Results

Table 7. Estimation of Perceived Interruptions at the Workplace

Question	Amount	Distribution
On average, how many times per day are you interrupted by information and communication technologies (e.g., an instant message via Slack or Microsoft Team) while completing an ongoing task?	Unable to estimate	29%
	<5	14%
	<10	14%
	<15	29%
	<20	14%
On average, how much time do you lose from your workday due to interruptions?	Unable to estimate	14%
	<30 min	29%
	<60 min	43%
	<90 min	14%
On average, how much time per week do you need to complete your work tasks outside of regular working hours due to interruptions?	Unable to estimate	14%
	<30 min	43%
	<60 min	29%
	<180 min	14%

Table 8. Insights into Influencing Factors for Immediate Management of Interruptions

Influencing Factor	Exemplary Statement
Current Workload of Interruption Receiver	"It depends on whether there is a lot or little to do now as well as how far the working day has progressed."
Interruption Consequences	"If the web store goes down, I would immediately take care of the interruption, otherwise we could lose sales."
Interruption Content	"When I am waiting for feedback that leads to the completion of an important activity, it is brought forward."
Interruption Duration	"In case of a short interruption that only slightly delays the schedule, I take care of the interruption and try to complete the planned work afterwards."
Interruption Task Type	"Operational activities with customers need to be handled immediately in the best case."
Status of Interruption Initiator	"When I receive instructions from my CEO, I immediately take care of the interruption."

Table 9. Insights into Personal Recommendation for Managing Interruptions

Recommendation	Exemplary Statement
Communication Rules	"Companies could establish communication rules that govern when an employee must be available."
Education and Training	"A company must support employees in creating an awareness of the resulting illnesses and, with appropriate support, ensure a certain vitality in everyday work."
Self-Management	"You have to decide for yourself which interruptions you pay attention to immediately and which ones you ignore for the time being."
Task Prioritization	"There are tasks that are more important and therefore should be prioritized in terms of interruptions."
Task-Oriented Use of ICT	"For important tasks, I use only the programs I need to perform the task and disable all other notifications."
Usage of "Do Not Disturb"	"I find MS Teams' "Do Not Disturb" mode or Microsoft's ability to disable notifications for a certain amount of time when you want to work in a focused manner very useful."

References

1. Carey, J., Kim, Y., Wildemuth, B.: The role of human computer interaction in management information systems curricula: A call to action. Commun. Assoc. Inf. Syst. **13**(23), 357–379 (2004). https://doi.org/10.17705/1CAIS.01323
2. Stephanidis, C., et al.: Seven HCI grand challenges. Int. J. Human-Comput. Interact. **35**(14), 1229–1269 (2019). https://doi.org/10.1080/10447318.2019.1619259
3. Brynjolfsson, E.: The contribution of information technology to consumer welfare. Inf. Syst. Res. **7**(3), 281–300 (1996). https://doi.org/10.1287/isre.7.3.281
4. Brynjolfsson, E., Hitt, L.M.: Beyond computation: Information technology, organizational transformation and business performance. J. Econ. Perspect. **14**(4), 23–48 (2000). https://doi.org/10.1257/jep.14.4.23
5. Keeney, R.L.: The value of Internet commerce to the customer. Manage. Sci. **45**(4), 533–542 (1999). https://doi.org/10.1287/mnsc.45.4.533
6. Hess, T., Benlian, A., Matt, C., Wiesböck, F.: Options for formulating a digital transformation strategy. MIS Q. Exec. **15**(2), 103–119 (2016).
7. Hacker, J., vom Brocke, J., Handali, J., Otto, M., Schneider, J.: Virtually in this together – How web-conferencing systems enabled a new virtual togetherness during the COVID-19 crisis. Eur. J. Inf. Syst. **29**(5), 563–584 (2020). https://doi.org/10.1080/0960085X.2020.1814680
8. Melville, N., Kraemer, K., Gurbaxani, V.: Information technology and organizational performance: An integrative model of IT business value. MIS Q. **28**(2), 283–322 (2004). https://doi.org/10.2307/25148636
9. Ganju, K.K., Pavlou, P.A., Banker, R.D.: Does information and communication technology lead to the well-being of nations? A country-level empirical investigation. MIS Q. **40**(2), 417–430 (2016). https://doi.org/10.25300/MISQ/2016/40.2.07
10. Bharadwaj, A., El Sawy, O.A., Pavlou, P.A., Venkatraman, N.: Digital business strategy: Toward a next generation of insights. MIS Q. **37**(2), 471–482 (2013). https://doi.org/10.25300/MISQ/2013/37:2.3

11. Caldeira, M.M., Ward, J.M.: Using resource-based theory to interpret the successful adoption and use of information systems and technology in manufacturing small and medium-sized enterprises. Eur. J. Inf. Syst. **12**(2), 127–141 (2003). https://doi.org/10.1057/palgrave.ejis.300 0454

12. Couffe, C., Michael, G.A.: Failures due to interruptions or distractions: A review and a new framework. Am. J. Psychol. **130**(2), 163–181 (2017). https://doi.org/10.5406/amerjpsyc.130. 2.0163

13. Addas, S., Pinsonneault, A.: The many faces of information technology interruptions: A taxonomy and preliminary investigation of their performance effects. Inf. Syst. J. **25**(3), 231–273 (2015). https://doi.org/10.1111/isj.12064

14. Cheng, X., Bao, Y., Zarifis, A.: Investigating the impact of IT-mediated information interruption on emotional exhaustion in the workplace. Inf. Process. Manage. **57**(6), 102281 (2020). https://doi.org/10.1016/j.ipm.2020.102281

15. Feldman, E., Greenway, D.: It's a matter of time: The role of temporal perceptions in emotional experiences of work interruptions. Group Org. Manag. **46**(1), 70–104 (2021). https://doi.org/ 10.1177/1059601120959288

16. Galluch, P.S., Grover, V., Thatcher, J.B.: Interrupting the workplace: Examining stressors in an information technology context. J. Assoc. Inf. Syst. **16**(1), 1–47 (2015). https://doi.org/10. 17705/1jais.00387

17. Addas, S., Pinsonneault, A.: E-mail interruptions and individual performance: Is there a silver lining? MIS Q. **42**(2), 381–405 (2018). https://doi.org/10.25300/MISQ/2018/13157

18. Addas, S., Pinsonneault, A.: Theorizing the multilevel effects of interruptions and the role of communication technology. J. Assoc. Inf. Syst. **19**(11), 1097–1129 (2018). https://doi.org/ 10.17705/1jais.00521

19. Mirhoseini, S., Hassanein, K., Head, M., Watter, S.: User performance in the face of IT interruptions: The role of executive functions. In: Davis, F.D., Riedl, R., vom Brocke, J., Léger, P.-M., Randolph, A.B., Fischer, T. (eds.) Information Systems and Neuroscience: NeuroIS Retreat 2019. LNISO, vol. 32, pp. 41–51. Springer, Cham (2020). https://doi.org/ 10.1007/978-3-030-28144-1_5

20. González, V.M., Mark, G.: "Constant, Constant, Multi-tasking Craziness": Managing multiple working spheres. In: Proceedings of the SIGCHI Conference on Human Factors in Computing Systems, pp. 113–120 (2004). https://doi.org/10.1145/985692.985707

21. Mark, G., González, V.M., Harris, J.: No task left behind? Examining the nature of fragmented work. In: Proceedings of the SIGCHI Conference on Human Factors in Computing Systems, pp. 321–330 (2005). https://doi.org/10.1145/1054972.1055017

22. Czerwinski, M., Horvitz, E., Wilhite, S.: A diary study of task switching and interruptions. In: Proceedings of the SIGCHI Conference on Human Factors in Computing Systems, pp. 175–182 (2004). https://doi.org/10.1145/985692.985715

23. O'Conaill, B., Frohlich, D.: Timespace in the workplace: Dealing with interruptions. In: Conference Companion on Human Factors in Computing Systems.,pp. 262–263 (1995). https:// doi.org/10.1145/223355.223665

24. Leroy, S., Glomb, T.M.: Tasks interrupted: How anticipating time pressure on resumption of an interrupted task causes attention residue and low performance on interrupting tasks and how a "Ready-to-Resume" plan mitigates the effects. Organ. Sci. **29**(3), 380–397 (2018). https://doi.org/10.1287/orsc.2017.1184

25. Russell, E., Jackson, T., Banks, A.: Classifying computer-mediated communication (CMC) interruptions at work using control as a key delineator. Behav. Inf. Technol. **40**(2), 191–205 (2021). https://doi.org/10.1080/0144929X.2019.1683606

26. Jia, L., Huang, L., Yan, Z., Hall, D., Song, J., Paradice, D.: The importance of policy to effective IM use and improved performance. Inf. Technol. People **33**(1), 180–197 (2019). https://doi.org/10.1108/ITP-09-2018-0409

27. McMurtry, K.: Managing email overload in the workplace. Perform. Improv. **53**(7), 31–37 (2014). https://doi.org/10.1002/pfi.21424

28. Tams, S., Ahuja, M., Thatcher, J.B., Grover, V.: Worker stress in the age of mobile technology: The combined effects of perceived interruption overload and worker control. J. Strat. Inf. Syst. **29**(1), 101595 (2020). https://doi.org/10.1016/j.jsis.2020.101595

29. Fischer, T., Riedl, R.: Technostress research: A nurturing ground for measurement pluralism? Commun. Assoc. Inf. Syst. **40**(17), 375–401 (2017). https://doi.org/10.17705/1CAIS.04017

30. Baham, C., Kalgotra, P., Nasirpouri Shadbad, F., Sharda, R.: Generational differences in handling technology interruptions: A qualitative study. Eur. J. Inf. Syst. 1–21 (2022). https://doi.org/10.1080/0960085X.2022.2070557

31. Woodard, C.J., Ramasubbu, N., Tschang, F.T., Sambamurthy, V.: Design capital and design moves: The logic of digital business strategy. MIS Q. **37**(2), 537–564 (2013). https://doi.org/10.25300/MISQ/2013/37.2.10

32. Vodanovich, S., Sundaram, D., Myers, M.: Research commentary—Digital natives and ubiquitous information systems. Inf. Syst. Res. **21**(4), 711–723 (2010). https://doi.org/10.1287/isre.1100.0324

33. Prensky, M.: Digital natives, digital immigrants part 1. On the Horizon **9**(5), 1–6 (2001). https://doi.org/10.1108/10748120110424816

34. Gerlach, J.P., Cenfetelli, R.T.: Constant checking is not addiction: A grounded theory of IT-mediated state-tracking. MIS Q. **44**(4), 1705–1732 (2020). https://doi.org/10.25300/MISQ/2020/15685

35. Barley, S.R., Meyerson, D.E., Grodal, S.: E-mail as a source and symbol of stress. Organ. Sci. **22**(4), 887–906 (2011). https://doi.org/10.1287/orsc.1100.0573

36. Mazmanian, M., Orlikowski, W.J., Yates, J.: The autonomy paradox: The implications of mobile email devices for knowledge professionals. Organ. Sci. **24**(5), 1337–1357 (2013). https://doi.org/10.1287/orsc.1120.0806

37. Ragu-Nathan, T.S., Tarafdar, M., Ragu-Nathan, B.S., Tu, Q.: The consequences of technostress for end users in organizations: Conceptual development and empirical validation. https://doi.org/10.1287/isre.1070.0165

38. Brod, C.: Technostress: The Human Cost of the Computer Revolution. Addison-Wesley Publishing, Reading (1984)

39. Puranik, H., Koopman, J., Vough, H.C.: Pardon the interruption: An integrative review and future research agenda for research on work interruptions. J. Manag. **46**(6), 806–842 (2020). https://doi.org/10.1177/0149206319887428

40. Stangl, F.J., Riedl, R.: Interruption science as a research field: Towards a taxonomy of interruptions as a foundation for the field. Front. Psychol. **14**, 1043426 (2023). https://doi.org/10.3389/fpsyg.2023.1043426

41. Lewin, K.T.: Vorsatz Wille und Bedürfnis: Mit Vorbemerkungen über die psychischen Kräfte und Energien und die Struktur der Seele. Springer, Heidelberg (1926). https://doi.org/10.1007/978-3-642-50826-4

42. Ovsiankina, M.A.: Die Wiederaufnahme unterbrochener Handlungen. Psychologische Forschung **11**(1), 302–379 (1928). https://doi.org/10.1007/BF00410261

43. Zeigarnik, B.W.: Das Behalten erledigter und unerledigter Handlungen. Psychol. Forsch. **9**(1), 1–85 (1927). https://doi.org/10.1007/BF02409755

44. Bailey, K.D.: Typologies and Taxonomies: An Introduction to Classification Techniques. Sage Publications, Thousand Oaks (1994)

45. Jett, Q.R., George, J.M.: Work interrupted: A closer look at the role of interruptions in organizational life. Acad. Manag. Rev. **28**(3), 494–507 (2003). https://doi.org/10.5465/amr.2003.10196791

46. Morgan, P.L., Patrick, J., Waldron, S.M., King, S.L., Patrick, T.: Improving memory after interruption: Exploiting soft constraints and manipulating information access cost. J. Exp. Psychol. Appl. **15**(4), 291–306 (2009). https://doi.org/10.1037/a0018008
47. Wan, M., Shaffer, M.A., Lau, T., Cheung, E.: The knife cuts on both sides: Examining the relationship between cross-domain communication and work–family interface. J. Occup. Organ. Psychol. **92**(4), 978–1019 (2019). https://doi.org/10.1111/joop.12284
48. Ragsdale, J.M., Hoover, C.S.: Cell phones during nonwork time: A source of job demands and resources. Comput. Hum. Behav. **57**, 54–60 (2016). https://doi.org/10.1016/j.chb.2015.12.017
49. Chen, A.J., Karahanna, E.: Life interrupted: The effects of technology-mediated work interruptions on work and nonwork outcomes. MIS Q. **42**(4), 1023–1042 (2018). https://doi.org/10.25300/MISQ/2018/13631
50. Laumer, S., Maier, C., Eckhardt, A., Weitzel, T.: Work routines as an object of resistance during information systems implementations: Theoretical foundation and empirical evidence. Eur. J. Inf. Syst. **25**(4), 317–343 (2016). https://doi.org/10.1057/ejis.2016.1
51. Schlachter, S., McDowall, A., Cropley, M., Inceoglu, I.: Voluntary work-related technology use during non-work time: A narrative synthesis of empirical research and research agenda. Int. J. Manag. Rev. **20**(4), 825–846 (2018). https://doi.org/10.1111/ijmr.12165
52. Derks, D., Bakker, A.B., Peters, P., van Wingerden, P.: Work-related smartphone use, work–family conflict and family role performance: The role of segmentation preference. Human Relations **69**(5), 1045–1068 (2016). https://doi.org/10.1177/0018726715601890
53. Keller, A.C., Meier, L.L., Elfering, A., Semmer, N.K.: Please wait until I am done! Longitudinal effects of work interruptions on employee well-being. Work Stress **34**(2), 148–167 (2020). https://doi.org/10.1080/02678373.2019.1579266
54. Zijlstra, F.R.H., Roe, R.A., Leonora, A.B., Krediet, I.: Temporal factors in mental work: Effects of interrupted activities. J. Occup. Organ. Psychol. **72**(2), 163–185 (1999). https://doi.org/10.1348/096317999166581
55. Sonnentag, S., Reinecke, L., Mata, J., Vorderer, P.: Feeling interrupted-Being responsive: How online messages relate to affect at work. J. Organ. Behav. **39**(3), 369–383 (2018). https://doi.org/10.1002/job.2239
56. Murray, S.L., Khan, Z.: Impact of interruptions on white collar workers. Eng. Manag. J. **26**(4), 23–28 (2014). https://doi.org/10.1080/10429247.2014.11432025
57. Kirmeyer, S.L.: Coping with competing demands: Interruption and the Type A pattern. J. Appl. Psychol. **73**(4), 621–629 (1988). https://doi.org/10.1037/0021-9010.73.4.621
58. McFarlane, D.C.: Interruption of people in human-computer interaction: A general unifying definition of human interruption and taxonomy. Naval Research Laboratory (NRL Formal Report NRL/FR/5510-97-9870), Washington (1997)
59. McFarlane, D.C.: Interruption of people in human-computer interaction. Doctoral Dissertation, George Washington University (1998)
60. McFarlane, D.C.: Coordinating the interruption of people in human-computer interaction. In: Sasse, M.A., Johnson, C. (eds.) Human-Computer Interaction, INTERACT '99: IFIP TC.13 International Conference on Human–Computer Interaction, 30th August–3rd September 1999, Edinburgh, UK. pp. 295–303. IOS Press, Amsterdam (1999)
61. McFarlane, D.C.: Comparison of four primary methods for coordinating the interruption of people in human-computer interaction. Human-Comput. Interact. **17**(1), 63–139 (2002). https://doi.org/10.1207/S15327051HCI1701_2
62. McFarlane, D.C., Latorella, K.A.: The scope and importance of human interruption in human-computer interaction design. Human-Comput. Interact. **17**(1), 1–61 (2002). https://doi.org/10.1207/S15327051HCI1701_1
63. Ayyagari, R., Grover, V., Purvis, R.: Technostress: Technological antecedents and implications. MIS Q. **35**(4), 831–858 (2011). https://doi.org/10.2307/41409963

64. Dey, I.: Qualitative Data Analysis: A User Friendly Guide for Social Scientists. Routledge, London and New York (1993)
65. Mayring, P.: Qualitative content analysis: Theoretical foundation, basic procedures and software solution. Klagenfurt (2014)
66. Russell, E., Purvis, L.M., Banks, A.: Describing the strategies used for dealing with email interruptions according to different situational parameters. Comput. Hum. Behav. **23**(4), 1820–1837 (2007). https://doi.org/10.1016/j.chb.2005.11.002
67. Cohen, J.: A coefficient of agreement for nominal scales. Educ. Psychol. Measur. **20**(1), 37–46 (1960). https://doi.org/10.1177/001316446002000104
68. Landis, J.R., Koch, G.G.: The measurement of observer agreement for categorical data. Biometrics **33**(1), 159–174 (1977). https://doi.org/10.2307/2529310
69. Brixey, J.J., Robinson, D.J., Turley, J.P., Zhang, J.: The roles of MDs and RNs as initiators and recipients of interruptions in workflow. Int. J. Med. Informatics **79**(6), e109–e115 (2010). https://doi.org/10.1016/j.ijmedinf.2008.08.007
70. Zellmer-Bruhn, M.E.: Interruptive events and team knowledge acquisition. Manage. Sci. **49**(4), 514–528 (2003). https://doi.org/10.1287/mnsc.49.4.514.14423
71. Csíkszentmihályi, M.: Flow: The Psychology of Optimal Experience. Harper & Row, New York (1990)
72. Tozman, T., Magdas, E.S., MacDougall, H.G., Vollmeyer, R.: Understanding the psychophysiology of flow: A driving simulator experiment to investigate the relationship between flow and heart rate variability. Comput. Hum. Behav. **52**, 408–418 (2015). https://doi.org/10.1016/j.chb.2015.06.023
73. Speier, C., Valacich, J.S., Vessey, I.: The effects of task interruption and information presentation on individual decision making. In: Proceedings of the 18th International Conference on Information Systems, pp. 21–35 (1997)
74. Basoglu, K.A., Fuller, M.A., Sweeney, J.T.: Investigating the effects of computer mediated interruptions: An analysis of task characteristics and interruption frequency on financial performance. Int. J. Account. Inf. Syst. **10**(4), 177–189 (2009). https://doi.org/10.1016/j.accinf.2009.10.003
75. Nadj, M., et al..: What disrupts flow in office work? A NeuroIS study on the impact of frequency and relevance of IT-mediated interruptions. MIS Q. (2023, in press)
76. Weil, M.M., Rosen, L.D.: TechnoStress: Coping with Technology @Work @Home @Play. Wiley, New York (1997)

Author Index

A

Alimu, Nurzat I-249
Amante, Daniel J. II-192
Anderson, Anthony I-381
Arbulú-Pérez Vargas, Carmen Graciela I-47
Assis, Luciana S. I-331
Auinger, Andreas I-170

B

Baer-Baldauf, Pascale I-3
Báez Martínez, Fernando II-155
Baier, Ralph II-86
Bajracharya, Adarsha S. II-192
Baldauf, Matthias I-3
Bao, Kaiwen I-260
Behery, Mohamed II-86
Bekiri, Valmir I-3
Belivanis, Dimitris II-377
Bettelli, Alice I-450
Brecher, Christian II-86
Buchler, Norbou II-66

C

Campanini, Maria Luisa I-450
Cao, Changlin II-130
Cao, Yan I-353
Carrasco-Espino, Danicsa Karina I-47
Chang Chien, Ya-Wen II-170
Chen, Chin-Yi I-150
Chen, Gang I-102, II-242
Chen, Hsi-His I-150
Chen, Langtao I-365
Chen, Po-Lin II-214
Chen, Wen-Hsin I-141
Chen, Yang-Cheng II-225
Cheng, Li-Chen II-170
Cheng, Yawei II-41
Chiang, Hao-Chun II-41
Chien, Shih-Yi II-41
Childress, Matt I-381
Chiu, Tseng-Ping I-314
Chiu, Yu-Jing I-150, II-3

Chu, Tsai-Hsin II-259
Chu, Wei-Hsin II-259
Chung, Pi Hui II-274
Cordeiro, Cheryl Marie II-334
Corves, Burkhard II-86

D

Dammers, Hannah II-86
Decouchant, Dominique I-421
Demaeght, Annebeth II-18
Denne, Bernhard II-364
Dieter, Sara I-381
Djamasbi, Soussan II-66, II-192
Du, Duo I-487, II-121
Du, Qianrui II-130

E

Eibl, Stefan I-170
Eschenbrenner, Brenda I-190
Evangelou, Semira Maria II-285

F

Facho-Cornejo, Jhoselit Lisset I-47
Figueroa-Montero, Manuel II-300
Fragoso González, Gilberto A. II-155
Freitas, Sergio A. A. I-331
Furman, Susanne M. I-14

G

Gailer, Carolin I-401
Gamberini, Luciano I-439, I-450
Ge, Jiao I-476, I-487, II-121
Ghoshal, Torumoy II-192
González, Víctor M. II-155
Gossen, Daniel II-86
Gries, Thomas II-86

H

Haney, Julie M. I-14
Hangrui, Cui I-202
Harberg, Max I-381
Henn, Thomas II-86

© Springer Nature Switzerland AG 2023
F. Fui-Hoon Nah and K. Siau (Eds.): HCII 2023, LNCS 14039, pp. 423–425, 2023.
https://doi.org/10.1007/978-3-031-36049-7

Hu, Yi-Chung II-3
Huang, Kai-Bin II-320
Huang, Lihua II-242
Hung, Li-Ting II-140

I
Israel, Kai I-461
Izumigawa, Christianne I-87

J
Jacobs, Jody L. I-14
Jiang, Hang I-74, II-104
Jin, Yong I-293

K
Kang, Lele I-260
Kowalewski, Stefan II-86
Krennhuber, Sarah I-219
Ku, Yi-Cheng I-141
Kwan, Ralston I-230

L
Lai, Yi-Pei II-41
Lakemeyer, Gerhard II-86
Lee, Yen-Hsien II-259
Li, Chien-Cheng II-3
Li, Guoxin II-30
Li, Hong I-249
Li, Keli II-30
Li, Zhi II-66
Liang, Siyi I-249
Liao, Qichen II-130
Lien, Shao-I. I-302
Lin, Jie I-102
Lin, Sung-Chien II-140
Lin, Szu-Yin II-41
Lin, Yuyang II-104
Lindström, Nataliya Berbyuk II-334
Liu, Xinhui I-260

M
Maguire, Martin I-271
Mark, Ben I-381
Martínez Pérez, Mayté V. II-155
Mashiro, Daisuke II-352
Megheirkouni, Majd II-53
Mendoza, Sonia I-421
Mi, Chuanmin I-249
Miclau, Christina I-401, II-364

Mihatsch, Alexander II-202
Misran I-114
Monaro, Merylin I-450
Monroy-Rodríguez, Guillermo I-421
Mower, Andrea I-381
Müller, Andrea I-401, II-18

N
Nakao, Yuri I-34
Nerb, Josef II-364
Nishihara, Yoko II-352
Nitsch, Verena II-86
Nurmandi, Achmad I-114

O
Oikonomaki, Eleni II-377
Olivas Martinez, Giorgio I-439
Orso, Valeria I-439, I-450
Osornio García, Fernando U. II-155
Ottersböck, Nicole II-181

P
Panitz, Adrian I-401
Panyawanich, Kwan I-271
Perry, Mark I-57
Peuker, Veronika I-401
Pierobon, Leonardo I-450
Polsri, Suyanee II-170
Portello, Giovanni I-450
Pradel, Patrick I-271

R
Reckter, Julian I-293
Rehman, Abdul I-114
Reyes-Perez, Moisés David I-47
Riedl, René II-400
Rojas-Palacios, Luis Eden I-47
Rusch, Tobias II-181

S
Saini, Vipin I-293
Salas Barraza, Mario H. II-155
Sánchez-Adame, Luis Martín I-421
Sancho-Chavarría, Lilliana II-300
Sankar, Gaayathri II-66, II-192
Satjawisate, Saiphit I-57
Sato, Jonathan I-87
Shan, Junjie II-352
Shaw, Norman I-190, I-230

Shi, Qiming   II-192
Shi, Yanqing   I-260
Si, Hanchi   I-249
Song, So-Ra   II-3
Stabauer, Martin   I-219
Stangl, Fabian J.   II-400
Stowasser, Sascha   II-181, II-202
Suchy, Oliver   II-202
Sun, Jianjun   I-260
Sun, Taipeng   I-74

T

Tang, Muh-Chyun   II-140
Taylor, Bethany   I-87
Telliel, Yunus Dogan   II-192
Terstegen, Sebastian   II-202
Trinh, Minh   II-86

V

Valdespin-Garcia, Ivan Giovanni   I-421

W

Wagner, Jenny   I-461
Walz, Natalie   II-18
Wang, Chou-Wen   II-214, II-225
Weir, David   II-53
Wu, Chin-Wen   II-214, II-225

Wu, I-Chin   I-302
Wu, Mingling   I-476

X

Xenos, Michalis   II-285
Xiao, Jing   II-66
Xiao, Shuaiyong   I-102, II-242
Xu, Michael   I-476

Y

Yang, Ya-Chun   I-314
Yang, Yu-Chen   I-293
Ye, Qiongwei   II-130
Yokota, Takuya   I-34
Younus, Muhammad   I-114
Yu, Hsin-Kai   I-302

Z

Zerres, Christopher   I-461
Zhai, Weimin   I-353
Zhai, Weiren   I-353
Zhang, Chenghong   I-102, II-242
Zhang, Xijie   I-74
Zhang, Yanling   I-487, II-121
Zhang, Zongxiang   I-102, II-242
Zheng, Yin   II-104
Zimmermann, Hans-Dieter   I-3

Printed in the United States
by Baker & Taylor Publisher Services